·The⚒Co·

A Manual of Determinative

Bacteriology

BY

FREDERICK D. CHESTER

*Bacteriologist of the Delaware College Agricultural Experiment Station, and
Director of the Laboratory of the State Board of Health of Delaware;
Member of the Society of American Bacteriologists; of the Society
for the Promotion of Agricultural Science, and of the
American Public Health Association.*

New York

THE MACMILLAN COMPANY

LONDON: MACMILLAN & CO., Ltd.

1901

Norwood Press
J. S. Cushing & Co. — Berwick & Smith
Norwood Mass. U.S.A.

PREFACE

PRELIMINARY to studies on the bacterial flora of cultivated soils, the writer undertook an arrangement of the several hundred species of bacteria already described, with the view of identifying the forms isolated, or at least of determining whether they were new to science. The labor involved in this arrangement has been so great that it was decided to embody the results in the present form that others might have the advantage of them. The writer does not claim that the system of arrangement is perfect or not open to criticism. The best use only could be made of the facts and material available. The present tables serve, therefore, only for purposes of identification, and not necessarily for those of classification.

For this reason the present book has been termed a Manual of Determinative, rather than one of Systematic, Bacteriology.

To the student working in the laboratory the determination of unknown bacteria has been almost impossible, except with the expenditure of an amount of labor which was impracticable. With the use of the present manual it is believed that the teacher can place a given culture in the hands of his pupil and expect him to determine it, as is done with other organic forms. It is therefore hoped that the present work will serve a useful purpose as a laboratory manual.

The chapter on morphology has been appended in order to make clearer the system of classification into orders and genera.

v

The chapter on terminology was necessary in order to make more intelligible the description of species.

The work does not claim to be a text-book on bacteriology, but aims only to supplement the latter.

It is evident that bacterial forms cannot be too carefully studied; the student is therefore urged to make use of the scheme for the study of species as given in Chapter III.

FREDERICK D. CHESTER.

DELAWARE COLLEGE,
AGRICULTURAL EXPERIMENT STATION,
NEWARK, DELAWARE,
December, 1900.

A MANUAL OF
DETERMINATIVE BACTERIOLOGY

CHAPTER I

THE MORPHOLOGY OF BACTERIA

1. THE STRUCTURE OF THE BACTERIAL CELL

BACTERIA are unicellular organisms of the simplest type. A bacterial cell consists of a *central body* which stains readily with the basic analine colors and other nuclear stains. This central body is surrounded by a *capsule* of variable form and thickness, but which in the majority of cases is thin and symmetrical. Such a typical cell is shown in Fig. 1, B.

A. The Central Body

The central body is generally homogeneous in structure, and is only in rare cases granulated. By special staining methods the protoplasm may show a number of deeply colored bodies known as *metachromatic granules*. These were formerly considered by certain authors as the initial elements of spores, and by others as nuclei. They are now regarded simply as denser aggregations of protoplasmic molecules which possess special staining properties. They are illustrated in Fig. 1, B–E.

According to Migula, in *Bacillus oxalaticus*, the central body may show the formation of vacuoles, Fig. 1, B. Vacuoles are, however, of phenomenal occurrence, and in the majority of bacteria the central protoplasm is homogeneous and dense. In certain species, however, the protoplasm may show unequal staining properties, giving the bacillus a beaded appearance with intervening unstained or feebly stained spaces; or again, the protoplasm may be aggregated at the poles, with a comparatively clear central portion.

B

In tubercle bacilli this beaded appearance is common, and the unstained spaces are considered by A. Coppen Jones to be of the nature of vacuoles. It is doubtful whether this is correct; on the contrary, it is more probable that the beaded structure represents a fragmentation of the protoplasm, which is a phase in all cellular degeneration.

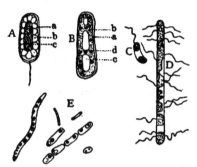

FIG. 1. — Showing structure of bacterial cell.

A. *Bact. lineola* after Bütschli, *a* capsule, *b* protoplasmic layer, *c* nuclear body. B. *Bact. oxylaticum* after Migula, *a* capsule, *b* central body, *c* vacuole, *d* metachromatic granules. C–D. plasmolysis of bacterial cell. E. bacilli, showing metachromatic granules.

According to Bütschli, in *Bact. lineola*, the central body is surrounded by an envelope of protoplasm corresponding to the cytoplasm of other cells. If such a cytoplasmic layer exists in other species of bacteria, it is too thin to be differentiated from the outer capsule. Whether the central body is a true nucleus or not cannot be positively decided, but there is no good reason to believe that it is.

That the central body is a distinct structure from the outer capsule is demonstrated by the phenomenon of plasmolysis. Thus when bacteria are placed in a 2.5 per cent potassium nitrate or a 1 per cent sodium chloride solution, the central body contracts, and separates itself in places from the capsule, as shown in Fig. 1, D.

The bacterial plasma in certain species may show the presence of granular bodies, as in *B. butyricus*, *Vibrio bugula*, and *Bact. Pasteurianum*, which stain bluish or violet-black with iodine; the so-called *granulose reaction*. The exact nature of granulose is not known. It may be identical with starch, or, at any rate, is a closely related carbohydrate.

In the sulphur bacteria the cell plasma may contain certain glistening, strongly refracting granules which are soluble in bisulphide of carbon, alcohol, xylol, and the alkalies. They consist of pure sulphur, and are products of the reduction of sulphur compounds found in the waters in which they abound.

Again, in certain bacteria the cell plasma may be tinged with color, either green by a chlorophyl-like substance, as in *Bact. viride*, and *B. virens*, Van Tieghem, and *Bact. chlorinum*, Engelmann; or by a violet-brown pigment known as *bacteriopurpurin*, as in *Chromatium Okenii*, *Rhabdochromatium fusiforme*, and other species described by Winogradsky.

B. The Capsule

The capsule consists of an inner tougher portion immediately surrounding the central body, and which gradually passes into a thinner and more watery outer portion which is uncolored by ordinary staining methods.

This outer portion of the capsule is furthermore so delicate in structure that it is easily destroyed or altered. Thus in the drying of films upon cover-glasses it shrinks to a fraction of its original thickness, or when in contact with water is subject to dissolution. Its failure to stain with the ordinary colors has caused it to be overlooked by most bacteriologists. It is well known that if bacteria are stained with aqueous analine dyes, only the central body is colored, and perhaps a portion of the denser part of the capsule. If, on the other hand, the same bacteria are stained by Löwit's method (see p. 6), it will be noted that the bacilli are as a rule larger and plumper, showing that a greater portion of the organism has been colored. If, again, such deeply stained preparations are partially decolorized with acid alcohol, the stain in the wall will be sufficiently removed to demonstrate it as distinct from the central body. This is shown in Fig. 2, E–F.

The usual method of incorporating upon the cover-glass a

small portion of an agar culture with a drop of water is faulty,
inasmuch as there is always danger of the water dissolving a
portion of the capsule. The better method is to touch the edge
of a square cover-glass to a portion of an agar culture; and
then draw this contaminated edge over the surface of a second
cover-glass, thus forming a thin film which dries instantly. The
cover-glass is then at once immersed in a 4 per cent forma-
line solution to fix the film. The latter can then be stained by
Löwit's method, which the author has found preferable to that

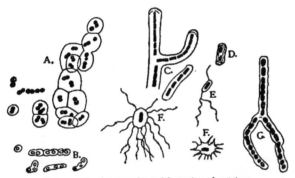

FIG. 2. — Showing capsules and formation of zoöglœa.

A. *Streptococcus mesenterioides* after Zopf. B. *Streptococcus capsulatus* after Binaghi.
C. *Bact. anthracis* after Babes. D. capsule bacillus, Babes. E. a bacillus simulating
typhoid, Babes. FF. *B. typhosus* after Babes and Löwit. G. *Bact. Pasteurianum* after
Hansen.

of Löffler (see p. 6). The form of the enveloping capsule
varies in different species: in Fig. 2, E, it is rather thick and
symmetrical with the central body; at F, of the same figure, the
capsule shows a number of angular processes to which individual
flagella are attached. The thickness of the capsule is governed
in certain cases by the chemical properties of the medium in
which the organism grows. Thus *Bact. pneumoniæ* when found
in the body fluids shows a much thicker capsule than when
grown on ordinary culture media. In certain species, as *Strep-*

tococcus mesenterioides, the capsules become enormously swollen by the imbibition of water. The capsules of adjacent individuals also coalesce, forming a common gelatinous envelope surrounding a number of individuals, as in A and D, Fig. 2. When these individuals are arranged in chains we have forms, as

FIG. 3. — Showing false branching.

A. in *Cladothrix dichotoma* after Fischer. B. a fungus filament. C. *Mycobact. tuberculosis* after A. Coppen Jones. D. *Mycobact. influenzæ* after Grassberger. E. bacteroid bodies of leguminous root tubercles after Beijerinck. F. an infecting filament of the latter after Atkinson.

in B, C, and G, of the same figure. In certain of the higher bacteria, as in *Cladothrix dichotoma,* Fig. 3, A, we have chains of individuals surrounded by a capsule, which at first sight is identical with that in *Bact. Pasteurianum,* Fig. 2, G, but in Cladothrix the capsule is firmer and of the nature of a membrane or sheath.

C. The Flagella

Certain genera of bacteria are provided with hairlike processes known as flagella. They are simply filamentous extensions of the capsule, and proceed from the latter and not from the central body. They vary in thickness from extremely delicate hairs, scarcely discernible with the highest powers of the microscope,

to thick sturdy filaments. They are of uniform thickness, or
may show slight nodular swellings, as in E–F, Fig. 2. They
may be continuous or branched. It is to the lashing movement
of these organs that the bacterium owes its motility. In the
genus Bacterium of Migula flagella are absent, and the organisms show no progressive motility, but as a rule only a vibratory
motion, *the Brownian movement.* In certain non-flagellated
forms a slow rotatory or squirming motion may result from successive dilation and contraction of the membrane. The flagella
vary as to their arrangement; and on this is based the classification of the genera Bacillus and Pseudomonas. In Bacillus the
flagella are *peritrichic;* that is, they may originate from any part
of the capsule, and frequently surround the organism, as in F,
Fig. 2. In Pseudomonas the flagella are polar or bipolar; that
is, a single flagellum may arise from one or both poles of the rod,
as in E, Fig. 2. Fischer distinguishes two types of polar flagella,
i.e. monotrichic, where they occur singly as in Pseudomonas,
and *lophotrichic,* where they occur in tufts of two or more, as in
Spirillum.

D. The Staining of Flagella

This is a matter requiring the greatest skill, and but few
bacteriologists are uniformly successful. Certain precautions
are essential to good results. In the first place the cover-glasses
must be absolutely clean and free from every trace of grease. If
an oese of water be placed on a cover-glass it should spread
evenly over the entire surface, and remain so; otherwise it is
not in a suitable condition. Four or five of these should be
placed in a row on a piece of black paper or tile, and on each an
oese-full of water should be deposited. A twenty-four hour
agar culture should be ready at hand. With a platinum wire,
remove a very small portion of the pure culture without touching
the underlying medium. With a single circular motion, no
more, since too much manipulation is apt to injure the delicate
flagella, mix the culture with the water on the first cover-glass,

which at the same time serves to spread the water. There should be a faint, only barely perceptible cloudiness. Then with an oese made of delicate platinum wire transfer an oese-full from the first cover-glass to the second, and with a single circular movement spread the drop as before into a thin layer. In the same manner transfer from the second to the third cover-glass, and so again to the fourth. On the third and fourth cover-glasses the bacteria will be sufficiently scattered and few in number to remain separated. The thin watery films dry quickly in the air. They are then fixed, and this is one of the most important operations, since most manipulators overheat in fixing. To do this, hold the two opposite edges of the cover-glass between the thumb and forefinger, and pass once through the flame so rapidly that the fingers feel no pain. The films are then ready for the mordant, which in Löwit's method is prepared as follows : Dissolve 5 g. of tannic acid in 20 c.c. of water, and filter twice ; to 10 cc. of this add 5 cc. of a saturated solution of copper sulphate, and 1 cc. of a saturated alcoholic solution of fuchsin. Filter the mixture twice. The necessary quantity of the mordant is then placed upon the films and allowed to act for 2–3 minutes in the cold. The films then are very thoroughly washed in water, and are ready for staining. The stain is the ordinary analine water gentian violet or fuchsin. The staining is done cold for 3–5 minutes. The films are then thoroughly washed in water, or for a few seconds in 50 per cent alcohol. They are then mounted and examined as usual. One important point is that all the solutions should be fresh and carefully filtered. This applies as well to the tannic acid solution. The copper sulphate solution is of course durable.

2. THE FORMS OF BACTERIA

Bacteria present a great variety of forms. There are, however, certain morphologic types to which the majority of them conform. The most common of these are *cocci* or spherical bac-

teria, Fig. 4, *a ; rods*, in which the length exceeds the diameter,
Fig. 4, *b–d ; filaments*, greatly elongated rods, Fig. 4, *e ; commas*
or curved rods, Fig. 4, *f ; spirals* or serpentine forms, Fig. 4, *g, h, t.*

Rods may be still further divided into *ovals*, in which the
length scarcely exceeds the diameter, Fig. 4, *b ; short rods*, in
which the length is 2–4 times the diameter, Fig. 4, *c ;* and *long
rods*, in which the length is 4–8 times the diameter, Fig. 4, *d.*

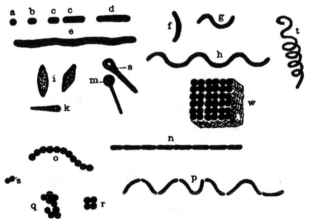

FIG. 4. — Showing forms and grouping of bacteria.

a cocci, *b* ovals, *c* short rods, *d* long rods, *e* filaments, *f* commas, *g* short spiral, *ht* long spiral, *i* clostridium forms, *k* cuneate forms, *s* clavate forms, *m* capitate forms, *n* streptobacilli, *o* streptococci, *s* diplococci, *q* staphylococci, *r* tetrads, *p* streptospirilli, *w* sarcina.

Spirals may also be divided in *short spirals*, the wave spiral,
Fig. 4, *g ;* and *long spirals*, which are multiples of the former,
Fig. 4, *h* and *t.*

Special forms may also be noted which commonly appear
during sporulation and as involution forms.   These are: *clostridium* forms, or rods swollen in the centre and attenuated at
both ends, Fig. 4, *i ; cuneate* forms, wedge-shaped, enlarged
at one end and gradually tapering toward the other, Fig. 4, *k ;
clavate* forms, as in Fig. 4, *s ;* and *capitate* forms, as in Fig. 4, *k.*

3. THE ARRANGEMENT OF BACTERIA

Bacteria may occur singly or be aggregated into groups. A unilateral arrangement of individual cells is called a chain. Chains of cocci are called *streptococci*, Fig. 4, *o;* chains of bacilli, *streptobacilli*, Fig. 4, *n;* and chains of spirals, *streptospirilli*, Fig. 4, *p.* Bacteria may be arranged in twos. Thus cocci in twos are *diplococci*, and bacilli in twos *diplobacilli.*

Cocci in irregular groups are designated as *staphylococci*, Fig. 4, *q;* in fours as *tetracocci;* in larger rectangular plates as *merismopedia;* and in cubical packets as *sarcina*, Fig. 4, *w.*

4. THE VEGETATING GROWTH OF BACTERIA

In the ordinary vegetative growth of bacteria multiplication is by fission. The exact process of cell division for the bacteria is not known, but is probably similar to that of the lower filamentous algæ, as illustrated by Strasburger in Cladophora, Fig. 5. Here, as seen at (*a*), a narrow ring of cellulose forms upon the cell wall, which gradually extends inward until it closes in the centre, a process which is nearly completed at (*b*). Bacteria multiply in geometric ratio, and under optimum

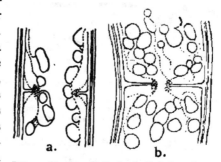

FIG. 5.—Showing cell division in *Cladophora fracta* after Strasburger.

conditions of temperature and nutrition a single cell of *B. subtilis* will divide in 30 minutes, and of the cholera *Microspira* in 20 minutes. At this rate a single cholera organism will in 24 hours produce a billion trillion individuals. In certain coccoid and rod-shaped bacteria division may take place in only one direction

of space, forming chains when the individuals remain adherent or
single individuals when they become separated. In others, as in
certain Coccaceæ, division may take place first in one direction,
forming two adherent hemispheres, as in the gonococci, followed
by a subsequent division of the two hemispheres in a direction at
right angles to the first, resulting in groups of four, *tetracocci*,
or in larger quadrangular plates as in Merismopedia. In Sarcina
division takes place at right angles to one another, forming
cubical packets, as in Fig. 4, *w*. In the Bacteriaceæ proper the
organisms are unicellular and unbranched. The union of individuals into groups, chains, or zoöglœa, even when there is apparent branching, as in Fig. 2, G, offers no morphologic difficulties,
inasmuch as each individual in the group is biologically distinct.
In the Mycobacteriaceæ the cellular protoplasm, on the other
hand, may show true dichotomous branching, as seen in tubercle,
diphtheria, and in influenza bacilli (see Fig. 3, *c, d, e, f*). Here
again the branching is quite distinct from the false branching
of Cladothrix, inasmuch as in the former the protoplasm in the
body and in its branches is continuous, while in Cladothrix,
Fig. 3, *a*, this continuity is absent. On the contrary, we have a
chain of individuals, one of which is thrust to one side and out
of line with the others, and which continues to elongate and
divide in the new direction, thus producing a false branching.
Furthermore, in the Mycobacteriaceæ the individual organisms,
whether simple, either rods or filaments, or branched, are continuous or unicellular. In the vegetating body of the true
fungi, which are most closely allied to the Mycobacteriaceæ,
while there is the same character of branching, the filaments
are longer and generally divided by septæ. The higher Leptothrix and the lowest Hypomycetes are, however, so closely
related that it is difficult to draw a hard and sharp line between the two groups.

5. THE FORMATION OF ENDOSPORES

Under unfavorable conditions of environment, as deficiency of food supply, unfavorable temperature or drying, bacterial cells may produce within their substance strongly refracting, roundish, or oval bodies known as endospores. The formation of endospores in *Bact. anthracis* consists in a contraction of the cellular protoplasm which collects as a naked body toward one pole of the rod, Fig. 6, H. This naked mass continues to contract through loss of water of imbibition, and becomes more strongly refracting and

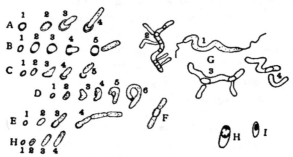

FIG. 6. — Showing methods of spore germination.

A. Polar germination of *B. butyricus* after Prazmowski. B. Equatorial germination of *B. subtilis* after Prazmowski. C–D. Equatorial germination of *B. tumescens* (c) and of *Bact. carotorum* after A. Koch. E–F. Polar germination of *Bact. sessile* after L. Klein. H. Germination by absorption of *B. anthracis* after De Bary. G. Endogermination in *Spirillum endoparagogicum* after Sorokin. I–K. Spore formation in *Bact. anthracis* after Migula.

symmetrically oval in form. Here it becomes invested with a capsule, where it remains enclosed in the empty cell until by the dissolution of the latter it is set free. In *B. subtilis* the cell plasma becomes granulated, and later a number of minute strongly refracting granules appear, which collect and aggregate themselves at one pole. These by their coalescence produce a dense mass which by later investing itself with a membrane becomes a mature spore. A spore consists of a central highly refracting, nearly water free, protoplasm, surrounded by a thick tough membrane, which, according to Burchard, is composed of two layers, an inner darker and denser portion and an outer bright delicate layer. The membrane is not equally thick throughout, but in the greater number of species, according to Migula, is thinnest at the poles, although in a number of species the polar portions of the membrane are thickened, and the equatorial portions thin. Since in germination the thinner portions constitute the *locus minoris resistentiæ*, the structure of the spore wall will determine the character of the germination, as will be explained in the next paragraph.

Spore Germination. Under favorable conditions of environment spores germinate and develop into vegetative forms. The method of germination varies in different species and groups of species, a point which is likely to have considerable taxonomic value. The first requisite of spore germination is the presence of moisture. It has been stated that the protoplasm of the spores is nearly or quite water free, but it is a common property of all protoplasm to absorb water, which causes it to swell. Thus the first process in spore germination is the enlargement of the spore to double its former dimensions. The tension thus produced causes the rupture of the wall of the spore at its thinnest portion, be this at one or both poles or at the equator. With the rupture of the wall comes the protrusion of its contents in the form of what may be called the *germinal rod*.

The methods of spore germination may be designated as follows : —

    1. Polar germination.

    2. Equatorial germination.

    3. Germination by absorption.

    4. Endo-germination.

Polar germination has been illustrated by Prazmowski in *Clostridium butyricum,* and by Klein in *Bact. sessile.* Here the

spore swells and becomes less strongly refracting; then one end
of the spore ruptures and an opening forms, as if by absorption
of the spore membrane at this point, and the germ rod emerges.
The latter then elongates, and the membrane either remains
attached to the rod or is cast off. The process is shown in
Fig. 6, A and E, and the figures 1–4 show the successive stages.
In *Bact. sessile*, as shown at F in the same figure, the germinal
rod proceeds from both poles of the spore. This may be designated *bipolar*, as distinguished from unipolar germination in
the former species. **Equatorial germination** is illustrated in *B.
subtilis* by Prazmowski, and in *B. tumescens* and in *B. carota-
rum* after A. Koch. Here the process is as before, except that
the germinal rod emerges in a direction at right angles, or ob-
liquely to the longer axis, or in a generally equatorial direction,
as shown in Fig. 6, B, C, and D. **Germination by absorption** is
illustrated in *B. leptosporus* by Klein, and in *Bact. anthracis* by
De Bary. In the former species the spores are long, strongly
refracting, and surrounded by a wide capsule. At the beginning
of the germination the spore thickens, then elongates, and becomes darker at the poles; the membrane gradually disappears
by absorption and the spore passes into a vegetating rod. A
similar process has been noted in *Bact. anthracis*, see Fig. 6, H,
1, 2, 3, 4. **Endo-germination.** According to Sorokin, *Spirillum
endoparagogicum* shows a unique method of germination. Here
the spores germinate within the body of the parent cell, as seen
in Fig. 6, G. The germinal rod then becomes detached, leaving
the empty capsule within the parent.

The Study of Spore Germination. Students should endeavor
to study the method of spore germination in all species which
they are led to investigate. A satisfactory method is to intro-
duce a quantity of spores into a small portion of bouillon, and then
make cover-glass preparations of the latter at intervals of 30
minutes, using Löffler's alkaline methylene blue, without heat,
as a stain.

6. THE FORMATION OF GONIDIA

In the higher bacteria, as in Mycobacteriaceæ, the cells or filaments may undergo multiple segmentation, resulting in the
formation of numerous short rods or coccoid forms, which par-
take of the nature of *gonidia*.

They may be termed resting bodies, inasmuch as they lie dor-
mant for a greater or less period until a favorable environment
causes them to elongate and produce the original vegetative form
from which they sprang.   Certain bacteriologists have considered
many of these so-called gonidia as degeneration forms; but it is
more likely that they are distinct morphologic elements, inasmuch as degenerative elements could not be expected to produce
new vegetative cells.   Thus, if one has ever searched in old
tuberculous lesions for the presence of tubercle bacilli, one must
have been struck with the complete absence of typical bacilli;
and yet it is well known that such tuberculous matter when
injected into guinea pigs will produce tuberculosis.   The only
explanation is that these granular particles are resting bodies of
the nature of gonidia, which are capable of reproducing the
species.   The great resistance of diphtheria germs to unfavorable conditions, as drying, conditions which rapidly destroy the
vegetative cell, makes it likely that the granular segments which
they often produce are of the nature of gonidia.   According to
A. Coppen Jones, tubercle bacilli produce gonidia.   These
stain more deeply than the vegetating portions, and more
strongly resist the decolorizing action of acid.   They are shown
in Fig. 7, E.   In Streptothrix we have frequent instances of the
formation of gonidia.   In *Streptothrix bovis* the filaments or
chains of filaments which are enclosed in a common capsule
undergo multiple segmentation, producing coccoid bodies which
at first remain enclosed within the sheath, and finally escape
from the dissolution of the latter as seen in Fig. 7, A.   In
Streptothrix chromogena, the branched filaments show multiple

constrictions and become elongated Monilia-like forms; later, the roundish gonidia become abstricted, and separate from the parent chain as seen in Fig. 7, B. In *Cladothrix dichotoma*, multiple segmentation occurs within the sheath, and oval gonidia are produced which escape from one pole of a fila‑ ment, either as motile swarm spores, or as simple non-motile

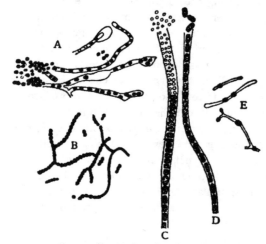

FIG. 7.— Showing formation of gonidia.

A. *Streptothrix bovis* after Lehmann-Newmann. B. *Streptothrix chromogena* after Macé. C–D. *Crenothrix Kühniana* after Zopf; *c* microgonidia, *d* macrogonidia. E. Gonidial bodies in *Mycobact. tuberculosis* after A. Coppen Jones.

bodies as seen in Fig. 3, A. In *Crenothrix Kühniana*, the pro‑ cess is the same as in the preceding, except that there are two classes of gonidia, the larger macrogonidia, as in Fig. 7, D, and the smaller microgonidia, as in Fig. 7, C. Gonidia germinate by the simple elongation of the cell, which by continued growth develops into a vegetative filament.

CHAPTER II

THE STUDY OF THE CULTURAL CHARACTERS AND BIOCHEMICAL FUNCTIONS OF BACTERIA

It is important that students shall study any new or old bacterial forms over a sufficiently long period to fully establish their characters. To publish a description of a bacterium after a study of one or two generations is the height of superficiality. An organism should be observed over a considerable period of time, at least until its characters become fixed and constant. Forms freshly isolated from a natural habitat, as soil or water, frequently show certain modifications of their characters after successive cultivation on artificial media ; so much so, that a description of a species in its early generations may differ rather widely from those of far later periods. Eventually the organism, conforming to its new environment, will establish characters which are reasonably constant. These variations, should they occur, need to be embodied in a description, or such a sufficient range given to the descriptive characters as to include said cultural variations. It is a familiar fact that slight differences in the condition of the medium will modify the macroscopic characters of a growth. Thus, in milk cultures, it may take very little to disturb the balance between an unchanged appearance of the medium and the formation of a coagulum. Indol formation may also become an uncertain factor in the biochemistry of an organism. Gelatin colonies in their microscopic characters are open to wide variations, so much so that it is a question whether they have any great value in species differentiation. In fact, even the macroscopic appearance of a gelatin colony may be

16

modified by rapidity of growth, a slowly growing organism forming a convex growth of small diameter, which becomes flatter
and more spreading with a more rapid development. Again, it
is important to decide as to those characters which shall possess
the greatest taxonomic value. The more profound chemical
changes induced in media probably have more value than the
microscopic appearances of growths. Among these are liquefaction or non-liquefaction of gelatin, proteolytic action, the
fermentation of carbohydrates, diastatic action, the reduction of
nitrates to nitrites, the formation of volatile and fixed organic
acids, together with the fermentation of definite carbohydrates
in the culture media; in the case of the production of lactic
acid, whether the latter is optically inactive, or active, and if the
latter, whether right or left handed in its action on the polarized ray. These and other investigations on the biochemistry
of bacterial species should demand greater attention. It means
that the bacteriologist must familiarize himself with chemical
methods, since in the future the study of the chemical functions
of bacteria will form a most important factor in species differentiation.

1. THE TERMINOLOGY OF DESCRIPTIVE BACTERIOLOGY

An important desideratum in descriptive bacteriology is the
adoption of a system of terminology. Many of the descriptions
of cultural characters are unnecessarily verbose. A few well-
defined terms will suffice to express as much as several sentences
of descriptive matter, and with greater exactness. Some examples will suffice. A description of *Bact. mycoides* reads : " In
gelatin stick cultures an outgrowth of branching filaments occurs
along the line of puncture, looking like a small fir tree turned
upside down." One term, *arborescent*, will express the phenomenon without this unnecessary verbosity, understanding an
arborescent growth to be one typically represented by this species.
Furthermore, the term arborescent is sufficiently elastic, and yet

c

sufficiently definite, to fit this and other related structures, for it
is a question whether every bacteriologist can always see a fir
tree turned upside down as he looks at a gelatin stab culture of
Bact. mycoides. Flügge compares the growth in gelatin stab of
the bacillus of the mouse septicæmia "to the brush bristles
used for cleaning test tubes." Such methods of description
may be realistic and wonderfully exact, but it would be better to
have a term, as *villous*, to express this particular type of structure whenever it occurs. Such terms would also possess a
certain elasticity of meaning, more generally applicable to
different cultures of the same organism than comparisons to fir
trees and test tube brushes. Furthermore, the appearance,
especially under the microscope, of colonies and growths is subject to such minor variations, to say the least, that the very exact
and detailed descriptions which we often read possess no value
except as a perfect word picture of the particular colonies or
what not, which the writer may happen to have observed at
some particular time; and while these exact descriptions are
useful in laboratory notes, with the view of eventually drawing
up a final average description, they are misleading to others,
who look for exact duplicates in their observations. A recent
description of the gelatin colonies of *Bact. mycoides* reads: "After
twenty-four hours the colonies appear as hazy, ill-defined spots,
with small, indistinct, slightly denser centres. On close in-
spection, they are seen to consist of a loose felt-work. The gelatin is liquefied in a short time. Under a low power, a network
is seen, formed of very long, hair-like filaments, which are sometimes straight and sometimes delicately undulating, running in
all directions and crossing one another at all angles. Toward
the centre of the spot the network is somewhat denser, and
here a dark, well-defined nucleus may be found. If the colonies
are few in number, they may very soon attain a diameter of a
centimetre or more." This is a fair sample of the verbosity
often found in bacterial descriptions. The author had in mind

a certain picture which he was anxious to paint as accurately as possible; but, with no system of terminology at hand, he found it necessary to write a whole composition. All that there is in this description can be expressed briefly as follows: Gelatin colonies, 24 hours, macroscopically, thin, emarginate, filamentous, approaching 1 cm. or more in diameter. Gelatin rapidly liquefied. Microscopically, centres dense, floccose; borders filamentous. Here the simple terms *floccose* and *filamentous*, each having definite meaning, express what was embodied in several sentences of descriptive matter. These illustrations might be multiplied indefinitely, but it will be sufficient to append a series of tables proposing certain terms which may be useful in descriptive bacteriology.

CHARACTERS OF BACTERIAL CULTURES

I. Gelatin Stab Cultures.
  A. Non-liquefying.
    Line of puncture.
      Filiform, uniform growth, without special characters. Fig. 8, 1 B.
      Nodose, consisting of closely aggregated colonies.
      Beaded, consisting of loosely placed or disjointed colonies. Fig. 8, 2 B.
      Papillate, beset with papillate extensions.
      Echinate, beset with acicular extensions. Fig. 8, 3 B.
      Villous, beset with short, undivided, hair-like extensions. Fig. 8, 5 B.
      Plumose, a delicate feathery growth.
      Arborescent, branched or tree-like, beset with branched hair-like extensions. Fig. 8, 4 B.
  B. Liquefying.
      Crateriform, a saucer-shaped liquefaction of the gelatin. Fig. 9, 1.
      Saccate, shape of an elongated sack, tubular, cylindrical. Fig. 9, 3.
      Infundibuliform, shape of a funnel, conical. Fig. 9, 4.
      Napiform, shape of a turnip. Fig. 9, 2.
      Fusiform, outline of a parsnip, narrow at either end, broadest below the surface.
      Stratiform, liquefaction extending to the walls of the tube and downward horizontally. Fig. 9, 5.

II. Stoke Culture (see plate cultural characters).

III. Plate Cultures, colonies.

    A. Form.

        Punctiform, dimensions too slight for defining form by naked eye,
           minute, raised, semi-spherical.

        Round, of a more or less circular outline.

        Irregular.

        Elliptical.

        Fusiform, spindle-shaped, tapering at each end.

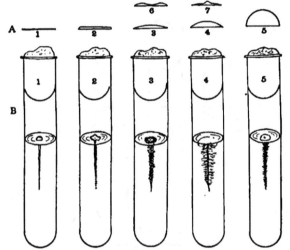

FIG. 8. — Showing characters of gelatin stab cultures.

A. Characters of surface elevation : 1 flat, 2 raised, 3 convex, 4 pulvinate, 5 capitate, 6 umbilicate, 7 umbonate. B. Characters of growth in depth : 1 filiform, 2 beaded, 3 tuber-
culate-ecinulate, 4 arborescent, 5 villous.

        Cochleate, spiral or twisted like a snail shell. Fig. 10, A.

        Ameboid, very irregular, streaming. Fig. 10, B.

        Mycelioid, a filamentous colony, with the radiate character of a
           mould. Fig. 11, D.

        Filamentous, an irregular mass of loosely woven filaments. Fig.
           11, E.

        Floccose, of a dense woolly structure.

Rhizoid, of an irregular branched, root-like character, as in *Bact. mycoides*. Fig. 10, C.

Conglomerate, an aggregate of colonies of similar size and form. Fig. 12, A.

Toruloid, an aggregate of colonies, like the budding of the yeast plant. Fig. 12, B.

Rosulate, shaped like a rosette.

B. Surface Elevation.

   1. General character of surface as a whole.

     Flat, thin, leafy, spreading over the surface. Fig. 8, A 1.

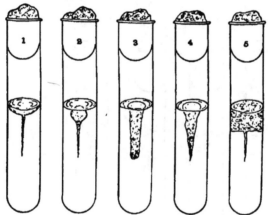

FIG. 9. — Showing types of liquefaction in gelatin stab cultures.

1 crateriform, 2 napiform, 3 saccate, 4 infundibuliform, 5 stratiform.

     Effused, spread over the surface as a thin, veilly layer, more delicate than the preceding.

     Raised, growth thick, with abrupt terraced edges. Fig. 8, A 2.

     Convex, surface the segment of a circle, but very flatly convex. Fig. 8, A 3.

     Pulvinate, surface the segment of a circle, but decidedly convex. Fig. 8, A 4.

     Capitate, surface hemispherical. Fig. 8, A 5.

   2. Detailed characters of surface.

     Smooth, surface even, without any of the following distinctive characters.

Alveolate, marked by depressions separated by thin walls, so as to resemble a honeycomb. Fig. 12, C.

Punctate, dotted with punctures like pin-pricks.

Bullate, like a blistered surface, rising in convex prominences, rather coarse.

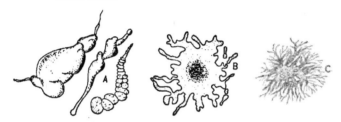

FIG. 10. — Types of colonies.

A. Cochleate. B. Ameboid. C. Rhizoid. F. Curled structure.

Vesicular, more or less covered with minute vesicles due to gas formation more minute than bullate.

Verrucose, wart-like, bearing wart-like prominences.

Squamose, scaly, covered with scales.

Echinate, beset with pointed prominences.

Papillate, beset with nipple or mamma-like processes.

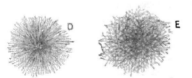

FIG. 11. — Types of colonies.

D. Mycelioid. E. Filamentous.

Rugose, short, irregular folds, due to shrinkage of surface growth.

Corrugated, in long folds. due to shrinkage.

Contoured, an irregular but smoothly undulating surface, like the surface of a relief map.

Rimmose, abounding in chinks, clefts. or cracks.

C. Internal structure of colony (microscopic).
  1. *Refraction weak*, outline and surface of relief not strongly defined.
  2. *Refraction strong*, outline and surface of relief strongly defined; dense, not filamentous colonies.
  General.
    Amorphous, without definite structure as below specified.
    Hyaline, clear and colorless.
    Homogeneous, structure uniform throughout all parts of the colony.
    Homochromous, colony uniform throughout.
  Granulations or *Blotchings.*
    Finely granular.
    Coarsely granular.
    Grumose, coarser than the preceding, a clotted appearance, particles in clustered grains. Fig. 12, D.
    Moruloid, having the character of a morula, segmented, by which the colony is divided in more or less regular segments. Fig. 12, E.
    Clouded, having a pale ground, with ill-defined patches of a deeper tint. Fig. 12, F.
  Colony Marking or *Striping.*
    Reticulate, in the form of a network, like the veins of a leaf. Fig. 12, G.
    Areolate, divided into rather irregular, or angular, spaces by more or less definite boundaries.
    Gyrose, marked by wavy lines, indefinitely placed. Fig. 12, I.
    Marmorated, showing faint, irregular stripes, or traversed by vein-like markings, as in marble. Fig. 12, H.
    Rivulose, marked by lines, like the rivers of a map.
    Rimmose, showing chinks, cracks, or clefts.
  Filamentous Colonies.
    Filamentous, as already defined. Fig. 11, E.
    Floccose, composed of filaments, densely placed.
    Curled, filaments in parallel strands, like locks or ringlets, as in agar colonies of *B. anthracis.* Fig. 10, F.
D. Edges of colonies.
    Entire, without toothing or division. Fig. 13, *a.*
    Undulate, wavy. Fig. 13, *b.*
    Repand, like the border of an open umbrella. Fig. 13, *c.*
    Erose, as if gnawed, irregularly toothed. Fig. 13, *i.*
    Lobate. Fig. 13, *d.*

Lobulate, minutely lobate. Fig. 13, *d*.
Auriculate, with ear-like lobes. Fig. 13, *e*.
Lacerate, irregularly cleft, as if torn. Fig. 13, *f*.

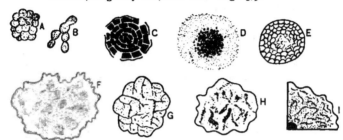

FIG. 12. — Structure of colonies.

A. Conglomerate colony. B. Toruloid colony. C. Alveolate structure. D. Grumose in centre. E. Moruloid. F. Clouded. G. Reticulate. H. Marmorated. I. Gyrose.

Fimbriate, fringed. Fig. 13, *g*.
Ciliate, hair-like extensions, radiately placed. Fig. 13, *h*.
Tufted.
Filamentous, as already defined.
Curled, as already defined.

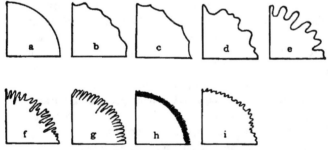

FIG. 13. — Showing characters of borders of colonies.

a entire, *b* undulate, *c* repand, *d* lobate-lobulate, *e* auriculate, *f* lacerate, *g* fimbriate, *h* ciliate, *i* erose.

E. Optical characters (after Shuttleworth)

Transparent, transmitting light.

Vitreous, transparent and colourless.

Oleaginous, transparent and yellow; olive to linseed-oil colored.

Resinous, transparent and brown; varnish or resin colored.

Translucent, faintly transparent.

Porcelaneous, translucent and white.

Opalescent, translucent, grayish white by reflected light, smoky brown by transmitted light.

Nacreous, translucent, grayish white, with pearly lustre.

Sebaceous, translucent, yellowish or grayish white.

Butyrous, translucent, and yellow.

Ceraceous, translucent, and wax-colored.

Opaque.

Cretaceous, opaque and white, chalky.

Dull, without lustre.

Glistening, shining.

Fluorescent.

Iridescent.

2. CULTURE MEDIA

The Reaction of Media

The reaction of media is a question of primary importance, since variations in their *titre* will often produce marked differences in the micro and macro characters of a growth. Hence it is important to work with media whose reaction is accurately known and uniformly the same. Phenolphthalein has been generally found to be the best indicator. A medium which is alkaline to litmus may be acid to phenolphthalein, showing that there are present in such media substances possessing an acid character which the litmus fails to indicate. ' These substances are weak organic acids and organic compounds, theoretically amphoteric, but in which an acid character predominates. Furthermore, the dibasic phosphates (Na_2HPO_4), present in considerable quantities in culture media, react alkaline to litmus, but neutral to phenolphthalein. Hence, if the acid phosphates are to be entirely neutralized, the medium must be made more than neutral to litmus. Exact neutralization can therefore only be determined with phenolphthalein. For the titration of media,

one-tenth normal sodium hydrate and hydrochloric acid solutions should be available; also, a 0.5 per cent solution of phenolphthalein in 50 per cent alcohol. Care should be taken to prevent the absorption of carbon dioxide by the soda solution, by arranging that all air which comes in contact with the latter, either in the stock bottle or in the burette, shall first pass through a strong solution of sodium or barium hydrate. The arrangement of the apparatus is described in any work on chemical analysis. The medium which has been previously boiled for at least several minutes to expel carbon dioxide is brought to the desired volume with water, and thoroughly mixed. Media are commonly warm or hot when measured, hence it must be remembered that true volumes cannot be thus obtained; for instance, a litre measured at, say 80° C., would be only 973 cc. if measured at 20° C., the temperature at which litre flasks are calibrated. Since many media cannot be cooled to 20° C. because of solidification, as in the case of agar or gelatin, it is a better plan to determine measures of volume by weight. For this, place a clean dry saucepan, in which the medium is to be prepared, upon one side of a trip scale, and counterbalance its weight exactly. The weight of a litre of bouillon, gelatin, or agar, having been determined once for all, the necessary weights added to the weight of the pan will give the amount which the pan and its contents must balance when the volume is exactly one litre. A portion of the medium brought to the exact volume is then taken and cooled to room temperature (20° C.), or to a point a few degrees above solidification, and 10 cc. withdrawn, placed in a small beaker, 50 cc. of distilled water and 1 cc. of the phenolphthalein solution added. If the medium is acid, the $\frac{N}{10}$ NaOH solution is then run in cautiously until a pale but decided pink color is obtained. The number of cubic centimetres of the solution used, multiplied by ten, will give the number of cubic centimetres of normal

sodium hydrate per litre necessary to effect complete neutralization.   The question as to what is the best reaction of media for
general work is not an easy one to settle, and one on which
bacteriologists differ.   Reactions are now commonly expressed
by plus or minus signs, the former representing an acid and the
latter an alkaline condition, the number following the sign
representing the percentage of normal acid or alkali present in
the medium.   Thus + 1.5 would indicate that the medium contained 1.5 parts per 100 or 1.5 per cent of free normal acid,
while − 1.5 would indicate that the medium contained an
equivalent quantity of free alkali.   The committee of the
American Public Health Association, in 1898, adopted a medium
whose titre was + 1.5 as the best for general work.   This reaction may be adopted if suitable, but it has been found by the
writer that many bacteria completely failed to grow in media of
this reaction.   This is especially true of a large number of soil
bacteria, which almost invariably require a neutral or slightly
alkaline medium.   I would therefore recommend a medium
whose reaction is 0.5 per cent acid to phenolphthalein as one of
more general applicability.   It cannot be too strongly impressed
upon the reader that whatever the reaction, its measure should be
stated in all descriptions of cultural characters.   To obtain uniform
results it is important that media should not only have identical
reactions, but that they should be prepared according to a fixed
and uniform rule.   The method here presented is that of the
Laboratory Committee of the American Public Health Association.   In the following formula it is noticed that Liebig's meat
extract is recommended instead of fresh meat infusion, not only
because it adds simplicity to the preparation of media, but
because media of more uniform composition can be obtained.
Furthermore, I have found the meat extract quite free from
muscle sugar, so objectionable when fresh meat infusions are
used.

Bouillon

In 1000 cc. of distilled water dissolve in the cold, Witte's peptone, 10 g.; common salt, 5 g.; and 5 g. of Liebig's extract of beef. When the ingredients are in solution, titrate, using phenolphthalein as an indicator. Make neutral to phenolphthalein with sodium hydrate. Boil for 15 minutes, using a rose burner so that the flame does not impinge against the bottom of the dish; restore to the original volume 1000 cc.; titrate with phenolphthalein; adjust reaction to the final point desired; boil 5 minutes; restore to original volume; filter, tube, and sterilize.

Nutrient Gelatin

In 1000 cc. of distilled water dissolve the ingredients used for bouillon. When the ingredients are in solution, add 100 g. of the best sheet gelatin, and warm until the latter is completely dissolved. Titrate; make neutral to phenolphthalein with sodium hydrate. Add the beaten whites of two eggs, and boil for 15 minutes with frequent stirring, when the medium should be perfectly clear. Restore to original volume; titrate; adjust reaction to final point desired. Boil for 5 minutes; restore to original volume; filter, tube, and sterilize.

Nutrient Agar

Boil 15 g. of shred agar in 500 cc. of distilled water for half an hour, or until the agar is completely dissolved; restore to original volume, cool and solidify. This constitutes the so-called *agar jelly*. In 500 cc. of distilled water dissolve 10 g. of Witte's peptone, 5 g. of common salt, and 5 g. of Liebig's extract. When ingredients are in solution, titrate, and make neutral to phenolphthalein with sodium hydrate. Add the 500 cc. of agar jelly previously prepared, breaking the latter into small fragments. Boil for 5 minutes to dissolve the agar; cool to 65° C. Add the beaten whites of two eggs; boil for

10 minutes longer, when the medium should be perfectly clear. Restore to original volume. Titrate with phenolphthalein; adjust reaction to final point desired. Boil for 5 minutes; restore to original volume. Filter, tube, and sterilize. In filtering agar or gelatin make a folded filter, and wet previous to filtration with distilled water. If the latter media are perfectly limpid they will rapidly run through the paper, and a hot water funnel is unnecessary. Often the latter portion of the agar will run through slowly. This can be hastened by placing the entire filtering apparatus in an autoclave with some pounds' steam pressure. In this case the funnel should be covered with a glass plate, to prevent condensed steam being added to the medium.

Milk

It is absolutely important where milk is used as a culture medium that it should be perfectly fresh. In stale milk certain fermentative changes have already taken place which render it unsuitable. It should be as free from fat as possible by running it through a centrifugal separator. The resulting skimmed milk is then immediately tubed and sterilized. Milk is best sterilized at 100° C. for several days, as higher temperatures are liable to discolor the medium. If the milk be acid in reaction, it should first be made exactly neutral to litmus paper before it is placed in the tubes, using the necessary quantity of normal NaOH.

Litmus Milk

Litmus milk is prepared from plain milk after the latter is sterilized, by adding sufficient sterilized litmus solution to give the medium a pale blue color. It is then distributed under aseptic conditions to sterile tubes. The best kind of vessel for holding the litmus milk is the well-known Lister flask, with a side tube placed just below the neck from which the medium can be poured with slight risk of contamination.

Saccharin Bouillon

Certain bacteria produce chemical changes in bouillon containing glucose, lactose, and saccharose, consisting either in the production of the gaseous products of fermentation, or of organic acids, or both. They are prepared by adding to ordinary bouillon 1 per cent by weight of one or the other of these sugars. In the preparation of lactose and saccharose bouillon the original bouillon must be free from muscle sugar. This can easily be tested by inoculating a fermentation tube, containing the plain bouillon, with a gas-producing organism like *B. coli*. If no gas is produced, the medium is free from muscle sugar.

Potato

The potatoes are cut into cylinders with a brass cork borer of the proper size for the tubes, and a couple of inches in length, and then cut across diagonally at an angle of about 30 degrees. The pieces are then washed in running water for 12 to 18 hours, placed in tubes, and sterilized.

For the preparation of other and special media, see various text-books on Bacteriology.

3. THE STAINING OF BACTERIA

In the examination of the staining properties of bacteria, the following solutions will be necessary : —

1. *Standard alcoholic solutions* of the anilin colors, notably fuchsin and gentian violet, made by dissolving 10 g. of the dry color in 100 cc. of 95 per cent alcohol, and filtering.

2. *Standard aqueous solutions* of fuchsin and gentian violet, composed of 10 cc. of distilled water and 1 cc. of the standard alcoholic solution. These solutions should be made fresh.

3. *Löffler's alkaline methylene blue solution.* To 100 cc. of distilled water add 1 cc. of a 1 per cent caustic potash solution,

and then 30 cc. of a saturated alcoholic solution of methylene blue.   Filter.

4. *Ehrlich's anilin-water fuchsin or gentian violet.*   To 10–15 cc. of distilled water add an excess of anilin oil.   Shake vigorously for several minutes and filter.   To 10 cc. of the filtrate add 1 cc. of the standard alcoholic solution (1) and filter again.

5. *Ziehl's carbol-fuchsin solution.*   To 10 cc. of a 5 per cent carbolic acid solution add 1 cc. of the standard alcoholic solution (1), and filter.

6. *Gabbet's methylene blue solution.*   To 75 cc. of water add cautiously 25 cc. of concentrated sulphuric acid; when cool, add 2 g. of methylene blue, with frequent stirring until dissolved, and filter.

7. *Gram's mixture.*   Dissolve 2 g. of potassium iodide in 300 cc. of distilled water, and add 1 g. of iodide.   Allow the mixture to stand with occasional stirring until the iodine is dissolved.   Filter and keep in a bottle protected from the light.

Gram's Method of Staining

1. Cover film with anilin gentian violet solution (4) for 30 seconds, cold.
2. Wash film in running water.
3. Immerse in Gram's mixture (7) for 30 seconds.
4. Immerse in 95 per cent alcohol until decolorized.
5. Wash in water, and mount.

Certain bacteria stain by Gram's method, others are decolorized.   This should be noted in each species under observation.

Capsule Staining, Welch's Method

1. Cover-slip preparations, made without water, see p. 4.
2. Flood film with glacial acetic acid, and at once allow to drain off.

3. Add anilin water gentian violet solution (4) repeatedly until the acid is removed.

4. Wash briskly in a 2 per cent solution of common salt.

5. Mount in salt solution and examine.

For further directions on the staining of bacteria see p. 6.

Spore Staining, Abbot's Method

1. Stain cover-glass preparation with Löffler's methylene blue (3), heating repeatedly until staining solution boils, but not continuously, for one minute.

2. Wash in water.

3. Wash in 95 per cent alcohol containing 0.2–0.3 per cent of hydrochloric acid.

4. Wash in water.

5. Stain 8–10 seconds, with anilin-water fuchsin (4).

6. Wash in water and mount.

Ehrlich's Method for Tubercle Bacilli

1. Stain cover-glass films, 5–10 minutes, cold, with Ehrlich's anilin water fuchsin, or gentian violet (4).

2. Wash in water.

3. Decolorize in 20 per cent nitric acid, one-half to one minute.

4. Wash in 70 per cent alcohol until no more color is given off ; dry, and mount.

Ziehl-Neelsen-Gabbet Method for Tubercle Bacilli

1. Stain cover-glass films, 5–10 minutes, cold, with Ziehl's carbol-fuchsin (5).

2. Decolorize for one minute with Gabbet's methylene blue solution (6).

4. Wash in water ; dry, and mount.

Löwit's Method of staining Flágella and the Capsules of Bacteria.

See p. 6.

All bacteria should be studied as to their ability —

1. To stain with standard aqueous solutions of fuchsin or gentian violet (2);

2. With Löffler's methylene blue (3).

3. If not stained by either of the preceding, try anilin-water fuchsin or gentian violet (4), and Ziehl's carbol-fuchsin (5).

4. Whether stained or decolorized by Gram's method.

5. Whether stained or decolorized by Ehrlich's or the Ziehl-Neelsen method as used for tubercle bacilli.

4. STUDY OF THE CHEMICAL FUNCTIONS OF BACTERIA

The Production of Indol and Phenol

Indol and phenol are products of putrefaction, and are frequently produced in bacterial cultures. Their presence or absence is therefore of value in species differentiation.

Indol production. Inoculate several tubes, each containing 10 cc. of bouillon free from glucose, and test for the presence of indol after 5 and 10 days' growth. To 10 cc. of the culture add 10 drops of chemically pure concentrated sulphuric acid, and then 1 cc. of a .02 per cent solution of sodium nitrite. If a pink color develops within 10 minutes, indol is present. In recording the production of indol it is necessary to state the age of the culture, since indol may be produced in 10 days and not in 5 days. The reaction may appear almost immediately after adding the reagents, or a faint reaction may appear after long standing; hence the necessity of a time limit for the reaction to manifest itself. Again, the reaction should be allowed to develop at room temperature, since a culture which may show no reaction in the cold may give one when heated.

Phenol production. A 100 cc. Erlenmeyer flask is connected with a condenser, and 50 cc. of a bouillon culture of the organ-

3

ism in question introduced, to which is added 5 cc. of concentrated hydrochloric acid. About 15 cc. are distilled, and the distillate divided into three portions, to be tested as follows: 1. To one portion is added a few drops of Millon's reagent, and the mixture heated to boiling; the development of a red color indicates the presence of phenol. 2. To another portion add a few drops of strong bromine water; a turbidity develops in the presence of phenol. 3. To the third portion add a few drops of a very dilute solution of ferric chloride; a violet color develops in the presence of phenol.

In recording the development of phenol in cultures, the age of the culture and temperature of growth should be stated. When indol and phenol occur together in the same culture, their separation is advisable before applying the tests. For this, the method as proposed by Hoppe-Seyler can be used. Distil 200 cc. of the culture with 50 cc. of concentrated HCl until 50–70 cc. passes over. The distillate will contain both indol and phenol. Render the distillate strongly alkaline with caustic potash and distil; the indol will be found in the distillate, the phenol in the residue. When the residue is cold, saturate with carbon dioxide, and distil; the phenol will pass over into the distillate.

The Reduction of Nitrates to Nitrites

For the study of the reduction of nitrates to nitrites a special medium is desirable, composed of Witte's peptone 10 g., nitrate of soda .02 g., and water 1000 cc. It is important to have the medium originally free from nitrites, and since distilled water frequently contains considerable quantities of nitrous acid, it is better to use well or spring water, which gives no reaction for nitrites. Furthermore, nitrous acid is present in the atmosphere, some of which will be absorbed by the culture during its growth. Hence it is necessary to have blank, un-inoculated tubes kept under the same conditions as those inoculated, which shall also be tested for nitrites. For the test two solutions are necessary.

I.   Naphthylamine . . . . 1.0 g.
     Distilled water . . . . 100 cc.

II. Sulphanilic acid . . . . 0.5 g.
     Dilute acetic acid . . . . 150 cc.

These solutions are kept separate in glass-stoppered bottles. Three cubic centimetres of each of solution I and II are placed in a test-tube, and mixed. Two cubic centimetres of this mixture are added to the cultures, and the same quantity to the blank tube. The tubes should be of the same capacity, and the fluid should be of the same height in both. The tubes are then allowed to stand for half an hour, keeping them closed with rubber stoppers. If a slight pink color develops in the blank tube, it may be due to nitrous acid originally present or absorbed. If appreciable amounts of nitrates have been reduced to nitrites, the pink color in the culture tubes will be deeper, and proportionate to the quantity of nitrites present. The absence of nitrites may not indicate non-reduction of nitrates to nitrites, since the nitrites previously formed may have already been reduced to free nitrogen, or to ammonia. In that case the nitrates will have disappeared. To test the presence or absence of nitrates, evaporate 10 cc. of the culture to dryness, and add to the residue 1 cc. of phenolsulphonic acid, composed of concentrated sulphuric acid 74 cc., water 6 cc., and phenol 12 g. Dilute with water, and transfer to a Nessler jar. Add enough concentrated caustic soda solution to make alkaline, and make up to 50 cc. If nitrates are present, the contents of the jar become a decided yellow. Since bouillon alone, when treated in this way, will give a slight yellow color, it is important, in case only a faint reaction for nitrates is obtained, to have a blank test made with a simple bouillon, and compare the color tints. A marked excess of color with the culture shows the presence of nitrates.

Ammonia Production

The cultures are made in bouillon in Erlenmeyer flasks. A 200 cc. flask is connected with a condenser, in which 100 cc. of the culture is placed, together with 2 g. of calcined magnesia. Fifty cubic centimetres of distillate are collected in a Nessler jar, and 1 cc. of Nessler reagent added. If ammonia is present, the distillate assumes a yellow color, whose depth of shade is proportionate to the amount.

Since bouillon alone, when distilled in the presence of magnesia, will with Nessler's reagent give a reaction for ammonia, an equal quantity of plain bouillon should be distilled over at the same time, and its depth of color, when treated with Nessler's reagent, compared with the distillate from the culture. Certain bacteria produce ammonia in bouillon cultures, others do not. This chemical function is therefore likely to possess value in differential diagnosis.

Acid Production in Saccharine Bouillon

In bouillon containing glucose, lactose, and saccharose, acids may or may not be generated. The writer's method is to grow the organism for 5 days in 2 per cent glucose bouillon. Ten cubic centimetres are taken, 50 cc. of water added, and the mixture titrated with $\frac{N}{10}$ NaOH, using phenolphthalein as an indicator. The original titre of the medium being known, the amount of acid produced in the culture can be estimated. The results are expressed in cubic centimetres of normal soda per 100 cc. of culture.

The products of the growth of bacteria in saccharine media may include the following : ethyl alcohol, aldehyde, acetone, formic, acetic, propionic, butyric, and lactic acids. For the study of these products the following system of analysis is proposed.

In a litre flask place 500 cc. of bouillon, containing 2 per cent of glucose or lactose. Sterilise, and add 10 g. of sterile

calcium carbonate. The flask is then inoculated with the culture, and allowed to incubate at the optimum temperature for 10 days. The culture is then filtered to remove undissolved calcium carbonate. The filtrate is then made slightly acid with HCl, and then slightly alkaline with Na_2CO_3. The precipitated $CaCO_3$ is filtered off. The filtrate is then introduced into a retort of the proper size, and distilled. The distillate, 50 cc. of which should be collected, may contain alcohol, acetone, or aldehyde. To test the presence of these bodies, add to 10 cc. of the distillate 5 or 6 drops of a 10 per cent solution of caustic potash, and warm the liquid to about 50° C. A solution of iodide of potassium, saturated with free iodine, is added drop by drop, until the liquid becomes a permanent yellowish brown color. It is then carefully decolorized by adding, drop by drop, the caustic potash solution. If any of the preceding bodies are present, iodoform is gradually deposited at the bottom of the tube as yellow crystals.

For the separate detection of alcohol, aceton, and aldehyde, consult Allen's "Commercial Organic Analysis," Vol. I.

The residue in the retort will contain the volatile and fixed organic acids. To separate the volatile acids, the contents of the retort is made strongly acid with sulphuric acid, and as much distilled over as is consistent with the safety of the retort. The distillate may contain acetic, formic, propionic, and butyric acids.

The quantity of the volatile acids is determined by titration. The preceding distillate is made alkaline with baryta water, and the solution evaporated to dryness. Then add 20 cc. of absolute alcohol, and let stand for 1–2 hours, with frequent stirring. Filter, and wash with alcohol. The residue will contain barium formate and acetate, and the filtrate mainly barium propionate and butyrate, and a little formate and acetate. Evaporate the filtrate to dryness; dissolve in 150 cc. of water; saturate with calcium chloride, and distil off the pure butyric acid. Dissolve the residue on the filter in hot water;

evaporate to a small volume, and test for formic and acetic acids as follows : —

To a small portion of the test add a few drops of ferric chloride solution, and then, drop by drop, dilute ammonia water, nearly to saturation.

A red color shows the presence of acetic or formic acid. Mercurous nitrate throws down formic and acetic acids as a white precipitate. With silver nitrate acetic acid salts are thrown down in the cold as a white precipitate, while the corresponding formic acid salts are precipitated only in concentrated solution, and upon boiling. A solution of formic acid salt heated with mercuric chloride gives a precipitate of mercurous chloride and a metallic mirror after a long time, while acetic acid salts give no precipitate with mercuric chloride.

To a small portion of the distillate containing butyric acid, add a small quantity of alcohol and several drops of strong sulphuric acid. If butyric acid be present, the fragrant odor of pineapple is evolved on heating.

The residue in the retort after the distillation of the volatile acids may contain lactic, oxalic, succinic, or glycocholic acid.

Evaporate the residue to a syrupy consistency, and extract with ether by agitation in a separatory funnel. The ether dissolves the fixed organic acids. Distil off the ether until the residue has a syrupy consistency ; add water, and boil with an excess of oxide of zinc. Filter. The filtrate contains zinc lactate in solution, the residue zinc oxalate and succinate. Evaporate the filtrate to dryness. The residue is nearly pure zinc lactate.

Nencki showed that different bacteria produce different isomeric forms of lactic acid, whose zinc salts behave differently in the polariscope. One, the inactive acid, has no effect upon the polarized ray ; another, the solution of whose zinc salt rotates the polarized ray to the left, is termed *right-handed acid;* and a third whose zinc salt rotates the ray to the right is termed *left-handed acid.*

Thus, when lactic acid is present, it is important to determine

the influence of a solution of its zinc salt upon the polarized ray. The residue of zinc salt is accordingly dissolved in only enough water to fill the shorter tube of the polariscope, and its action upon the polarized ray determined in the usual manner.

The system of analysis for the organic acids is given in outline in the following diagram:—

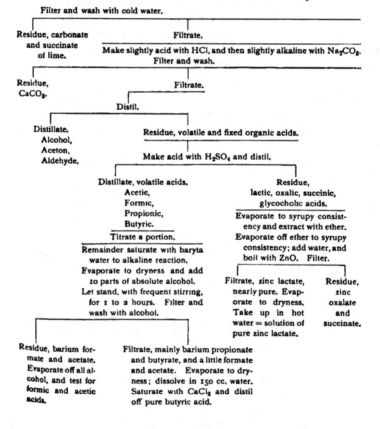

CULTURE MADE IN A ONE-LITRE FLASK CONTAINING 500 CC. OF 2 PER CENT GLUCOSE BOUILLON AND 10 GR. OF STERILE CaCO$_3$.

Filter and wash with cold water.

Residue, carbonate and succinate of lime.

Filtrate.
Make slightly acid with HCl, and then slightly alkaline with Na$_2$CO$_3$. Filter and wash.

Residue, CaCO$_3$.

Filtrate.
Distil.

Distillate. Alcohol, Aceton, Aldehyde.

Residue, volatile and fixed organic acids.
Make acid with H$_2$SO$_4$ and distil.

Distillate, volatile acids. Acetic, Formic, Propionic, Butyric.
Titrate a portion.
Remainder saturate with baryta water to alkaline reaction. Evaporate to dryness and add 10 parts of absolute alcohol. Let stand, with frequent stirring, for 1 to 2 hours. Filter and wash with alcohol.

Residue, lactic, oxalic, succinic, glycocholic acids.
Evaporate to syrupy consistency and extract with ether. Evaporate off ether to syrupy consistency; add water, and boil with ZnO. Filter.

Filtrate, zinc lactate, nearly pure. Evaporate to dryness. Take up in hot water = solution of pure zinc lactate.

Residue, zinc oxalate and succinate.

Residue, barium formate and acetate. Evaporate off all alcohol, and test for formic and acetic acids.

Filtrate, mainly barium propionate and butyrate, and a little formate and acetate. Evaporate to dryness; dissolve in 150 cc. water. Saturate with CaCl$_2$ and distil off pure butyric acid.

The Production of Proteolytic Enzymes

Certain bacteria produce enzymes which have the power of converting proteids into propeptones and peptones. Such a process takes place in the liquefaction of gelatin and blood serum, and in the peptonization of milk. To study the process, use as a culture medium either blood serum or milk which has been freed of its fat by passing through an unglazed porcelain filter. Since serum is a more difficult material to obtain in an aseptic condition, milk will serve as a better medium. This should be perfectly fresh. The porcelain filter should be sterilized for one-half an hour in an autoclave under a pressure of 5–10 pounds. The milk serum or filtrate can thus be obtained perfectly aseptic. Fifty cubic centimetres of this serum placed in a sterile flask is then inoculated with a culture and kept at the optimum temperature for 10 days. To 40 cc. of the culture add 60 g. of ammonium sulphate, and warm to 50° C. for one-half an hour. This will precipitate all the proteid bodies except the peptones. Filter; the filtrate will contain the peptones and propeptones. To test their presence add to a portion of the filtrate enough caustic potash solution to make strongly alkaline, and then a few drops of a 1 per cent solution of copper sulphate. A violet color indicates the presence of peptones.

The Production of Diastatic Ferments

Certain bacteria produce enzymes which have the property of converting starch into sugar. To test their presence, the following method is recommended : Inoculate a few tubes of bouillon, free from sugar, with the organism to be tested, and incubate for 10 days. Prepare a thin starch paste, to which is added 2 per cent of thymol. The latter must be free from sugar. Mix equal parts of the broth culture and the above paste, and place in a thermostat for 6–8 hours. Filter, and test filtrate with Fehling's solution for sugar.

The Production of Invertin Ferments

A certain few bacteria, according to Fermi, produce enzymes which have the power of converting cane sugar into glucose.

To a 2 per cent solution of cane sugar, to which has been added 2 per cent of carbolic acid, add an equal quantity of a 10 days' old culture of the organism, and allow the mixture to stand for several hours. It is then tested with Fehling's solution for the presence of glucose.

SCHEME FOR THE STUDY OF BACTERIA
Morphology

On agar, grown . . . days at . . . ° C. (standard 24 hours).

In bouillon, grown . . . days at . . . ° C.

Cocci ; single, staphlococci, diplococci, streptococci, tetracocci, sarcina.

Bacilli ; ovals, short rods, long rods, single, diplobacilli, streptobacilli, filaments.

Spirilli ; commas, short spirals, long spirals.

Size ; breadth . . . μ, extreme length from . . . μ to . . . μ, average length . . . μ.

Stain with standard watery fuchsin, easily or with difficulty, uniformly or irregularly. Record size as above. Stain by Gram's method. Stain by Löwit's method, and record size as above. Note *capsule*, form and size. Make drawings. Note presence or absence of *flagella* ; length, structure, arrangement, numbers ; monotrichic, lophotrichic, peritrichic.

Motility ; sluggish, active, rotatory, progressive.

Spores, presence or absence ; form, size, location, effect on mother cell ; clostridium, clavate, or capitate forms. Stain by Abbot's method. Heat a small portion of an old agar or potato culture, suspended in bouillon, for 10 minutes at 80° C., and determine vitality.

Spore germination ; polar, equatorial, by absorption.

Pleomorphism on media of different reaction, and of different kind.

Relative growth at 20° and 37° C. Thermal death point in . . . minutes.

Optimum temperature . . . ° C. Growth limits, maximum and minimum.

Relative growth in + 1.5, − 1.5, and 0 bouillon.

Gelatin Colonies

Deep colonies ; form, edges, size, color, internal structure.

Surface colonies ; form, surface elevation, internal structure, edges, optical characters.

Agar Colonies

Characters as before.

Stroke Cultures on Agar, Gelatin, Blood Serum, Potato

(See II, p. 20.)

Bouillon

Reaction of medium. Liquid remains clear, or becomes opalescent or turbid. Surface growth, presence or absence; membranous, coriaceous, farinaceous, gelatinous. Deposit, granular, flocculent, viscid, coherent. Odor.

Milk

Coagulated or not coagulated in . . . days at . . . ° C. Curd hard or soft, in one mass or in fragments; presence or absence of gas. Whey, presence or absence, clear or turbid. If not coagulated, does or does not curdle on boiling.

Change of consistency or color; presence or absence of peptones.

Reaction in . . . days at . . . ° C.

Litmus Milk '

Changes in color, at daily intervals of observation.

Chemical Relations

Fermentation tubes containing glucose, lactose, and saccharose bouillon. Relative growth in both arms. Presence or absence of gas. Amount of gas, measured daily. Total gas formed, in mm. of tube length. Amount after absorption with NaOH. The H CO_2 ratio.

Neutral bouillon containing 2 per cent of glucose. Determine titre at end of 5 days at . . . ° C.

Neutral bouillon containing 2 per cent of lactose. Determination as before.

Neutral bouillon free from muscle sugar and containing 0.05 per cent of nitrate of soda. Determine titre at end of 5 days at . . . ° C. Test for ammonia. Presence of nitrites, indol, phenol, diastatic, and invertin ferments.

In neutral bouillon containing 2 per cent of glucose, and an excess of CaCO₃, incubated for two weeks; make analysis of organic acids in accordance with scheme proposed on p. 39.

Pathogenesis

Inoculation of fresh bouillon cultures into mice, guinea pigs, and rabbits.

CHAPTER III

THE CLASSIFICATION OF BACTERIA

1. THE POSITION OF BACTERIA IN THE CLASSIFICATION OF ORGANIC FORMS

BACTERIA occupy a unique place in the classification of organic forms. The early view that bacteria were animal structures is no longer seriously held, since their relation to plant forms is so much closer as to leave no further doubt as to their claim to a position in the vegetable kingdom. Their evident relationship to both plants and animals places them, however, on the border between certain forms clearly on one side or the other. In certain respects bacteria are related to the Flagellata. Here the organism consists of a protoplasmic body surrounded by a thin cuticle, and provided with elongated appendages or flagella. The protoplasm contains a nucleus and a pulsating vacuole. The organism multiplies by the fission of the parent and by the endogenous formation of resting bodies or cysts. The main difference between the Flagellata and the bacteria lies in the character of the membrane. In the former it is continuous with the plasma, differing only in its physical structure; while in the bacteria the capsule, as seen in the phenomena of plasmolysis, is a distinct structure, separable from the central body.

In the Flagellata, furthermore, the flagella are definite in number in the different species, while in the bacteria their number is variable, especially when peritrichic. The Flagellata also contain a distinct nucleus and vacuole, which is not certainly the case in any of the bacteria. There is, therefore, little reason

to believe that there is any more than a remote, or at least
superficial, relationship between bacteria and their animal con-
geners, the Flagellata.   Bacteria are closely related to the
lower fission algæ, *the Cyanophyceæ*.   The Cyanophyceæ consist of a homogeneous colorless central body, which, according
to Hegler, has the structure of a true nucleus and shows indirect
karyokinetic division.   Surrounding the nucleus is a colored
peripheral layer which contains a blue-green pigment, *phycochrome*,
consisting of true chlorophyl and *phycocyanin*, or a modification
thereof.   Within the latter colored peripheral layer are small
granular bodies, *cyanophycin grains*.   Vacuoles also occasionally
occur within the cells.   The cell wall consists of cellulose, and
in certain cases may undergo a mucilaginous modification of its
outer layer.   Endogenous spores are not found, but certain cells
have their walls thickened, producing resting bodies or arthro-
spores.   The Cyanophyceæ are not progressively motile, but
show only a slow vibratory or rotatory motion due to an undulating membrane.   Flagella are not known.   From the above
description it is seen that the *Cyanophyceæ* are distinctly higher
structures than the bacteria, in possessing a distinct nucleus,
and in the function of carbon assimilation.   The greatest
morphologic difference lies in the character of the cell wall,
which, in the Cyanophyceæ, is composed of cellulose and in the
bacteria of a proteid body.   On the other hand, certain pigmented
bacteria approach closely to some of the lower Cyanophyceæ;
thus, *Bact. viride*, *B. virens* Van Tieghem, and *Bact. chlorinum*
Englemann are tinted a faint green by a substance which, with
some doubts, may be regarded as chorophyl.   Another class of
bacteria studied by Winogradsky contains a red-violet-brownish
pigment known as *bacteriopurpurin*.   This latter pigment is easily
soluble in absolute alcohol; and, according to Bütschli, if
Chromatium cells be treated with the latter, the violet pigment,
bacteriopurpurin, is removed, and there remains a greenish color
which is due to chlorophyl, or a related body.   Solutions of

bacteriopurpurin, according to Lankester, show, when examined with the spectroscope, absorption bands, as in chlorophyl and other cell pigments. Whether certain bacterial cells contain chlorophyl or related bodies with similar physiologic relationships has not been positively decided, but there is reason to believe that such is the case, and that in this is to be found a close relationship between the bacteria and the lower Cyanophyceæ. In the highest filamentous Schizomycetes there exist close relationships with the filamentous Cyanophyceæ. Thus, Beggiatoa are morphologically quite identical with Oscillatoria, except in the absence of phycocyanin in the former and its presence in the latter. In *Beggiatoa rosea-persicina*, which contain bacteriopurpurin, the relationship to *Oscillatoria* is still closer. Equally close relationship exists between Spirochæta and the fission algæ, Spirulina. In the same way Streptothrix is related to Lyngbya or Chamæsiphon; Cladothrix to Glaucothrix or Tolypothrix. Crenothrix and Phragmidiothrix, on the other hand, are unique in having no near relatives among the Cyanophyceæ. In cell-grouping the analogy between bacteria and the lower algæ is a striking one. Thus Streptococci, in the arrangement of the cells, are similar to Anabæna; Micrococcus has its analogy in Chroococcus, and more especially in the tetrad arrangement of its cells. *Micrococcus tetragenus* has its prototype in *Chroococcus turgidus*. It has been stated that the cell membrane of bacteria is a proteid body of indefinite composition, but that it may contain in certain cases, dependent upon the composition of the medium, cellulose or other carbohydrate molecules. Thus Brown believes cellulose to exist in the membrane of *Bact. xylinum*. From their investigations on *Bact. aceti*, Nageli and Löw conclude that the cell membrane of that species contains 84 per cent of ash-free cellulose. Similar observations on the presence of cellulose within the capsule have been made by Bovet on the bacterium of *Erythema nodusum*, and by Hammerschlag on *Mycobact. tuberculosis*. Thus it is seen that, in the case of certain bacteria, the cell membrane may approach in character that of the fission algæ, thus making the relationship of the two groups closer than was at first indicated. The relation between bacteria and the lower fungi, the Hypomycetes, is quite marked. Between Bacillus and Mycobacterium it is extremely close, and certain branched filamentous forms of the latter run by indistinct transitional stages into very much elongated Streptothrix forms, which are indistinguishable from certain Hypomycetes (Oidium). The cell wall of the Hypomycetes is, according to the researches of Gilson, composed of mycosin, a body related to animal chitin, which, according to Nishimura, is also represented in the capsules of bacteria. Thus, in the nature of the cell wall the bacteria are more closely related to the Hypomycetes than to the Cyanophyceæ. The true position of the bacteria, however, is one intermediate between the two groups, and this phylogenetic relationship can best be illustrated in the accompanying diagram : —

TABLE SHOWING THE RELATIONSHIPS OF BACTERIA

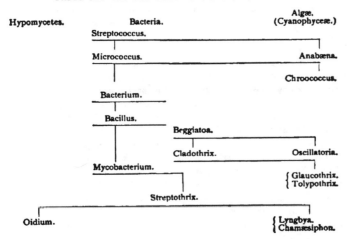

2. THE CLASSIFICATION OF THE SCHIZOMYCETES IN FAMILIES

In the present work, the sytsem of classification as first proposed by Migula in Engler and Prantl's Die Näturlichen Pflanzenfamilien, I 1 a, 1896, has been adopted, with minor modifications. The synopsis of the families is given in the following table, and of the genera under the separate family headings.

SCHIZOMYCETES

BACTERIA

Minute, one-celled, chlorophyl-free, colorless, rarely violet-red or green colored, organisms, which divide in one, two, or three directions of space, and are thus united into filamentous, flat, or cubical aggregates. Filamentous species often surrounded by a common sheath. Capsule or sheath composed in the main of proteid matter. The cell plasma generally homogeneous without a nucleus. Sexual reproduction absent. In many species resting bodies are produced either endospores or gonidia.

SYNOPSIS OF THE FAMILIES

I. Cells unbranched, or show only a false branching in Cladothrix.
   A. Cells in their free condition globular, becoming slightly elongated before division. Cell division in one, two, or three directions of space.   COCCACEAE, p. 55.
   B. Cells short or long, cylindrical, straight, curved, or spiral. Without a sheath surrounding the chains of individuals; motile or non-motile; endospores present or absent.   BACTERIACEAE, p. 117.
   C. Cells surrounded by a sheath and arranged in elongated filaments.
          CHLAMYDOBACTERIACEAE, p. 369.
   D. Cells not surrounded by a sheath, arranged in elongated filaments, and motile by means of an undulating membrane.
          BEGGIATOACEAE, p. 379.
II. Cells short or long, cylindrical or filaments, often clavate; cuneate or irregular in form. Without endospores, but with the formation of gonidia-like bodies due to the segmentation of the cells. Without flagella. Division at right angles to the axis of a rod or filament. Filaments not surrounded by a sheath as in Clamydobacteriaceae. With true dichotomous branching.   MYCOBACTERIACEAE, p. 349.

3. THE NOMENCLATURE OF SPECIES

A matter requiring more careful attention is the nomenclature of species. Little or no regard has, in many instances, been paid to the most ordinary rules of botanical nomenclature. The common rule is to express a species as a binomial, and yet in bacteriologic nomenclature we find almost as many trinomials as binomials; and quadrinomials are not infrequent. The Pneumococcus, Fränkel's bacillus, *Diplococcus pneumoniæ*, and *Micrococcus lanceolatus* are used indiscriminately, and many think it makes little difference what the organism is called, provided it is understood what is meant. There are certain rules governing the naming of species, and these should be observed. Each bacillus should be given its proper name, as determined by these rules, and it should become the practice to use such names only, and not one of its various synonyms indiscriminately. The rules of bacteriologic nomenclature should be those of the Paris code of 1867, together with those of the Botanical Club of the American Association for the Advancement of Science, adopted at the recent Rochester and Madison meetings. The writer has endeavored to apply these rules to all of the better-known species of bacteria. This has involved a careful study of the synonymy of each species, a rather laborious task, but the fulfilment of which, we trust, will result in placing bacteriologic nomenclature on a better basis. The rule of priority must be the guiding one in the naming of species, and custom or preference must yield to this inexorable law. Article 59 of the Paris code reads, "Nobody is authorized to change a name because it is badly chosen or disagreeable, or another is preferable or better known, or for any other motive, either contestable or of little import." This will require us to apply some new names to a number of familiar species. Thus Koch's bacillus of mouse septicæmia becomes *Bacterium insidiosum* (Trevisan) Migula. The reason for this is seen in the following synonymy of the species:

Bacillus der Mauseseptikämie Koch: Mitteilungen a. d. Kaiserl. Gesund-
heitsamte, I, 1881, 80.
Bacillus insidiosus Trevisan: Nouv. gen. di Batter, 1885, 10.
Bacillus murisepticus Flügge: Die Mikroorganismen, 1886.
Bacillus murinus Schröter: Die Pilze von Schlessian, 1886.
Bacterium murisepticum Migula: Engler u. Prantl, Die Natürlichen Pflanzen-
familien, 1895.

The hog cholera bacillus likewise becomes *Bacillus Salmoni*
(Trevisan), notwithstanding the fact that it is more generally
known as *Bacillus suipestifer* Kruse. The reason for this is
shown in the following synonymy : —

Bacillus of swine plague or swine fever, Klein: Report of the Local Govern-
ment Board, England, 1877-78.
Hog Cholera Bacillus, Salmon-Smith: Report U. S. Dept. Ag., 1885.
Bacillus der Schweinepest, Bang-Selander: Centralblatt f. Bakteriologie, III,
1888, 361.
Pasteurella Salmoni Trevisan: Genera, 1889, 21.
·Amerikanischen Schweineseuche, Frosch: Zeitsch. f. Hygiene, IX, 1890, 235.
Bacterium cholerae-suum Lehmann-Neumann: Bakt. Diagnostik, 1896, 233.
Bacillus suipestifer Kruse: Flügge, Die Mikroorganismen, 1896, 233.

Attention has already been called to the frequent use of trinomials in the naming of species. This is, of course, entirely
contrary to the most ordinary rule of botanical nomenclature.
A trinomial, properly speaking, is an expression for a varietal
form. Thus, *Bacillus fluorescens-undulatus* (Ravenel), if strictly
interpreted, would mean variety *undulatus*, of the species *Bacillus fluorescens*. The fact is, it stands for nothing of the sort.
Other forms of a similar character occur, as *B. fluorescens-aureus,
fluorescens-crassus, fluorescens-immobilis, fluorescens-liquefaciens,
fluorescens-longus, fluorescens-minutissimus, fluorescens-nivalis,
fluorescens non-liquefaciens, fluorescens-putridus, fluorescens-
tenuis*, and so on, in a dozen or more instances. Bacillus,
or, more properly speaking, *Pseudomonas fluorescens-undulata*
(Ravenel), should be changed to *Pseudomonas undulata* (Rave-

E

nel), since it cannot be called *Pseudomonas fluorescens*, owing to
the fact that the specific name *fluorescens* is claimed by a previously described species of this same genus. Furthermore,
Bacillus (Pseudomonas) fluorescens-liquefaciens Flügge becomes
Pseudomonas fluorescens (Flügge) Migula, on the ground that it
was probably the first member of the genus to receive the
specific appellation fluorescens. In the same way, *Pseudomonas
fluorescens-capsulata* (Pottier) becomes *Pseudomonas capsulata*
(Pottier), and *Pseudomonas fluorescens-Schuylkilliensis* (Wright)
becomes *Pseudomonas Schuylkilliensis* (Wright). Should it,
however, be demonstrated that *Pseudomonas Schuylkilliensis*
(Wright) is only a variety of *Pseudomonas fluorescens*, then, of
course, the original name becomes again applicable. Again, the
second term in the trinomial should be retained, provided no
earlier species with the same name is known. Thus, *Micrococcus
cumulatus-tenuis* v. Besser becomes *Micrococcus cumulatus*. Instances such as these might be multiplied indefinitely, but the
few examples cited will serve to illustrate the principles involved.

4. THE ARRANGEMENT OF BACTERIAL SPECIES IN GROUPS

In the following pages the writer has undertaken to arrange,
for purposes of identification, all those forms of bacteria which
have been already described with any degree of detail. The
task has proved a laborious one, and has occupied several years
of close work. The final outcome has been more or less unsatis-
factory, and yet an entirely satisfactory system of bacterial classification cannot be expected until our knowledge is greatly
extended beyond its present limits.

This much, however, has been accomplished in a provisional
way, *i.e.* the ability to cluster a considerable number of imperfectly described forms around a few reasonably defined species,
and thus to establish groups which may become the basis of
future comparative studies.

In the majority of cases the facts regarding certain species were too few to differentiate between the greater number of the members of these groups. In fact, it is highly probable that many of these so-called species which are members of any one group are not true species at all, and that at best they are only varieties of some central group organism.

The conclusion is permanent that probably nine-tenths of the forms of bacteria already described might as well be forgotten or given a respectful burial. This will then leave comparatively few well-defined species to form the nuclei of groups, in one or another of which we shall be able to place all new and sufficiently described forms.

The question of what is a species in bacteriology I shall not undertake to settle. Bacteria are so subject to morphologic, cultural, and pathologic variations that one form appears to merge into another, making distinctions often difficult; and yet a typical form — an ideal species — may exist.

Bacillus coli and *Bacillus typhosus* may present to the mind certain characters which we commonly attribute to these two types. These characters determine the species. Variation from these characters may establish varieties of one or the other *ad libitum*, forming a chain of types or races connecting the two ideal species. Nevertheless, we retain in mind the typical *B. coli* and the typical *B. typhosus* as distinct identities.

That typical forms or species of bacteria do exist no one can deny. These typical forms furthermore present certain definite morphologic, biologic, cultural, and perhaps pathogenic characters which establish the types, independent of minor variations.

The most marked of these types we select to become the centres of groups, around which are gathered all related species or varieties. The division of the bacteria into groups, so far as said grouping was possible, is outlined in the following tables : —

A PROPOSED SYNOPSIS OF GROUPS OF BACTERIA

Bacterium

I. Without endospores.
   A. Aerobic and facultative anaerobic.
      a. Gelatin not liquefied.
         * Decolorized by Gram's method.
            † Obligate aerobic. ACETIC FERMENT GROUP.
            †† Aerobic and facultative anaerobic.
               Gas generated in glucose bouillon.
                  Gas generated in lactose bouillon.
                                BACT. AEROGENES GROUP.
                  Little or no gas generated in lactose bouillon.
                                FRIEDLANDER GROUP.
               No gas generated in glucose bouillon.
                  Milk coagulated. FOWL CHOLERA GROUP.
                  Milk not coagulated. SWINE PLAGUE GROUP.
        ** Stained by Gram's method.
            † Gas generated in glucose bouillon. LACTIC FERMENT GROUP.
      b. Gelatin liquefied.
         * Colonies on gelatin ameboid or proteus-like.
                              BACT. RADIATUM GROUP.
        ** Colonies on gelatin round, not ameboid.
                              BACT. AMBIGUUM GROUP.
II. Produce endospores.
   1. No growth at room temperature, or below 22°-25° C.
                         THERMOPHILLIC GROUP.
   2. Grow at room temperatures.
      a. Gelatin liquefied. ANTHRAX GROUP.
      b. Gelatin not liquefied. BACT. FÆCALIS GROUP.

Bacillus

I. Without endospores.
   A. Aerobic and facultative anaerobic.
      a. Gelatin colonies roundish, not distinctly ameboid.
         * Gelatin not liquefied.
            † Decolorized by Gram's method.
               Gas generated in glucose bouillon.
                  Milk coagulated. COLON GROUP.
                  Milk not coagulated. HOG CHOLERA GROUP.
               No gas generated in glucose bouillon. TYPHOID GROUP.

†† Stained by Gram's method. B. MURIPESTIFER GROUP.
** Gelatin liquefied.
  † Gas generated in glucose bouillon. B. CLOACA GROUP.
  †† No gas generated in glucose bouillon. Include a large number of bacteria not sufficiently described to arrange in groups.
  b. Gelatin colonies ameboid, cochleate, or otherwise irregular.
    * Gelatin liquefied. PROTEUS VULGARIS GROUP.
    ** Gelatin not liquefied. B. ZOPFI GROUP.
II. Produce endospores.
  A. Aerobic and facultative anaerobic.
    1. Rods not swollen at sporulation.
      a. Gelatin liquefied.
        * Liquefaction of the gelatin takes place slowly. Ferment urea, with strong production of ammonia.
            URO-BACILLUS GROUP OF MIQUEL.
        ** Gelatin liquefied rather quickly.
        † Potato cultures rugose. POTATO BACILLUS GROUP.
        †† Potato cultures not distinctly rugose. B. SUBTILIS GROUP.
      b. Gelatin not liquefied. B. SOLI GROUP.
    2. Rods spindle-shaped at sporulation. B. LICHENIFORMIS GROUP.
    3. Rods clavate at sporulation. B. SUBLANATUS GROUP.
  B. Obligate anaerobic.
    1. Rods not swollen at sporulation. MALIGNANT ŒDEMA GROUP.
    2. Rods spindle-shaped at sporulation. CLOSTRIDIUM GROUP.
    3. Rods clavate-capitate at sporulation. TETANUS GROUP.

Pseudomonas Migula

I. Cells colorless, without a red-colored plasma and without sulphur granules.
  A. Grow in ordinary culture media.
    1. Without endospores.
      a. Aerobic and facultative anaerobic.
        * Without pigment.
        † Gelatin not liquefied.
          Gas generated in glucose bouillon.
            PS. MONADIFORMIS GROUP.
          No gas generated in glucose bouillon.
            PS. AMBIGUA GROUP.

†† Gelatine liquefied.
　　Gas generated in glucose bouillon.

                                      Ps. COADUNATA GROUP.
　　No gas generated in glucose bouillon.

                                      Ps. FAIRMONTENSIS GROUP.
* Produce pigment on gelatin or agar.
　† Pigment yellowish.
　　Gelatin liquefied. Ps. OCHRACEA GROUP.
　　Gelatin not liquefied. Ps. TURCOSA GROUP.
　†† Pigment blue-violet.
　　Gelatin liquefied. Ps. JANTHINA GROUP.
　　Gelatin not liquefied. Ps. BEROLINENSIS GROUP.
** Produce a greenish-bluish fluorescence in culture media.
　† Gelatin liquefied. Ps. PYOCYANEA GROUP.
　†† Gelatin not liquefied. Ps. SYNCYANEA GROUP.
2. With endospores, aerobic and facultative anaerobic.
　a. Non-chromogenic.
　　* Rods not swollen at sporulation. Ps. ROSEA GROUP.
　　** Rods swollen at one end at sporulation.

                                      Ps. TROMELSCALÄGEL GROUP.
　b. Produce a greenish-bluish fluorescence in culture media.
　　* Gelatin liquefied. Ps. VIRIDESCENS GROUP.
　　** Gelatin not liquefied. Ps. UNDULATA GROUP.
B. Do not grow in nurient gelatin or other organic media.
                                      NITROMONAS GROUP.
II. Cell plasma with a reddish tint, also with sulphur granules.
                                      CHROMATIUM GROUP.

Microspira Migula

I. Cultures show a bluish-silvery phosphorescence.
                                      PHOSPHORESCENT GROUP.
II. Cultures not phosphorescent.
　A. Gelatin liquefied.
　　1. Cultures show the nitro-indol reaction.
　　　a. Very pathogenic to pigeons. MSP. METSCHNIKOVI GROUP.
　　　b. Not distinctly pathogenic to pigeons. CHOLERA GROUP.
　　2. Nitro-indol reaction negative or very weak, at least after 24 hours.
                                      CHOLERA NOSTRAS GROUP.
　B. Gelatin not liquefied or only slightly so. MSP. SAPROPHILA GROUP.

Mycobacterium Lehmann-Neumann

I. Stain with basic aniline dyes, and easily decolorized by mineral acids when stained with carbol fuchsin.

    A. Grow well on nutrient gelatin. Gelatin liquefied very slowly or merely softened.

        1. Stain by Gram's method. SWINE ERYSIPELAS GROUP.

        2. Not stained by Gram's method. GLANDERS GROUP.

    B. Little or no growth in ordinary nutrient gelatin.

        1. Grow well in nutrient bouillon at body temperatures.

            a. Stained by Gram's method. Rods cuneate — clavate — irregularly swollen. DIPHTHERIA GROUP.

        2. No growth in nutrient bouillon or on ordinary culture media. Rods slender, tubercle-like.

            a. Stain by Gram's method. LEPROSY GROUP.

            b. Do not stain by Gram's method. INFLUENZA GROUP.

        3. No growth in nutrient bouillon or on ordinary culture media. Rods variable. ROOT-TUBERCLE GROUP.

II. Not stained with aqueous solutions of basic aniline dyes; not easily decolorized by acids. TUBERCLE GROUP.

COCCACEAE

Cells in their free condition globular, becoming slightly elongated before division. Cell division in one, two, or three directions of space.

    A. Cells without flagella.

        1. Division in only one direction of space. *Streptococcus* Billroth, p. 55.

        2. Division in two directions of space. *Micrococcus* Hallier, p. 71.

        3. Division in three directions of space. *Sarcina* Goodsir, p. 109.

    B. Cells with flagella.

        1. Division in two directions of space. *Planococcus* Migula, p. 114.

        2. Division in three directions of space. *Planosarcina* Migula, p. 115.

STREPTOCOCCUS (Billroth)

Untersuch. über die Vegetationsformen von Coccobacteria septica, 1874.
Leuconostoc, Van Tieghem: Traité de Botanique, 1883.

Cells round to slightly elongated. In short chains of only a few elements, or in long chains containing several hundred. Never in threes, tetrads, or irregular clusters. Division in only one direction of space. Non-motile. Single chains or cocci may be surrounded by a capsule.

SYNOPSIS OF THE GENUS

I. Chains on all culture media without a thick gelatinous envelope, or at most only a thin capsule.

  A. Non-chromogenic; no pigment on gelatin or agar.

    1. Do not grow at room temperatures, or at best but poorly.

CLASS I, p. 56.

    2. Grow at room temperatures, and in nutrient gelatine.

      a. Gelatin liquefied. CLASS II, p. 58.

      b. Gelatin not liquefied. CLASS III, p. 62.

  B. Chromogenic; produce a pigment on gelatin or agar. CLASS IV, p. 68.

II. Chains at least when grown in saccharine media surrounded by a gelatinous envelope. See Nos. 35, 36.

CLASS I. NON-CHROMOGENIC. DO NOT GROW AT ROOM TEMPERATURES, OR AT BEST BUT POORLY

I. On the surface of agar a scarcely discernible growth.

    1. *Str. giganteus* Lustgarten.

    2. *Str. enteritis* Hirsh.

II. On the surface of agar a visible growth.

  A. Stain by Gram's method.

    3. *Str. capsulatus* Binaghi.

    4. *Str. Cappelletti.*

  B. Decolorized by Gram's method.

    5. *Str. Kirchneri.*

1. **Str. giganteus** Lustgarten

Str. giganteus-urethræ Lustgarten.
Vierteljahresber. f. Dermatol. u. Syph., 1887, 918.

Morphology. Round, 0.8–1.0, in chains, often of many hundreds of elements, in thick tangled masses.

Agar colonies. Thin, flat, conglobate; easily overlooked, iridescent by transmitted light.

Agar slant. Growth mostly in the water of condensation.

Habitat. In the normal human urethra.

2. **Str. enteritis** Hirsh

Centralblatt f. Bakteriologie, XXII, 1897, 369.

Morphology. Similar to *Str. involutus* Kurth.

Agar colonies. Macroscopically no growth. Microscopically the colonies are small, brown, and coarsely granular.

Gelatin stab. Very little growth either in depth or on the surface.

Bouillon. Without sugar, very slight growth.

Glucose bouillon. In 12 hours at 37° C. a general turbidity; medium rendered acid. No gas produced.

Milk. No fermentation.

Pathogenesis. With white mice results variable. One cc. of culture injected intraperitoneally may give a marked diarrhœa. The organism in blood and stools.

Habitat. Isolated from stools in cases of infant diarrhœa.

3. Str. capsulatus Binaghi

Centralblatt f. Bakteriologie, XXII, 1897, 273.

Morphology. Cocci in long chains. In animal body with a capsule. Stain by Gram's method.

On gelatin and other media, no development.

Agar slant. At 37° C., transparent dewy colonies.

Pathogenesis. Guinea pigs inoculated subcutaneously with purulent matter containing the cocci show at point of inoculation a gelatinous hemorrhagic diffuse œdema. Spleen, liver, and kidneys enlarged. Inoculations with cultures give negative results.

Habitat. Isolated from a case of peribronchial pneumonia, with multiple abscesses.

4. Str. Cappelletti

Str. equi Capelletti-Vivaldi: Centralblatt f. Bakteriologie, XXV, 1899, 251.

Morphology. Cocci round-oval; occur singly, in twos and in short chains. Facultative anaerobic. Grow poorly at 20° C., and well at 24°–30° C.

Gelatin colonies. Round, delicate, yellow, granular.

Agar colonies. Watery.

Blood serum. A thin, grayish transparent layer.

Bouillon in 24 hours, 37° C., shows a sediment as gray-white flocculi.

Pathogenic. For mice and rabbits.

Habitat. Isolated from pneumonic exudate, and the spleen of three horses which died of an epidemic form of lymphatic inflammation, *gourme.*

5. Str. Kirchneri

No name, Kirchner: Zeitsch. f. Hygiene, IX, 1891, 528.

Morphology. Similar to *Str. lanceolatus,* but elements round and smaller; with a capsule.

Decolorized by Gram's method.

Agar colonies. Small, grayish white, round, transparent; becoming round grayish plaques.

Pathogenesis. Guinea pigs. One cc. of culture inoculated into pleural cavity is followed by death in 24 hours. Lungs hyperæmic; spleen not enlarged; cocci in the blood. Rabbits and mice immune.

Habitat. Isolated from sputum in influenza.

CLASS II. NON–CHROMOGENIC. GROW AT ROOM TEMPERATURES. GELATIN LIQUEFIED

I. Produce only a viscid softening of the gelatin after several weeks.
   6. *Str. Brightii* Trevisan.
II. Grow rather poorly in nutrient gelatin.
   7. *Str. enteritidis* Escherich.
   8. *Str. Bonvicini.*
   9. *Str. Fischeli.*
III. Grow well in nutrient gelatin, but slowly liquefy the medium.
   10. *St. carnis.*
IV. Grow well in gelatin at room temperatures, and rapidly liquefy the medium.
   A. Liquefaction of the gelatin takes place along the line of stab; funnel-formed — cylindrical.
      11. *Str. coli* Escherich.
      12. *Str. septicus* Babes.
      13. *Str. liquefaciens* Sternberg.
   B. Liquefaction of the gelatin takes place only on the surface; shallow, crateriform.
      14. *Str. albus* Sternberg.
      15. *Str. lactis* Sternberg.

6. Str. Brightii Trevisan

Streptococcus bei Morbus Brightii, Mannaberg: Centralblatt f. Bakteriologie, V, 1889, 93.
Str. Brightii, Trevisan: Saccardo, Syllog. Fungorum, VIII, 1057.

Morphology. Cocci 0.9 μ; solitary — pairs — short chains of 6–10 elements. Stain by Gram's method.

Gelatin stab. A beaded growth along line of stab; after 3–4 weeks a funnel containing a very viscid gelatin, with brush-like outgrowths.

Agar slant. Growth like *Str. pyogenes.*

Potato. In 4–5 days at 37° C., white colonies, 0.5 mm., becoming a confluent slimy growth.

Milk. Coagulated, acid.

Pathogenesis. Dogs and rabbits inoculated with 0.7–1.0 cc. of a culture subcutaneously show a local abscess. By intravenous injections there result inflammatory changes in kidneys. At the end of 3 days urine contains red corpuscles, renal epithelium, blood clots, and albumen.

Habitat. Isolated by Mannaberg from urine in acute Bright's disease.

7. Str. enteritidis Escherich

Wiener klin. Wochensch., 1887, No. 42.

Morphology. Diplococci — chains of 8 or more elements. In cultures similar to *M. lanceolatus.* Stain by Gram's method. Grow best in glucose bouillon, with addition of human blood serum.

On agar, white colonies. *On gelatin,* a feeble growth, with slight liquefaction.

Pathogenesis. Mice die in 2–3 days, with diarrhœa.

Habitat. Isolated from acute enteritis, in stools, organs, and lymphatics of the intestines.

8. Str. Bonvicini

Streptococcus della leucemia nel câne e nel bue Bonvicini : Centralblatt f. Bakteriologie, XXI, 1897, 211.

Morphology. Diplococci, and chains in old cultures ; elements 0.9–1.0 μ. Stain by Gram's method.

Agar colonies. In 24 hours at 37° C., colonies small, round, white.

Gelatin slant. In 40 hours, small white colonies.

Bouillon. In 24 hours at 37° C., medium turbid, becoming clear, with flocculi on the walls of the tube.

Pathogenesis. Dogs ; inoculations cause some disturbance with swelling of ganglia. Guinea pigs and mice, leucæmia. Rabbits, negative.

Habitat. Isolated from a dog in a case of leucæmia.

9. Str. Fischeli

Micrococcus No. 2 of Fischel : Zeitsch. f. Heilkunde, XII, 1891 ; ref. Centralblatt f. Bakteriologie, IX, 611.

Morphology. Cocci 1.0–1.2 ; in pairs and chains. Stain by Gram's method.

Gelatin colonies. Milk-white ; liquefaction commences in 4 days, and progresses slowly.

Agar colonies. At 37° C., milk-white.

Potato. 37° C., 8 days, thin, glistening, yellowish white ; at 20° C., no growth.

Milk. No growth.

Pathogenesis. Dogs inoculated intravenously with 3-4 cc. of culture show symptoms resembling distemper, in other cases keratitis, pyrexia, catarrhal conjunctivitis, etc.

Habitat. Isolated from the blood in two cases of influenza.

10. Str. canis

Ueber einen bei der bakteriologischen Fleischbeschau aufgefundenen Diplococcus.
Harrevelt: Centralblatt f. Bakteriologie, XXVI, 1899, 121.

Morphology. Diplococci, or as chains of 4 elements, rather smaller than *Str. pyogenes.* Stains with ordinary aniline stains. Decolorized by Gram's method.

Gelatin colonies. No deep colonies. Surface colonies small, similar to those of *Str. pyogenes,* but liquefy the gelatin extremely slowly. Quite flat, dirty white, with a brownish centre, round-oval, granular, with a clear border.

Gelatin slant. On 8 per cent gelatin a plumose growth along the line of inoculation. In 6 days the gelatin is softened, and later entirely liquefied.

Agar slant. In 48 hours, at 36° C., a white transparent spreading layer with an iridescent lustre.

Glycerine agar and *blood serum.* As on agar.

Gelatin stab. In 48 hours spreading irregularly over one-third of the surface, flat ; liquefaction infundibuliform.

Bouillon. Turbid ; slimy sediment.

Milk. 36°, 3 × 24 hours, coagulated ; coagulum slimy ; little acid.

Glucose and lactose bouillon. No gas ; rendered acid.

Starch paste. 36° C., 24 hours, starch liquefied; no sugar, no diastatic ferment. Grows in an acid medium. A slight growth in media containing glycerine. Young cultures have an odor of cooked crabs. Reaction of media without sugar rendered acid. H₂S negative. Indol and nitrites negative.

Pathogenesis. Inoculations of rabbits may or may not cause a fatal infection, with peritonitis, inflammation of liver, spleen, kidneys, and lymph spaces.

Habitat. Isolated from meat.

11. Str. coli Escherich

Str. coli-gracilis Escherich: Die Darmbakterien des Säuglings, Stuttgart, 1886.

Morphology. Cocci, 0.2-0.4 μ, usually in S-shaped chains of 6-20 elements.

Gelatin colonies. Small, round, becoming punctiform, and lying in funnels of liquefied gelatin.

Gelatin stab. Liquefaction cylindrical; sediment white; gelatin rendered acid.

Agar slant. A scanty growth.

Blood serum. Scaly colonies; medium not liquefied.

Potato. Thin white colonies, no growth on old potato.

Milk. Not coagulated; rendered acid after some time.

Habitat. Fæces of healthy children fed on a flesh diet.

12. Str. septicus Babes

Str. septicus-liquefaciens Babes: Septic Prozesse Kindesalters, 1889.

Morphology. Cocci 0.3-0.4; in pairs—short chains. Stain by Gram's method.

Gelatin stab. Along line of stab, in 24 hours, a thin, granular whitish growth. On the surface, growth depressed, becoming funnel-formed. Liquefied gelatin but slightly clouded; upon walls of funnel flat white leafy jagged colonies.

Agar slant. Small white glistening transparent colonies.

Blood serum. A scarcely visible granular layer.

Pathogenesis. Mice and rabbits inoculated subcutaneously show a local inflammation, with œdema and death in about 6 days. Cocci in effused serum, in blood and spleen.

Habitat. Isolated from the blood of a child who died of septicæmia.

13. Str. liquefaciens Sternberg

Manual of Bacteriology, 1892, 613.

Morphology. Cocci round-oval, 0.4-0.6; solitary—pairs—short chains.

Gelatin stab. Liquefaction occurs rapidly along the entire line of puncture, and in 7 days tube completely liquefied.

Agar stab. A beaded growth along line of puncture, with but scanty growth on the surface.

Potato. Growth thin, limited, dry, white.

Pathogenesis. Rabbits and guinea pigs, negative.

Habitat. Isolated from liver of yellow fever cadaver, and from contents of the intestine.

14. Str. albus Sternberg

Weisser Streptococcus Maschek: Jahresber. der Oberrealschule zu Leitmeritz, 1887.
St. albus Sternberg: Manual of Bacteriology, 1892.

Morphology. Streptococci, which show independent movements only during period of division (?).

Gelatin colonies. Flat, round, with white margins; liquefaction crateriform.

Gelatin stab. Develop chiefly on the surface; liquefaction rapid; a white sediment.

Potato. A slimy white growth.

Habitat. Water.

15. **Str. lactis** (Sternberg)

Micrococcus lactis-viscosus Sternberg: Manual of Bacteriology, 1892.
Micrococcus of bitter milk Conn: Centralblatt f. Bakteriologie, IX, 1891, No. 20.

Morphology. Cocci medium-sized, frequently in pairs. In agar colonies form short chains. Growth at 35° C.

Gelatin colonies. Small, round, becoming liquefied and spreading; also thin granular.

Gelatin stab. A shallow crateriform liquefaction; liquefaction progresses rapidly; liquefied gelatin viscous.

Agar slant. A glistening white growth.

Potato. Glistening white discrete masses.

Bouillon. Good growth; a thin film on surface; the medium becomes viscous.

Milk. Rendered bitter. In 24 hours, at 37° C., coagulated; acid, peptonized; butyric acid produced.

Habitat. Isolated from cream which had a bitter taste.

CLASS III. NON–CHROMOGENIC. GROW AT ROOM TEMPERATURES.
GELATIN NOT LIQUEFIED

I. Streptococci, at least in the animal body, with a capsule.

                                    PNEUMOCOCCUS GROUP.

  A. Elements frequently elongated.

        16. *Str. pneumoniæ* (Weichselbaum) Gamaleïa.

  B. Elements spherical.

        17. *Str. Weichselbaumii* (Trev.) L. and N.

II. Streptococci without a capsule.

  A. Cocci actively motile.

        18. *Str. sanguinis* Pitfield.

  B. Cocci non-motile.

    1. Aerobic and facultative anaerobic.

      a. Gelatin colonies floccose-filamentous.

        19. *Str. mirabilis* Roscoe-Lunt.

      b. Gelatin colonies not as above specified.

        * In bouillon, elements in conglomerate masses, arranged in chains; free chains uncommon.

        20. *Str. conglomeratus* Kurth.

      ** Streptococci in free chains.

       † But little growth in the depth of the gelatin, aerobic.

        21. *Str. acidi-lactici* (Marpmann).

†† Growth in depth of gelatin in stab cultures.

STREPTOCOCCUS PYOGENES GROUP.

§ Species pathogenic to animals.

22. *Str. pyogenes* Rosenbach.

23. *Str. enteritis var. Libmanni.* (See No. 2.)

§§ Pathogenic to plants.

24. *Str. phytophthorus* (Frank).

§§§ Non-pathogenic (?) milk species.

25. *Str. Grotenfeldtii.*

26. *Str. proteus.*

††† Growth in gelatin stab cultures not described. Cultural characters imperfectly known.

27. *Str. agalactiæ* Adametz.

28. *Str. Weisii* Trev.

29. *Str. equi* Sand-Jensen.

16. Str. pneumoniæ (Weichselbaum) Gamaleïa

Microbe de la salive Pasteur: Acad. Med. Paris, 1881.
Pneumococcus Fränkel: Verhandl. d. III Congress f. inn. Med., 1884.
Diplococcus pneumoniæ Weichselbaum: Centralblatt f. Bakteriologie, 1887, 588.
Bacillus salivarius-septicus Biondi: Zeitsch. f. Hygiene, II, 1887, 196.
Diplococcus lanceolatus-capsultatus-pneumonicus Foà-Bord: Zeitsch. f. Hygiene, IV, 1888, Heft. 1.
Klebsiella salavaris Trevisan: Gen., 1889, 26.
Streptococcus lanceolatus Gamaleïa: Ann. Inst. Pasteur, 1888, 440.

Morphology. Cocci in twos or chains of 4–6 elements; round-lenticular; in body with a capsule. Stain by Gram's method.

Gelatin colonies. Deep colonies: macroscopically small, round, white or gray; microscopically round, entire, slightly granular. *Surface colonies:* macroscopically round, gray, in 4 days 1–2 mm.; microscopically round, entire, finely granular.

Agar colonies. Deep colonies: microscopically round-lenticular, entire, opaque, grayish black, coarsely punctate. *Surface colonies:* macroscopically like gelatin colonies; microscopically round, entire, translucent, finely punctate.

Glycerin agar slant. A feeble growth, small grayish colonies.

Blood serum. Growth slimy, transparent.

Bouillon. Faintly turbid.

Milk. Coagulation variable.

Litmus milk. Sometimes pink and coagulated.

Potato. No growth.

Pathogenesis. For pathogenesis see various text-books.

Habitat. Pulmonary exudate in croupous pneumonia; frequently associated with bronchopneumonia, peri- and endocarditis, pleuritis, meningitis, otitis-media, etc. Also in saliva of healthy man.

17. Str. Weichselbaumii (Trev.) L. and N.

Diplococcus intracellularis-meningitidis Weichselbaum: Fortschritte d. Medizin, 1887, No. 18.

Neisseria Weichselbaumii Trevisan: Gen., 1889, 32.

Streptococcus intracellularis-meningitidis Lehmann-Neumann: Bak. Diag., 1896.

Morphology. Cocci in twos, with a capsule. Stain by Gram's method.

Blood serum. In 48 hours colonies 2–3 mm., round, flat, convex, colorless, moist, glistening; becoming confluent.

Agar slant. Growth round flat grayish translucent colonies.

Bouillon. Slightly turbid; a whitish sediment.

Potato. A scanty growth, or invisible.

Litmus milk. Unchanged.

In gelatin. Only feeble growth; no liquefaction.

Pathogenesis. Mice and guinea pigs only feebly or doubtfully affected.

Habitat. In meningeal exudate in an epidemic of cerebrospinal meningitis.

VARIETY.—*Streptococcus of Bonome* Sternberg: Manual of Bacteriology, 1892, 325.

Not sufficiently described to distinguish from the above, except that it does not grow on blood serum, and is more pathogenic to mice and rabbits.

Habitat. Obtained by Bonome from the exudations in the cerebrospinal meninges in cerebrospinal meningitis.

18. Str. sanguinis Pitfield

Str. sanguinis-canis Pitfield: Queen's Microscopic Bulletin, Philadelphia, 1897, 44.

Morphology. Cocci actively motile, usually occur singly, but occasionally in pairs. On agar or in bouillon they form long chains. Stain by Gram's method. Flagella were not stained, but were seen occasionally. Growth on media white.

Pathogenesis. Pathogenic to guinea pigs and rabbits. In dogs a local abscess.

Habitat. Found in blood of dogs, in healthy and diseased animals.

19. Str. mirabilis Roscoe-Lunt

Phil. Trans. Roy. Soc., London, CLXXXIII, 1892, 648.

Morphology. Cocci forming very long chains; elements 0.4 μ.

Gelatin colonies. Deep colonies: in 4 days, minute, gnarled-convoluted thread-like masses. *Surface colonies:* faint transparent expansions, 2 mm. Microscopically, masses of fine long threads, filamentous.

Gelatin stab. On the surface a thin transparent film, 3–5 mm.

Agar slant. Like growth on gelatin.

Potato. Growth inappreciable.

Bouillon. A fine sediment like masses of threads of cotton wool. Grow in an atmosphere of hydrogen.

Habitat. Isolated from sewage.

20. Str. conglomeratus Kurth

Trans. 9th International Med. Congress, Berlin, 1891, 335.

Morphology. Cocci in chains as conglomerate masses. Differs from *Str. pyogenes* in that in bouillon cultures at 37° C. there form at the bottom of the tube smooth, round, very firm white scales, or a single white layer, which is not disintegrated when the tube is slightly agitated.

Pathogenesis. Very pathogenic for mice.

Habitat. Associated with scarlet fever; etiological relation undetermined.

21. Str. acidi-lactici (Marpmann)

Sphærococcus acidi-lactici Marpmann: Ergänzungshefte des Centralblatt f. allgemeine Gesundheitspflege, II, 121.

Morphology. Cocci very small; oval, in pairs or short chains.

Gelatin colonies. Porcelain-white, the size of a pin's head.

Gelatin stab. In depth but slight growth; surface growth elevated, white, with dentate margins, becoming in 6 weeks light yellow.

Milk. Becomes reddish, coagulated, acid.

Habitat. Fresh cow's milk.

22. Str. erysipelatos Fehleisen

Str. erysipelatos Fehleisen : Aetiol des Erysipels, Berlin, 1883.

Str. pyogenes Rosenbach: Mikroorganismen bei den Wundinfectionskrankheiten des Menschen, Wiesbaden, 1884.

Str. puerperalis Arloing: Recherches sur les Septicémies, 1884.

Morphology. Cocci about 1.0 μ. Chains best seen in bouillon cultures. Stain by Gram's method. Optimum temperature 37° C.

F

Gelatin colonies. Small, round, white, flat to slightly convex. Microscopically, round, entire — lobed — laciniate, grayish-yellowish, punctiform — finely granular.

Gelatin stab. Along stab filiform — tuberculate; on the surface as in plate cultures.

Agar colonies. As in gelatin. Microscopically round, translucent, grayish yellow, punctiform — granular.

Agar slant. Minute grayish translucent colonies.

Bouillon. Growth variable, turbid, becoming clear, with sediment.

Milk. Coagulated.

Potato. Growth invisible. Indol negative. H_2S slight.

Glucose bouillon. No gas, acid.

Pathogenesis. Virulence variable. Inoculations of mice and rabbits cause local suppuration, septicæmia, etc.; bacilli in internal organs.

Habitat. Associated with phlegmonous inflammation and suppurative processes; septicæmia, erysipelas, bronchopneumonia, puerperal septicæmia, angina, etc.

VARIETIES. — V. Lingelsheim, Zeitsch. Hygiene, X, 1891, 331, makes the two following principal varieties of the species: —

1. Var. *brevis.* In bouillon form short chains, causing a turbidity of the medium. Growth on potato evident. Grow at a temperature of 10°–12° C. Generally non-virulent.

2. Var. *longus.* In bouillon form long chains of 40 and more elements, the medium remaining clear, with a heavy slimy or flocculent sediment. Growth on potato fails. Does not grow below 14–16° C. Generally notably virulent.

The following forms are also probably identical with one or more of the varieties of Str. pyogenes : —

a. *Str. cadaveris* Sternberg: Manual of Bacteriology, 1892.

b. *Str. septicus* Flügge: Die Mikroorganismen, 1886.

c. *Str. of Mannaberg:* Centralblatt f. Klin. Med., 1888, No. 30.

23. Str. enteritis var. Libmanii (See No. 2)

Str. enteritis of Libman : Centralblatt f. Bakteriologie, XXII, 1887, 380

L. isolated from cases of enteritis a streptococcus which he thinks identical with *Str. enteritis* Hirsh, with additional characters.

Morphology. Cocci 0.7–0.9 µ.

Gelatin colonies. In 48 hours, minute points. Microscopically irregular, dark granular in centre, o.1 mm.

Agar colonies. Deep colonies: irregular — oval, coarsely granular, dark yellowish in centre, with a clear erose border. *Surface colonies:* 24 hours, minute points. Do not increase in size.

Potato. A moist glistening thin growth composed of small white colonies.

Milk. In 2–3 days coagulated, acid.

Agar slant. A thick glistening layer.

Habitat. Isolated from a case of enteritis.

24. Str. phytophthorus Frank

Die Bakterienkrankheiten der Kartoffeln: Centralblatt f. Bakteriologie, Zweite Abt., V. 1899.

Morphology. Cocci 0.5 μ; occur singly, in twos, and short chains.

Gelatin colonies. Flat, roundish, lobate, white, rosette-like, sunken in the centre.

Habitat. Associated with a rot and blight of the potato.

25. Str. Grotenfeltii .

Str. acidi-lactici, Grotenfelt: Fortschritte d. Medizin, VII, 124.

Morphology. Cocci round-ellipsoidal, 0.5–1.0 : 0.3–0.6 μ.

Gelatin colonies. Round, white.

Gelatin stab. A good growth in depth.

Milk. Coagulated, acid; no gas; lactic acid.

Habitat. Milk.

26. Str. proteus

Str. No. 52 Conn: Report Storr's Ag. Expt. Sta., 1894, 81.

Morphology. Cocci 0.3 μ, forming chains which have a tendency to arrange themselves in rings. Grow at 35° C.

Gelatin colonies. Smooth, raised, very white, becoming 1.0 mm.; not spreading.

Gelatin stab. In depth a good growth; on the surface growth snow-white, raised.

Agar slant. Growth smooth, thick, white, glistening.

Milk. Coagulated, acid.

Bouillon. Turbid; a slight sediment.

Habitat. Milk.

27. Str. agalactiæ Adametz

Str. mastitidis-sporadicæ, Guillebeau; Landw. Jahrb. d. Schweiz, Bd. IV, 1890, 27.
Str. mastitis-epidemicæ, Guill.: Centralblatt f. Bakteriologie, XVIII, 209.
Str. agalactiæ Adametz: Milch Zeitung, 1893.

Morphology. Cocci in short-long chains, like *Str. pyogenes.*
Milk. Yellowish flocculent coagulæ, and often much gas. Glucose and milk
   sugar converted into lactic acid, CO_2, and traces of volatile fatty acids and
   alcohol.
Habitat. Associated with mastitis of cows and goats.
Str. of mastitis in cows Nocard-Mollereau: Ann. Inst. Pasteur 1887, 109.
   Insufficiently described to distinguish from the above.

28. Str. Weissii Trevisan

Atti d. Acc. Fis. Med. Stat. Mil., Ser. IV, Vol. III, 119.
Mikrokokus der Lungenseuche des Rindes, Pocls-Nolen: Fortschritte d.
   Medizin, 1886, 217.

Morphology. Cocci round, 0.8–1.0 μ, singly and in chains of six.
Gelatin colonies. Round white; like Friedlander's bacillus.
Pathogenesis. Direct inoculation of cultures into the lungs of cattle, dogs,
   cats, rabbits, and guinea pigs produce pneumonia, with abundant cocci in
   exudate. Subcutaneous inoculations negative.
Habitat. Isolated from lung exudate in pleuropneumonia of cattle.

29. Str. equi Sand-Jensen

Streptococcus der Druse der Pferde Sand-Jensen: Archiv f. wissen. u. prakt. Thierheilkunde, XIV, 1887.
Str. equi Sand-Jensen: l. c.

Morphology. Like *Str. pyogenes.*
Pathogenesis. Mice inoculated subcutaneously show an abscess at the point
   of inoculation, and metastatic suppuration of the lymphatics.
Habitat. Associated with an inflammation of the upper air passages of horses,
   and neighboring lymphatic spaces.

CLASS IV. CHROMOGENIC; PRODUCE PIGMENT ON GELATIN OR AGAR

I. Pigment yellowish.
   A. Gelatin liquefied.
      30. *Str. vermiformis* Sternberg.

B. Gelatin not liquefied.
    31. *Str. aurantiacus* (Bruyning).
    32. *Str. ochroleucus* (Prove) Trev.
II. Pigment violet.
  A. Gelatin liquefied.
    33. *Str. fluorescens* (Klamann).
  B. Gelatin not liquefied.
    34. *Str. violaceus* (Schröter) Trev.

30. Str. vermiformis Sternberg

Wurmförmiger Streptococcus Maschek: Jahresber. Oberrealschule zu Leitmeritz, 1887.
Str. vermiformis Sternberg: Manual Bacteriology, 1892.

Morphology. Streptococci, which show slow vermiform movements. Chains resemble filaments.
Gelatin colonies. Yellowish white, becoming liquefied; concentric structure. Microscopically the colonies show radiate margins.
Gelatin stab. Liquefaction rapid; a dirty yellow sediment.
Potato. A dirty yellow growth.
Habitat. Water.

31. Str. aurantiacus (Bruyning)

Micrococcus aurantiacus-sorghi Bruyning: Archiv. Néerland Sci. Exact. et Nat., I, 1898, 297.

Morphology. Cocci round-oval, 0.7–0.9 μ, which form chains.
Bouillon. Turbid; medium acid.
Potato. A bright yellow growth.
Milk. Coagulated in 7 days.
On gelatin. A yellow growth.
Pigment. Soluble in alcohol.
Habitat. Isolated from blighted sorghum.

32. Str. ochroleucus (Prove) Trevisan

Micrococcus ochroleucus Prove: Cohn, Beiträge Biol., IV, 409.
Str. ochroleucus Trevisan: Gen., 1889, 31.

Morphology. Cocci 0.5–0.8 μ; solitary—pairs—chains.
Gelatin colonies. In 24 hours small, colorless, with elevated and wavy borders.

Gelatin stab. On the surface growth, thin, colorless, becoming in 3–4 days sulphur-yellow.

Pigment. Soluble in alcohol. Old gelatin cultures give off a peculiar odor.

Potato. Growth scarcely visible before the fifth day.

Habitat. Isolated from urine of man.

33. Str. fluorescens (Klamann)

Diplococcus fluorescens-fœtidus Klamann : Allgemeine medizin Centralzeitung, 1887, 1347.

Morphology. Cocci 1.4 μ; diplococci, chains of 6–10. Grow at 37° C.

Gelatin colonies. Surface colonies round, gray brownish, becoming a crateriform liquefaction, with a brownish gray sediment. The surrounding gelatin becomes grassy green — violet.

Gelatin stab. On the surface a round, shallow, crateriform liquefaction, with a glistening iridescent film, and a greenish gray sediment. Film on surface violet; later a saccate liquefaction of the gelatin.

Agar slant. A granular brownish gray growth.

Potato. A greenish growth, becoming bluish green. Potato colored blue.

Habitat. Isolated from posterior nares of man.

34. Str. violaceus (Schröter) Trevisan

Bacteridium violaceum Schröter: Cohn Beiträge Biol., 1870, 122.
Micrococcus violaceus Cohn: Beiträge, 1870, 157.
Streptococcus violaceus Trevisan: Gen., 1889, 31.

Morphology. Cocci elliptical, larger than *B. prodigiosus*; in chains.

Gelatin colonies. Slimy, raised, violet.

Potato. Slimy spots, deep violet, becoming coalescent, spreading violet.

Habitat. Water.

SPECIES WITH A THICK GELATINOUS ENVELOPE

35. Str. involutus Kurth

Mitteilungen a. d. Kaiserl. Gesundheitsamte, 1893, 439.
Streptococcus d. Maul und Klauenseuche.

In fluid blood-serum growth in upper part of medium, as a bright creamy layer, composed of zoöglœa of streptococci, with gelatinous envelopes.

Habitat. Associated with Maul and Klauenseuche of cattle and sheep.

36. Str. mesenterioides (Cienkowski) Migula

Ascococcus mesenterioides Cienkowski: Arbeit. d. Naturforsch. Gesellsch. Univ.
  Charkow, XII, 1878.
Leuconostoc mesenterioides Van Tieghem: Ann. Sci. Nat., 1878.
Str. mesenterioides Migula: Engler-Prantl. Natürlichen Pflanzenfam., 1895.

Morphology. In media free from glucose or cane sugar, like *Str. pyogenes.*
  In saccharine media, as thick white gelatinous clumps, containing chains.
  In the latter gas production and acid.
Milk. Coagulated.
Habitat. Causes a frog-spawn alteration of saccharine solutions.

MICROCOCCUS

Monas Ehrenberg: Die Infusionstierchen als volkommene Organismen, 1838.
Mikrokokkus Hallier, 1866–68.
Micrococcus Cohn: Ueber Bacterien, Beiträge zur Biol. der Pfl., I, 2, 1872, 127.
Ascococcus Billroth: Untersuch. über die Vegetationsformen von Coccobacteria
  septica, 1874.
Bacteridium Schröter: Die Pilze, in Kryptogamenflora von Schlesien, 1886.
Lampropedia Schröter: l. c.
Hyalococcus Schröter: l. c.
Leucocystis Schröter: l. c.

Cells round; occur singly, in twos, threes, tetrads, and in irregular groups.
  Division in two directions of space. Non-motile. Spores absent, at least
  never certainly demonstrated.

SYNOPSIS OF THE GENUS

I. Aerobic and facultative anaerobic.
  A. Without pigment on gelatin or agar.
    1. Do not grow on ordinary culture media, or, at best, but very poorly.
                                                          CLASS I, p. 72.
    2. Grow on ordinary culture media.
      a. Grow best at body temperatures, and not at 20° C.
                                                          CLASS II, p. 73.
      b. Grow at ordinary room temperatures, 20° C.
        * Gelatin liquefied. CLASS III, p. 74.
        ** Gelatin not liquefied. CLASS IV, p. 82.

B. Doubtfully chromogenic; growth on gelatin or agar light yellow — yellowish white.

    1. Gelatin liquefied. CLASS V, p. 91.
    2. Gelatin not liquefied. CLASS VI, p. 94.
    3. Do not grow on gelatin. No. 61.

C. Distinctly chromogenic; form a pigment on gelatin or agar.

    1. Pigment yellowish orange.
        a. Gelatin liquefied. CLASS VII, p. 96.
        b. Gelatin not liquefied. CLASS VIII, p. 102.
    2. Pigment reddish, pinkish, flesh-colored. CLASS IX, p. 105.
    3. Pigment bluish black. CLASS X, p. 108.

CLASS I. WITHOUT PIGMENT ON GELATIN OR AGAR. DO NOT GROW ON ORDINARY CULTURE MEDIA, OR, AT BEST, BUT VERY POORLY.

I. No growth on ordinary culture media.
    1. *M. gonorrhœæ* (Bumm) Flügge.
II. Very feeble growth on gelatin or agar.
    2. *M. catarrhalis* Frosch-Kolle.

1. M. gonorrhœæ (Bumm) Flügge

Gonococcus Neisser: Centralblatt f. Med. Wissensch., 1879, No. 28.
Diplococcus gonorrhœæ Bumm: Der Mikroorganismen der gonorrh. Schleimhauterkrankung, Wiesbaden, 1885.
Micrococcus gonorrhœæ Flügge: Die Mikroorganismen, 1886.

Morphology. Coccio .8–1.0: 0.6–0.8 μ; in pairs, flattened at points of contact, or in tetrads. Decolorized by Gram's method. Grow only in specially prepared media, best in serum agar, one part of human ascites fluid and two parts of melted agar.

Differential diagnosis. Examination of pus, cover-glass preparations, shows gonococci within the cells, when stained with alkaline methyl blue: or stain by Gram's method, and then with aqueous solution of Bismarck brown.

Pathogenesis. Not pathogenic for smaller laboratory animals. Inoculation experiments upon man give positive results.

Habitat. In gonorrheal pus, and associated with lesions of subsequent infection.

2. M. catarrhalis Frosch-Kolle

Kokkus bei infektioser Bronchitis Seifert: Volkmann's Klin. Vortr., No. 240.
Micrococcus catarrhalis Frosch-Kolle: Flügge, Die Mikroorganismen, 1896, 154.

Morphology. Cocci 1.0–2.0 μ, slightly oval, mostly in twos, with a clear space between, like gonococci.
Decolorized by Gram's method.
Gelatin stab. A slight growth along line of stab; medium not liquefied.
Agar slant. White colonies like *M. pyogenes var. albus.* Grow best on blood agar, as white discrete colonies. Cultures on artificial media die in 3–4 days.
Pathogenesis. Inoculations of smaller animals negative.
Habitat. Associated with an infectious bronchitis: found in the sputum and nasal secretions. According to Pfeiffer, perhaps an identical species is found in the purulent secretions of cases of bronchopneumonia in children.

CLASS II. WITHOUT PIGMENT. GROW ON ORDINARY CULTURE MEDIA. DO NOT GROW AT 20° C., OR POORLY IN NO. 3.

I. Gonococcoid forms.
    3. *M. bovis* (Babes).
    4. *M. Demmei.*
II. Staphlococcus forms.
    5. *M. endocarditis* Weichselbaum.

3. M. bovis (Babes).

Hæmatococcus bovis Babes: Virchow's Archiv, CXV, 1889.

Morphology. Gonococcoid-oblong forms; isolated or united in groups. Free cocci surrounded by a pale yellowish shining aureole. Decolorized by Gram's method. Facultative anaerobic.
Gelatin stab. Along the line of stab a feeble growth, composed of small white colonies.
Agar slant. Small transparent colonies.
Potato. At 37° C., growth thin, spreading, yellowish, glistening, scarcely visible.
Blood Serum. Small, moist, transparent colonies.
Pathogenesis. Inoculations of rabbits and rats positive, with death in 6–8 days; spleen enlarged, lungs hyperæmic, bloody serum in peritoneal cavity, cocci in blood. Guinea pigs immune.
Habitat. Isolated from blood and organs of cattle in an epidemic malady with hæmaglobinuria.

4. M. Demmei

Diplococcus of pemphigus acuta Demme : Verhandl. des Vet. Congress für innere Med. in
Wiesbaden, 1886.

Morphology. Cocci 0.8–1.4 μ, usually in pairs like gonococcus. No capsule ;
usually in irregular masses.

Aerobic. Minimum temperature ± 32° C.

Agar colonies. In 36–48 hours at 37° C., round, raised, milk-white, becoming
rosette-like forms, with outgrowths, or moruloid ; creamy white.

Agar slant. Growth creamy white, with clavate and stalactite elevations.

Potato. At 37° C. a slow growth.

Pathogenesis. Injections into lungs of guinea pigs cause emaciation and de-
bility, with foci of bronchopneumonia.

Habitat. Isolated from contents of bullæ in a case of pemphigus.

5. M. endocarditis Weichselbaum

M. endocarditis rugatus Weichselbaum : Ziegler's Beiträge, IV, 1889, 127.

Morphology. Similar to *M. pyogenes.* Aerobic.

Agar Colonies. *Deep colonies :* irregular, granular ; a large central, yellowish
brown nucleus, surrounded by a narrow grayish brown periphery. *Surface
colonies :* a small brown central nucleus, with a translucent granular,
grayish brown periphery, becoming wrinkled, with a stearine lustre, and
viscid.

Potato. At 37° C. a feeble growth, as small, dry, pale brownish masses.

Blood serum. Discrete, confluent, colorless colonies.

Pathogenesis. Rabbits inoculated subcutaneously in ear show a tumification
and œdema. With guinea pigs, a local suppuration. Dogs, inoculated
intravenously, endocarditis.

Habitat. Isolated from affected cardiac valves in fatal cases of ulcerative
endocarditis.

CLASS III. WITHOUT PIGMENT. GROW ON ORDINARY CULTURE
MEDIA AND AT ROOM TEMPERATURES, 20° C. GELATIN
LIQUEFIED.

I. Liquefaction of the gelatin proceeds more or less rapidly.

   A. Grow on potato.

      1. Growth on potato white.

         a. Liquefaction of the gelatin takes place along the entire length
of the stab ; *i.e.* infundibuliform — cylindrical.

MICROCOCCUS PYOGENES GROUP.

* Species more or less pathogenic.
  6. *M. pyogenes var. albus* Rosenbach.
  7. *M. mastitis.*
  8. *M. polymyositis* Martinotti.
** Saprophytic species.
  † Gelatin colonies compact; border entire — undulate.
  9. *M. aethebius* (Trev.) Flügge.
  †† Gelatin colonies with lacerate-fimbriate borders.
  10. *M. coronatus* Flügge.
  ††† Gelatin colonies not described.
  11. *M. liquefaciens.*
b. Liquefaction of the gelatin takes place near the surface; *i.e.* crateriform becoming stratiform.
  * Milk coagulated.
  12. *M. simplex* Wright.
  13. *M. acidi-lactis* Krüger.
  ** Milk not coagulated.
  14. *M. xanthogenicus* (Freire) Sternberg.
2. Growth on potato yellowish — yellowish brown.
  15. *M. Freudenreichii* Guillebeau.
  16. *M. radiatus* Flügge.
3. No growth on potato.
  17. *M. dissimilis* Dyar.
II. Liquefaction of the gelatin proceeds very slowly.
  A. Liquefaction takes place only on the surface, crateriform.
  1. Gelatin colonies very minute.
  18. *M. ascoformans* Johne.
  2. Growth on gelatin normal.
  19. *M. albicans* (Trev.) Bumm.
  20. *M. aerogenes* Miller.
  21. *M. fœtidus* Klammen.
B. Liquefaction takes place in depth, infundibuliform.
  22. *M. alvi.*
  23. *M. Rheni.*

6. M. pyogenes var. albus Rosenbach

Wundinfectionskrankh. des Menschen, Wiesbaden, 1884.

Morphology. Cocci ± 0.8 μ; occur singly and in clumps.

Gelatin colonies. Deep colonies: round-oval, yellow to brown, entire, finely granular. *Surface colonies:* round, irregular, white, becoming 1.5 mm., and sunken in the liquefied gelatin; microscopically, yellowish — brownish with a transparent border; slightly granular.

Gelatin stab. Liquefaction begins in 2–3 days, becoming saccate; gelatin turbid, with a heavy sediment.

Agar colonies. Deep colonies: round-oval, grayish yellow, opaque with a granular border. *Surface colonies:* round, moist, glistening, becoming several millimetres in diameter, white; microscopically, round, entire with a transparent granular zone, and with a gray homogeneous centre.

Bouillon. Turbid, a delicate pellicle, and a heavy white sediment.

Milk. Coagulated.

Potato. Growth limited, white, thick; cultures have a pasty odor; H_2S produced; indol in small amount.

VARIETIES

Staphlococcus epidermis-albus Welch: Am. Jour. Med. Sci., Phila., 1891, 439.

Differs from No. 6 in the following points: Liquefies gelatin more slowly; does not quickly cause coagulation of milk, and is much less virulent when injected into the circulation of rabbits.

Habitat. Common on surface of body, and in the deeper parts of the epidermis.

Staphlococcus pyosepticus Héricourt-Richet: Compt. rend. CVII, 1888, 690.

Probably a variety of No. 6, but more pathogenic to rabbits, subcutaneous inoculations causing extensive inflammatory œdema, with death in 12–24 hours.

7. M. mastitis

Micrococcus of Gangrenous Mastitis in Sheep, Nocard:
Annales Inst. Pasteur I, 1887, 417.

Morphology. Cocci solitary, in pairs and irregular clusters. Stain by Gram's method. Facultative anaerobic.

Gelatin colonies (surface). Round, white; microscopically, round, homogeneous, brown, surrounded by a translucent aureole.

Gelatin stab. Liquefaction infundibuliform.

Agar stab. A good growth in depth; on the surface a thick white spreading growth.

Blood Serum. Liquefied.

Potato. Growth thin, viscid, grayish; central portion yellowish.

Pathogenesis. Sheep inoculated subcutaneously in the mammary gland show inflammatory œdema and death in 24–48 hours. Rabbits by subcutaneous inoculation, show a local abscess. Guinea pigs immune.

8. M. polymyositis Martinotti

Centralblatt f. Bakteriologie, XXIII, 1898, 877.

Morphology. Cocci solitary, in pairs and groups. Stain by Gram's method.

Gelatin colonies. In 36 hours white points, becoming in 3 days sunken in the liquefied gelatin.

Agar colonies. In 24 hours, at 37° C. *Deep colonies:* convex, grayish white, with sharp thin edges.

Agar slant. Growth thick, grayish white, becoming spreading.

Gelatin stab. In depth a white stripe. On the surface growth quite spreading. In three days liquefaction begins on the surface, becoming infundibuliform. Gelatin turbid, with a white sediment.

Bouillon. In 24 hours, at 37° C., turbid.

Potato. Growth white-gray, becoming deeper-colored.

Milk. Coagulated.

Litmus milk. Reddened.

Pathogenesis. Rabbits inoculated subcutaneously show œdema, redness, and a gangrenous condition. Animals become greatly emaciated, but generally recover. Guinea pigs immune.

Habitat. Isolated from an abscess in kidneys in a case of polymyositis acuta.

9. M. aethebius (Trev.) Flügge

M. ureæ-liquefaciens Flügge: Die Mikroorganismen, 1886.
Str. æthebius Trevisan: Gen. e Spec. Batteriaceæ, 1889.

Morphology. Cocci 1.2–2.0 µ occur singly and in chains of 3–10 elements, or in irregular groups.

Gelatin colonies. In 2 days small white points; microscopically, round, dark gray, entire, becoming yellowish brown granular with an undulate border.

Gelatin stab. Growth along needle track beaded, becoming a liquefied funnel.

Habitat. Air.

10. M. coronatus Flügge

Die Mikroorganismen, 1886.

Morphology. Cocci 0.8–1.6 µ; in chains and clumps.

Gelatin colonies. White; microscopically, gray, coarsely granular, with lacerate-fimbriate borders, becoming liquefied funnels.

Agar slant and potato. Growth gray-white, rather dry, spreading.

Bouillon. Feeble growth, turbid. Indol negative. H$_2$S slight.

Milk. In 10 days gelatinous, in 14 days curdled and acid.

Habitat. Air.

11. M. liquefaciens v. Besser

M. albus-liquefaciens v. Besser: Beiträge path. Anat., VI, 1889, 46.

Morphology. Cocci round-elliptical; in twos, clumps, and chains. Facultative anaerobic.

Gelatin stab. A saccate liquefaction.

Agar colonies. Elevated, glistening, 0.5 mm. Microscopically, the colonies show a brownish centre, with an outer concentric zoning.

Potato. Growth white, glistening.

Habitat. Isolated from normal nasal mucus.

12. M. simplex Wright

Memoirs National Academy of Sciences (U. S. A.), VII, 1895, 32.

Morphology. Cocci medium-sized; in pairs, tetrads, and clumps.

Gelatin colonies. Deep colonies: dark brownish, granular; entire — irregular margins. *Surface colonies:* in 3 days round, glistening, white, translucent, 1.0 mm. Microscopically, dense-opaque, granular; later immersed in a clear liquid and somewhat fragmented.

Gelatin stab. In depth a filiform growth; on the surface liquefaction crateriform, becoming stratiform. Gelatin rendered alkaline.

Agar slant. Growth milk-white, glistening, limited.

Bouillon. Turbid, alkaline, with flocculi on the surface and a white flaky sediment.

Potato. Growth feeble, composed of whitish discrete colonies.

Litmus milk. In 2 weeks viscid; pink above, decolorized below, becoming coagulated and acid.

Pepton-rosolic acid solution. Unchanged. *Indol?* No nitrites. Grow at 36° C.

Habitat. Water.

13. M. acidi-lactis Krüger.

Centralblatt f. Bakteriologie, VII, 1890, 495.

Morphology. Cocci oval, 1.0–1.5 μ; in twos and tetrads. Facultative anaerobic.

Gelatin colonies. Round, white, border lacerate.

Gelatin stab. In depth growth beaded; on the surface growth round, white, becoming sunken in the liquefied gelatin.

Milk. Coagulated, becoming slimy, peptonized, and has a pasty odor. Lactic acid produced.

Habitat. Isolated from cheesy butter.

14. M. xanthogenicus (Freire) Sternberg

Cryptococcus xanthogenicus Freire.
Micrococcus of Freire Sternberg : Manual of Bacteriology, 1892.

Morphology. Cocci 0.5–0.8 μ ; occur singly, in twos and in groups. Stain by Gram's method.

Gelatin stab. A crateriform liquefaction, becoming stratiform, with a milk-white sediment.

Agar stab. In depth growth white-opaque ; on the surface soft, milk-white.

Potato. Growth milk-white, limited.

Pathogenesis. Guinea pigs, negative. According to Freire the organism is pathogenic in summer, and for small birds.

Habitat. Isolated from cases of yellow fever and supposed by Freire to be the cause of the disease.

15. M. Freudenreichii Guillebeau

Schweizer Archiv. f. Thierheilkunde, XXXIV, 1892, 128.

Morphology. Cocci large, 2.0 μ and more ; occur singly, rarely in chains.

Gelatin colonies. White, entire, granular ; liquefied in 2 days.

Agar slant. Growth white.

Potato. Growth yellowish — yellowish brown.

Bouillon. Turbid, becoming clear, with a flocculent sediment.

Milk. Coagulated, viscid, acid.

Habitat. Milk and cheese.

16. M. radiatus Flügge

Die Mikroorganismen, 1886.

Morphology. Cocci small, 0.8–1.0 μ ; occur singly, in short chains and clumps.

Gelatin colonies. Large white ; in 2 days slightly liquefied, with a yellowish green fluorescence. Microscopically brownish yellow, granular ; border fimbriate — ciliate. Colonies become 10–15 mm. in diameter, and show a concentrically zoned structure.

Gelatin stab. Growth along line of stab arborescent, becoming an infundibuliform liquefaction.

Potato. Growth yellowish brown.

Habitat. Air and water.

17. M. dissimilis Dyar

Trans. N. Y. Acad. of Sci., VIII, 1895, 353.

Morphology. Cocci 1.0 μ; in masses.
Gelatin colonies. Round, opaque, surrounded by an obscure granular veil.
Agar slant. Growth white, opaque.
Milk. Coagulated. Nitrates reduced to nitrites after 28 days.
Litmus milk. Color unchanged.
Pepton-rosolic acid solution. Unchanged. Gelatin rapidly liquefied.
Habitat. Air.

18. M. ascoformans Johne

Berichte Veterinärwesen in K. Sachsen, 1885, 27.
M. botryogenus Rabe: Deutsche Zeitsch. f. Thiermedizin, 1886, 137.

Morphology. Diplococci and tetrads, similar to *M. luteus.*
Gelatin colonies. Very minute, like granules; microscopically round, entire, without special characters.
Gelatin stab. In depth growth white, filiform; on the surface a slow crateriform liquefaction.
Potato. Growth yellowish, with a fruity odor.
Agar slant. Growth scarcely visible.
Pathogenesis. Guinea pigs die of septicæmia. Mice immune. Rabe and Kitt by inoculations produced true fibroma in horses. Inoculations of sheep cause inflammation, œdema, and necrosis.
Habitat. Associated with botryomycosis in horses.

19. M. albicans (Trev.) Bumm

Gray white micrococcus Bumm: Der Mikroorg. Schleimhauterkrank., 1885, 25.
Neissera albicans Trevisan: Gen. e Spec. Batteriaceæ, 1889.
M. albicans-amplus Kruse: Flügge, Die Mikroorganismen, 1896, 186.

Morphology. Diplococci like gonococcus, but larger, 2-2.8 μ, or in groups of 3-4 elements.
Gelatin colonies. Grayish white, slightly elevated.
Gelatin stab. In depth a grayish white stripe; on the surface a grayish white growth; after a time a liquefaction takes place under the surface growth.
Habitat. Isolated from vaginal secretions.

20. **M. aerogenes** Miller

Deutsche medizinische Wochenschrift, 1883, No. 3.

Morphology. Cocci large, oval.

Gelatine colonies. Round, smooth.

Gelatin stab. In depth a brownish yellow line; on the surface a flat grayish white button; a slight liquefaction after some days.

Agar slant. Growth yellowish white.

Potato. Growth yellowish white.

Habitat. Found in the alimentary canal.

21. **M. fœtidus** Klamman

Allg. med. Centralzeitung, 1887, 344.

Morphology. Cocci of irregular sizes; occur singly, in pairs, short chains, and irregular groups. Single diplococci 1.4 μ. At 37° C. only slight growth.

Gelatin colonies. White, round-oval.

Gelatin stab. Growth milk-white, glistening, with an elevated knobby surface, becoming concentric and brownish. Liquefaction occurs slowly. Cultures have a disagreeable odor of ozæna.

Agar slant. Growth irregular, white, spreading.

Potato. Growth irregular, slimy, pale reddish gray, with a knobby surface. Odor of ozæna.

Habitat. Isolated from the posterior nares of man.

22. **M. alvi**

M. albus liquefaciens Sternberg: Manual of Bacteriology, 1892.

Morphology. Cocci 0.8–1.2, often oval, and 3.0 μ long; in irregular groups. Considerably larger than *M. pyogenes*.

Gelatin colonies. Round, white; after some time a liquefaction of the surrounding gelatin takes place.

Gelatin stab. In depth a feeble growth, later a funnel of liquefaction, with a scum on the liquefied gelatin.

Agar slant and *blood serum.* Growth white; latter not liquefied.

Potato. Growth very scanty, thin, colorless, becoming a collection of white button-like masses.

Habitat. Isolated from alvine discharges of healthy infants.

G

23. M. Rheni

Rhine Water Micrococcus Burri: Archiv f. Hygiene, XIX, 1893, 1.

Morphology. Cocci variable 0.5–1.25 μ; scarcely ever quite round, mostly flattened; in tetrads.

Gelatin Colonies. Deep colonies: Lenticular, granular near the periphery. *Surface colonies:* In 2–3 days, white; microscopically round; border slightly irregular, granular; after 7 days still small, when they begin to sink in the softened gelatin.

Gelatin stab. In 2 days a slight depression, which later resembles an air bubble, becoming in 7 days a narrow funnel, with a white granular sediment, and a surface scum. Liquefaction takes place slowly.

Glycerin agar colonies. Deep colonies: Like gelatin colonies. *Surface colonies:* In 2 days, white, glistening, slightly convex; microscopically granular. Growth viscid.

Bouillon. Turbid, not clearing; finally a granular white sediment.

Milk. No appreciable growth, and no coagulation.

Glucose bouillon. No gas.

Pathogenesis. Guinea pigs, negative.

Habitat. Rhine water.

CLASS IV. WITHOUT PIGMENT. GROW ON ORDINARY CULTURE MEDIA AND AT ROOM TEMPERATURES 20° C. GELATIN NOT LIQUEFIED.

I. Gelatin colonies distinctly filamentous or arborescent.

        24. *M. viticulosus* Flügge.

        25. *M. stellatus* Maschek.

II. Gelatin colonies compact.

  A. Cocci in tetrads, surrounded by a capsule.

        26. *M. tetragenus* Koch-Gaffky.

        27. *M. Mendozæ* (Trevisan) Mendoza.

  B. Cocci morphologically identical with gonococcus.

        28. *M. tardissimus* (Trev.) Bumm.

        29. *M. magnus* (Rosenthal).

  C. Cocci united in zoöglœa masses.

        30. *M. candidus* Cohn.

  D. Cocci not as before specified, staphlococci.

    1. Colonies concentrically zoned.

      a. Growth on gelatin dirty white, yellowish white.

        31. *M. eczemæ.*

        32. *M. Sornthalii* Adametz.

b. Growth on gelatin thin bluish gray.

      33. *M. concentricus* Zimmerman.

2. Colonies not concentrically zoned.

  a. Growth on potato invisible, or very scanty.

    * From animal habitats, or pathogenic.

      34. *M. cumulatus* v. Besser.

      35. *M. salivarius* Biondi.

      36. *M. beta.*

    ** Terrestrial species; in water, milk, etc.

      37. *M. aquatilis* Vaughan.

      38. *M. tenacatis.*

      39. *M. acidi-lactici* (Lindner).

      40. *M. cerevisæ* (Balcke).

  b. Growth on potato abundant.

    * Gelatin colonies distinctly lobed.

      41. *M. fervidosus* Adametz.

      42. *M. rosettaceus* (Zimmerman).

    ** Gelatin colonies not distinctly lobed.

    † Milk coagulated.

      43. *M. ovalis* Escherich.

    †† Milk not coagulated.

      § Potato cultures yellowish-brownish.

      44. *M. lactis.*

      §§ Potato cultures whitish-grayish.

      45. *M. candicans* Flügge.

      46. *M. nivalis.*

24. M. viticulosus Flügge

Die Mikroorganismen, 1886, 178.

Morphology. Cocci oval, 1.0–1.2 μ; occur as zoöglœa masses.

Gelatin colonies. Deep colonies: Show a network of hairy ramifications, like *B. Zopfi,* consisting of zoöglœa. *Surface colonies:* Thin, cloudy, opaque, whitish, from which fine threads penetrate into the gelatin.

Gelatin stab. Along line of puncture a villous-feathery growth; on the surface like surface colonies.

Potato. A dirty white growth.

Habitat. Air and water.

25. M. stellatus Maschek

Jahresbericht der Oberrealschule zu Leitmeritz, 1887.

Morphology. Cocci never in chains.

Gelatin colonies. Star-shaped, due to numerous extensions, the ends of which are branched.

Gelatin stab. Along line of stab an arborescent growth.

Potato. A brownish yellow, slimy growth.

Habitat. Water.

26. M. tetragenus Koch-Gaffky

Mitteilungen a. d. Kaiserl. Gesundheitsamte, II, 1884, 42.

Morphology. Cocci round-oval, 1.0 μ, variable; in animal body as tetrads, the latter surrounded by a capsule.

Aerobic, only slight growth under anaerobic conditions.

Gelatin colonies. *Deep colonies:* Irregular, entire, opaque, granular — moruloid — nodular. *Surface colonies:* Small, irregular, entire, white, moist, slightly raised. Microscopically round, entire, becoming lobed-lacerate, gray.

Gelatin stab. A white beaded growth along line of stab; on the surface in 10 days a white, thick growth.

Agar slant. A thick, slimy, grayish yellow, moist, glistening growth.

Bouillon. Medium clear, with a sediment.

Milk. May or may not be coagulated.

Potato. Growth thick, white, viscid, limited.

Glucose bouillon. Rendered acid. Indol negative. H_2S negative.

Pathogenesis. Mice inoculated subcutaneously show somnolence in 2 days, and die in 3-6 days. Bacilli in spleen, liver, kidneys, and lung. Guinea pigs show a local abscess or septicæmia. Intraperitoneal inoculations cause a purulent peritonitis.

Habitat. Found in phthisical cavities, and in sputum.

27. M. Mendozæ (Trev.) Mendoza

M. tetragenus-mobilis Mendoza: Centralblatt f. Bakteriologie, VI, 1889, 506.
Gaffkya Mendozæ Trevisan: Saccardo, Sylog. Fungorum, VIII, 1043.

Morphology. Cocci in tetrads, surrounded by a capsule.

Gelatin colonies. Round, dirty white, finely granular.

Gelatin stab. On the surface a dirty white growth. Old cultures give off an odor of skatol.

Agar slant. Growth dirty white.

Habitat. Isolated from the contents of the stomach.

28. M. tardissimus (Trev.) Bumm

Milk-white Micrococcus Bumm : Mikroorganismus der gon. Schleimhauterk., 1885.
Diplococcus albicans-tardissimus Flügge: Die Mikroorganismen, 1886.
Neisseria tardissima Trevisan: Gen. e spec. Batteriaceæ, 1889.

Morphology. Gonococcoid forms.
Gelatin colonies. Small, punctiform ; microscopically round, opaque, brownish
  gray ; colonies in 2 weeks become 2 mm. in diameter.
Gelatin stab. A grayish white beaded growth along the line of inoculation ;
  on the surface, growth thin, white, stearine-like, with dentate margins.
Habitat. Found in vaginal secretions, especially those of puerperal women.

29. M. magnus (Rosenthal)

Diplococcus magnus Rosenthal: Centralblatt f. Bakteriologie, XXV, 1899, 1.

Morphology. A diplococcus with a distinct capsule, also gonococcoid-like
  forms. Non-motile. Decolorized by Gram's method.
Gelatin colonies. After some days, yellowish white, convex.
Glycerin agar colonies. In 24 hours, convex, grayish white, with a fatty
  lustre, radially striped and concentrically ringed, with a sharp indented
  border ; opalescent.
Gelatin stab. Very small colonies along line of stab, and a small segmented
  colony on the surface.
Agar slant. In 18 hours at 37° C. a growth composed of punctiform colonies.
Glycerin agar slant. A scanty growth composed of minute colonies, scarcely
  discernible to the naked eye.
Agar stab. In depth a slight growth ; on the surface growth grayish white,
  segmented and radially striped.
Gelatin slant. As on glycerin agar.
Bouillon. In 24 hours, at 16–18°, turbid ; little sediment.
Glucose bouillon. No gas.
Potato. An abundant growth, as orange-yellow points.
Habitat. Isolated from the air.

30. M. candidus Cohn

Beiträge Biologie, I, 1875, 160.

Morphology. Cocci round, 0.5–0.7 μ.
Gelatin colonies. Snow-white ; margins irregular, granular.

Gelatin stab. A scanty growth along line of stab; a flat, milky-white growth on the surface.

Agar slant. A milk-white growth.

Habitat. Water.

31. M. eczemæ

Diplococcus albicans-tardus Unna: Tomasoli, Monatsch. prakt. Dermatol, IX, 1889, 54.

Morphology. Oval diplococci, often in short chains or irregular groups.

Gelatin colonies. *Deep:* oval, dark yellow. *Surface:* round, elevated, grayish yellow; in 8 days 2 mm. Microscopically granular, grayish yellow.

Gelatin stab. Scanty growth in depth; on the surface growth thin, waxy, yellowish white.

Agar slant. A yellowish gray, dull streak.

Habitat. Upon the surface of the body in eczema.

32. M. Sornthalii Adametz

Centralblatt f. Bakteriologie, 2d. abt., 1895, 465.

Morphology. Cocci round-oval, 0.7 μ; in twos, fours, clumps, and short chains of 6-8 elements.

Gelatin colonies. *Surface:* white, flat, slimy, in 8-10 days, 4-5 mm.; microscopically, dirty — yellowish white, concentric; centre brownish opaque; border thin erose.

Lactose gelatin stab. A beaded growth in depth; on surface growth dirty-yellowish white, slimy.

Milk. Coagulated, acid; gas.

Lactose bouillon. Gas $\dfrac{H}{CO_2} = \dfrac{1}{3}$.

Habitat. Milk, cheese.

33. M. concentricus Zimmerman

Bak. Trink u. Nutzwässer, Chemnitz, 1890.

Morphology. Cocci 0.9 μ; in irregular clumps.

Gelatin colonies. *Deep:* round, light brown — grayish yellow, granular, concentrically zoned. *Surface:* small, bluish gray disks, becoming larger and more irregular, with lobate periphery. Microscopically, a grayish brown irregular-edged and lobate centre, with a lighter granular ring and a whitish glistening border.

Gelatin stab. On surface growth thin bluish gray, concentric.
Agar slant. Growth thin, spreading, bluish gray — whitish.
Potato. Growth thin, yellowish gray, slimy.
Habitat. Water.

34. **M. cumulatus** v. Besser

M. cumulatus-tenuis v. Besser: Beiträge Path. Anat., VI, 347.

Morphology. Cocci large, oval; in masses.
Gelatin stab. In depth growth white, beaded; on the surface elevated slightly, or flat, transparent.
Agar colonies. Surface: elevated, punctiform, 0.2 mm., becoming 5 mm., flat and transparent; microscopically, shows a brownish nucleus with a grayish brown border; wrinkled.
Potato. Only a scanty growth.
Bouillon. Remains clear, with a sediment.
Habitat. Common in nasal mucus in man.

35. **M. salivarius** Biondi

M. salivarius-septicus Biondi: Zeitsch. f. Hygiene, II, 1887, 194.

Morphology. Cocci round — slightly oval. Stain by Gram's method.
Gelatin colonies. Surface: round, grayish white, which may become darker.
Gelatin stab. In depth beaded, white.
Potato. Growth scanty.
Pathogenesis. Inoculations of mice, guinea pigs, and rabbits cause death in 4–6 days. Cocci in organs. No inflammatory reactions in tissues.
Habitat. Saliva of man.

36. **M. beta** Foutin

Coccus B Foutin: Centralblatt f. Bakteriologie, VI, 1890, 372.

Morphology. Cocci large, round, 1.0 μ; in twos, threes, and short chains.
Gelatin colonies. Surface: 6 days, round, 1.0 mm., white, slightly raised; microscopically, gray — yellowish gray, slightly granular towards periphery, smooth, and sometimes lobular.
Gelatin stab. The growth in depth shows lateral extensions; surface colony flat.
Agar slant. Growth white, smooth, glistening, rimmed.
Potato. A thin, scanty, almost transparent, whitish growth.
Pathogenesis. White rats inoculated intraperitoneally died in 5–6 days; cocci in blood, liver, and spleen.
Habitat. Isolated from hail.

37. M. aquatilis Vaughan

M. aquatilis-invisibilis Vaughan: Am. Jour. Med. Sci., 1892.

Morphology. Cocci oval.
Gelatin colonies. Spreading irregularly; microscopically deep brown.
Gelatin stab. Only a slight growth in depth; on surface growth spreading.
Agar slant. Growth thin, white.
Potato. Growth invisible.
Pathogenesis. Negative.
Habitat. Water.

38. M. tenacatis

M. No. 43 Conn: Conn. (Storrs) Ag. Expt. Sta. Report, 1894, 78.

Morphology. Cocci oval, 0.7–0.9 μ.
Gelatin colonies. Round, smooth, white; slightly raised.
Gelatin stab. In depth a good beaded growth; on surface colony raised, transparent, becoming thick, white.
Agar slant. Growth thin, transparent, tenacious, yellow.
Potato. No visible growth.
Milk. Unchanged, becoming slightly acid.
Bouillon. Turbid; dense yellowish sediment, becoming clear.
Habitat. Milk.

39. M. acidi-lactici (Lindner)

Pediococcus acidi-lactici Lindner: Centralblatt f. Bakteriologie, II, 1887, 342.

Morphology. Cocci 0.6–1.0, solitary, pairs, and tetrads.   Optimum temperature 41° C.
Gelatin colonies. Small, colorless, becoming yellowish brown.
Agar stab. On surface growth thin, colorless, moist, glistening.
Potato. Growth scanty, scarcely visible.   Lactic acid is produced in saccharine media.
Habitat. In mash from malt, and in hay infusions.

40. M. cerevisiæ (Balcke)

Pasteur: Études sur la bière, 1876; no name.
Pediococcus cerevisiæ Balcke · Wochenschrift f. Brauerei, 1884, 183.

Morphology. Cocci, single, in twos and as tetrads; involution forms.
Gelatin colonies. Small, colorless, becoming yellowish brown.
Gelatin stab. In depth a grayish stripe; on the surface growth white, leafy, spreading.

Agar slant. Growth moist, grayish white, iridescent, with smooth border.

Potato. A scanty, scarcely visible growth. Optimum temperature 25° C. In hop beerwort a good growth, with a pellicle, and production of lactic acid.

Habitat. In beer, air of breweries, water. A common cause of turbidity in beer.

41. M. fervidosus Adametz

Bak. Nutz u. Trinkwässer, 1888.

Morphology. Cocci 0.6 μ, diplococci, and in clumps.

Gelatin colonies. Deep: entire, light yellow, strongly refracting. *Surface:* 5-6 days, transparent, yellow, serrate-lobate; later colonies show a brownish granular centre, and a yellowish, slightly folded border.

Gelatin stab. In depth a beaded growth; on surface growth round, thin, and finely serrate.

Glycerin gelatin stab. Bubbles of gas.

Agar slant. Growth milk-white, slimy, iridescent.

Potato. Growth dirty white.

Habitat. Water.

42. M. rosettaceus Zimmerman

Bak. Nutz u. Trinkwässer, Chemnitz, 1899.

Morphology. Cocci round-elliptical, 0.7-1.0 μ; in clumps, like a bunch of grapes.

Gelatin colonies. Deep: round, entire. *Surface:* grayish yellow, glistening, drop-like expansions, with irregular contours.

Gelatin stab. In depth but slight growth; on surface colony round, spreading, with rosette-shaped expansions.

Agar slant. Growth smooth, gray, glistening, spreading.

Potato. Growth yellowish gray.

Habitat. Water.

43. M. ovalis Escherich

Die Darmbakterien des Säuglings, 1886, 90.

Morphology. Cocci oval, 0.2-0.3 μ; sometimes as short rods with a fission line.

Gelatin colonies. Small, not characteristic.

Gelatin stab. In depth a thin whitish growth; on surface little or no growth.

Potato. Growth rather abundant; white.

Milk. Coagulated, acid.

Habitat. In menconium and fæces of milk-fed infants.

44. M. lactis

M. No. 44, Conn: l.c. 1894, 79.

Morphology. Cocci round, very large, 1.5 μ.

Gelatin colonies. Round, smooth, raised, gray white.

Gelatin stab. A good growth in depth; on surface growth thin, transparent, and slightly spreading.

Agar slant. Growth thin, dry, whitish, iridescent, tough.

Potato. Growth yellowish-brownish.

Milk. Unchanged.

Bouillon. Turbid; no pellicle.

Habitat. Milk.

45. M. candicans Flügge

Die Mikroorganismen, 1886, 173.

Morphology. Cocci 1.0–1.2 μ; occur singly and in clumps, with generally a septum in centre. Grows at 37° C.

Gelatin colonies. *Deep:* round-lenticular, opaque, dark, entire. *Surface:* round, glistening, porcelain-white, slightly raised (convex), thinner on margin. In 8 days colonies 2–3 mm.

Gelatin stab. In depth growth white, filiform — beaded; on surface growth white, glistening, raised.

Agar slant. Growth white, limited, slightly raised, greasy.

Bouillon. Turbid with a pellicle, becoming clear.

Milk. Not coagulated; in 14 days slightly acid.

Potato. Growth thick, white, greasy, with wavy edges.

Glucose bouillon. No gas. Indol negative. Sulphuretted hydrogen negative.

Habitat. Air, water, milk, urine, etc.

46. M. nivalis

M. No. 47, Conn : l.c., 1894, 80.

Morphology. Cocci round, 0.4 μ. Grow at 35° C.

Gelatin colonies. Small, round, smooth, raised, whitish.

Gelatin stab. A good growth in depth; on surface growth thick, moist, slightly spreading.

Agar slant. Growth moist, snow-white, slightly spreading.

Potato. Growth very watery, spreading, snow-white.

Milk. Unchanged, becoming in 3 weeks slightly alkaline.

Bouillon. Turbid, in 4 weeks clear.

Habitat. From the air.

CLASS V. DOUBTFULLY CHROMOGENIC ; *i.e.* GROWTH ON GELATIN OR AGAR LIGHT YELLOW TO YELLOWISH WHITE. GELATIN LIQUEFIED.

I. Cocci surrounded by a capsule.
        47. *M. Heydenreichii*.
II. Cocci not surrounded by a capsule.
    A. Gonococcoid forms.
        48. *M. hæmorrhagicus* Klein.
    B. Cocci only as tetrads, merismopedia forms.
        49. *M. expositionis*.
    C. Cocci in irregular groups.
        1. Liquefaction of the gelatin takes place only at the surface — crateriform.
            50. *M. Finlayensis* Sternberg.
            51. *M. descidens* Flügge.
            52. *M. alpha* Foutin.
            53. *M. tetragenus-pallidus* Dyar.
        2. Gelatin liquefied along the length of the needle track.
            54. *M. cremoides* Zimmerman.

47. M. Heydenreichii

M. of Briska-button Heydenreich : Centralblatt f. Bakteriologie, V, 1888, 163.

Morphology. Diplococci 0.8–1.0 μ, or tetrads, surrounded by a capsule.

Gelatin stab. In 48 hours a grayish white beaded growth in depth ; on the surface growth thin, yellowish white ; in 3 days a liquefied funnel.

Agar slant. Growth grayish white — yellowish, — with a varnish-like lustre.

Potato. Growth white, yellow.

Pathogenesis. Inoculations of rabbits, dogs, chickens, horses, and sheep produce a skin affection identical with briska-button.

Habitat. Found by H. in pus and serous fluid from tumors and ulcers in an Oriental skin affection — briska-button.

48. M. hæmorrhagicus E. Klein

Centralblatt f. Bakteriologie, XXXII, 1897, 81.

Morphology. Cocci 0.4–0.6 μ, of gonococcoid type. Stain by Gram's method.

Gelatin colonies. Surface: 24 hours, small round gray points ; microscopically, with thick centres and thin borders. In 48 hours the gelatin softens, and colonies are surrounded by a zone of fluid gelatin ; the central colony slightly yellowish.

Agar colonies. In 24 hours, at 37° C., surface colonies round, white, and slightly raised; microscopically, brown, granular, light yellowish, thicker in the middle and with a narrow thin translucent border, 1–1.5 mm.; radially striped.

Agar slant. Growth flat, only slightly raised; white, becoming yellowish toward the centre; with a thin border.

Gelatin slant. A whitish gray streak, becoming slowly liquefied, with yellowish pulverent masses in liquid gelatin.

Gelatin stab. Growth in depth yellowish brown, beaded; on surface growth white, flat, spreading, becoming sunken; later a funnel with grayish white — yellowish sediment.

Glucose gelatin stab. No gas; a growth in depth.

Litmus bouillon. Reduced.

Milk. In 8–9 days at 37° C.; coagulated.

Litmus milk. In 4 days, red.

Potato. At 37° C.; growth grayish-yellowish.

Blood serum. Not liquefied; growth yellowish where thickest.

Habitat. Associated with an erythema of the skin, simulating anthrax.

49. M. expositionis

M. No. 34, Conn.: l.c., 1894. 77.

Morphology. Cocci 0.4 μ, with characteristics of merismopedia. Grow at 35° C.

Gelatin colonies. Surface: A liquid pit, with a nucleus, and cloudy edges.

Gelatin stab. A good growth in depth; on the surface a slight growth, not spreading, rather thick, and does not very thoroughly liquefy the gelatin.

Agar slant. Growth thick, moist, with thin edges tinged with yellow.

Potato. No visible growth.

Bouillon. Slightly turbid, with a tough yellowish sediment, becoming in 4 weeks clear.

Habitat. Milk.

50. M. Finlayensis Sternberg

Report on Etiology and Prevention of Yellow Fever, Washington, 1891, 219.

Morphology. Cocci 0.5–0.7 μ, solitary — pairs — tetrads — irregular masses.

Gelatin stab. A good growth in depth; on the surface a crateriform depression, lined with a viscid, light yellow layer of cocci.

Agar slant. Growth light yellow, viscid.

Pathogenesis. Negative for guinea pigs and rabbits.

Habitat. In cultures from liver and spleen of yellow fever cadaver.

51. M. descidens Flügge

Die Mikroorganismen, 1886.

Morphology. Small cocci and diplococci, or in threes and short chains.

Gelatin colonies. *Deep:* 2 days, small, white-yellowish; microscopically oval, yellowish brown, granular. *Surface:* 4 days, round-lobular, 5–10 mm., light yellowish-brownish smooth slimy expansions, not at all elevated; finally colonies sink in a flat circular depression.

Gelatin stab. In depth growth white, filiform; on surface growth yellowish brown, slimy, becoming sunken.

Potato. Growth yellowish brown, thick, slimy.

Habitat. Air and water.

52. M. alpha Foutin

Coccus A Foutin: Centralblatt f. Bakteriologie, VII, 1890, 372.

Morphology. Cocci round. Stain by Gram's method.

Gelatin colonies. In 4 days, elevated white disks; microscopically, darker in centre; periphery lighter and slightly granular.

Gelatin stab. But scanty growth in depth; on surface a yellowish nailhead; liquefaction begins in 5–6 days, and proceeds slowly.

Agar slant. Growth light rose-colored, smooth, glistening.

Potato. Growth like *B. typhosus.*

Pathogenesis. Negative for mice and guinea pigs.

Habitat. Found in hail.

53. M. tetragenus-pallidus Dyar

L.c., p. 354. According to Dyar, a variety of *M. tetragenus-vividus*, No. 74.

54. M. cremoides Zimmerman

Bak. Nutz u. Trinkwässer, Chemnitz, 1890.

Morphology. Cocci 0.8 μ, in clumps.

Gelatin colonies. *Deep:* round, entire, yellowish — grayish brown, granular. *Surface:* lobular-denticulate, concentric, crateriform; the yellowish white colony lines the bottom; microscopically, a brownish yellow granular centre, and a less dense granular border, with radial extensions.

Gelatin stab. In 3 to 4 days liquefaction extends along the line of the needle; on the surface, a bubble of gas lined with a yellowish white growth; a yellowish white sediment in the liquefied gelatin, and a pellicle on the surface.

Agar slant. Growth yellowish white, glistening, granulated.
Potato. Growth raised, creamy.
Habitat. Water.

CLASS VI. DOUBTFULLY CHROMOGENIC, *i.e.* GROWTH ON GELA-
TIN OR AGAR LIGHT YELLOW OR YELLOWISH WHITE. GEL-
ATIN NOT LIQUEFIED.

I. Cocci very large, 1.5–2.0 microns in diameter.
    55. *M. citreus* Sternberg.
II. Cocci smaller than the preceding.
  A. Growth in gelatin stab beset with radiate acicular extensions.
    56. *M. plumosus* Bräutigam.
  B. Gelatin stab cultures not as above specified.
    1. Grow well in nutrient gelatin.
      57. *M. viridis* (Guttmann).
    2. Growth in gelatin slight, colonies very small.
      58. *M. versicolor* Flügge.
      59. *M. tardigradus* Flügge.
      60. *M. Jongii.*
    3. No growth in gelatin.
      61. *M. subflavus* v. Besser.

55. **M. citreus** Sternberg

Cremfarbiger micrococcus List: Adametz, Bak. Nutz u. Trinkwässer, 1888.
M. citreus Sternberg: Manual of Bacteriology, 1892, 599.

Morphology. Cocci very large, round, 1.5–2.2, occur singly, as diplococci, or
    as chains of 8 or more elements.
Gelatin colonies. Elevated, irregular, moist, glistening, dirty yellow — cream-
    colored.
Gelatin stab. Only a slight growth in depth.
Agar slant. Growth light yellow.
Potato. At 37°, a yellowish growth.
Habitat. Water.

56. **M. plumosus** Bräutigam

Adametz, Bak. Nutz u. Trinkwässer, 1888.

Morphology. Cocci round, 0.8 μ, forming zoöglœa.
Gelatin colonies. Raised, lobed, yellowish white.

Gelatin stab. The growth in depth is beset with echinulate-acicular extensions; on surface, growth slimy, with radiating acicular outgrowths.

Potato. Growth yellowish white, with lobular offshoots.

Habitat. Water.

57. M. viridis (Guttmann)

Staphlococcus viridis-flavescens Guttmann: Virchow's Archiv, CVI, 1887, 259.

Morphology. Cocci of irregular sizes, solitary, in pairs and clumps.

Gelatin colonies. Both deep and surface colonies grayish yellow.

Agar stab. At 37° C., 24 hours, growth grayish yellow.

Potato. Growth at 37° C. abundant.

Habitat. Water (?).

58. M. versicolor Flügge

Die Mikroorganismen, 1886, 177.

Morphology. Cocci in twos and clumps.

Gelatin colonies. *Deep:* round, entire, yellowish gray, opaque, granular, 1 mm. *Surface:* in 4 days 10 mm., irregular—quadrangular—lobed, slimy, glistening, yellowish gray, iridescent, often punctiform in centre.

Gelatin stab. In depth, yellowish, beaded; on the surface, like gelatin colonies.

Habitat. Air.

59. M. tardigradus Flügge

Die Mikroorganismen, 1886, 175.

Morphology. Cocci large, round, in clumps, with often darker poles.

Gelatin colonies. *Deep:* round-oval, entire, dark olive-green, 0.4–0.6 mm. *Surface:* slightly elevated, smooth, glistening, 0.5–1.0 mm.; microscopically, grayish yellow, lighter toward the periphery.

Gelatin stab. In depth, after 6–8 days, a beaded growth, composed of yellowish colonies.

Habitat. Air.

60. M. Jongii

Staphlococcus pyogenes bovis Jong: Centralblatt f. Bakteriologie, XXV, 1899, 67.

Morphology. Cocci 0.6–1.0 μ, variable in size, singly, and in clumps. Stain with ordinary colors. Stain by Gram's method.

Gelatin colonies. In 24–48 hours, minute, round-oval, whitish yellow to yellow; microscopically, light — dark brown.

Gelatin slant. Yellow round colonies, or a yellow layer.

Gelatin stab. In depth, white — yellowish white round-oval colonies; on the surface, a minute golden yellow growth.

Agar slant and *glycerine agar slant.*   A good growth, which at 37° C. is yellow, and at 22° C. a whitish yellow.

Blood serum.   Growth whitish yellow.

Potato.   Growth glistening, whitish yellow.

Milk.   Not coagulated.

Bouillon.   Turbid, stringy sediment.   Media first rendered alkaline, then acid.   *Indol* negative.

Glucose, lactose, and *saccharose bouillon.*   No gas.

Pathogenesis.   Dogs, rabbits, and guinea pigs negative by subcutaneous, intravenous, and intraperitoneal injections.

Habitat.   Associated with suppurating processes in cattle.

61. **M. subflavus** v. Besser

M. tetragenus-subflavus v. Besser : Ziegler's Beiträge path. Anat., 1889, 331.

Morphology.   Cocci round-oval, medium-sized, in tetrads.   Do not grow in nutrient gelatin.

Agar colonies.   Flat, dirty white, glistening, 0.5 mm., with wrinkled margins ; microscopically, a small brown nucleus, a grayish brown irregular-striped zone, and a wrinkled outer margin.

Agar slant.   Growth flat, spreading, grayish white ; later, the color of *M. pyogenes aureus.*

Potato.   Growth pale brown.

Habitat.   Isolated from nasal mucus.

CLASS VII. DISTINCTLY CHROMOGENIC. FORMS A PIGMENT ON GELATIN OR AGAR. PIGMENT YELLOWISH-ORANGE. GELATIN LIQUEFIED.

  I. Typical gonococcoid forms.

        62. *M. epidermis.*

 II. Sarcina, or typical tetrad grouping of the elements.

        63. *M. albus* (Lindner).

III. Staphlococcus grouping of the elements.

   A. Gelatin liquefied along the length of the stab, or to a considerable depth.

      1. Gelatin rapidly liquefied.

          64. *M. pyogenes* var. *aureus* Rosenbach.

          65. *M. aureus* Dyar.

          66. *M. flavus* Flügge.

2. Gelatin liquefied slowly, or imperfectly.
>67. *M. luteus* Cohn.
>68. *M. Uruguæ.*

B. Gelatin liquefied only on the surface; shallow, crateriform.
>1. Colonies with distinct lobular projections from the central body.
>>69. *M. conglomeratus* (Bumm) Flügge.
>>70. *M. flavus* Flügge.
>2. Colonies not specified as before.
>>*a.* Potato cultures salmon-colored — brown; rough — rugose.
>>>71. *M. rugosus.*
>>*b.* Potato cultures yellow, not rugose.
>>>72. *M. orbicularis* Ravenel.
>>>73. *M. Tommasoli.*
>>>74. *M. versatilis* Sternberg.

62. M. epidermis

Diplococcus flavus-liquefaciens Unna: Tommasoli, Monatshefte f. prakt. Dermatol., IX, 56.

Morphology. Cocci 0.5–0.8 μ, like gonococci.

Gelatin colonies. *Deep:* small, round, opaque, olive-brownish yellow. *Surface:* in 8 days, very small, round, glistening, light-grayish yellow; later, 3 weeks, chrome yellow — greenish yellow, surrounded by a zone of liquefied gelatin.

Gelatin stab. Growth in depth, thin, yellowish; on surface, thin, yellowish white, slimy; in 3 weeks 3–4 mm.; later, depressed, due to liquefaction; in 8 weeks one-half of the tube is liquefied. Gelatin yellowish, with a yellowish sediment.

Agar slant. Growth thick, slimy, yellowish white — greenish yellow.

Potato. Growth sulphur-yellow.

Habitat. From the skin in eczema.

63. M. albus (Lindner)

Pediococcus albus Lindner: Die Sarcina Organismen der Gährungsgewerbe, Berlin, 1888; ref. Bot. Centralblatt, 1888, 99.

Morphology. Cocci solitary, in pairs and tetrads, frequently in pseudo-sarcina.

Gelatin colonies. Round, becoming liquefied.

Gelatin stab. In 24 hours, a deep channel of liquefied gelatin, with a whitish sediment, becoming pale orange.

Agar slant. Growth broad, dry; old cultures have an orange color.

Potato. Growth dirty white. An acid reaction of the medium.

Habitat. From water and Weiss beer, causing a slight acidity of the latter.

H

64. M. pyogenes var. aureus Rosenbach

l.c., No. 6.

Morphology. Cocci 1.0 μ, in irregular clumps. Stain by Gram's method.

Gelatin colonies. *Deep:* round-oval, granular, brownish. *Surface:* whitish yellow points, becoming more yellowish; not increasing in size, but becoming sunken in the liquefied gelatin. Microscopically, round; border undulate — erose — lacerate; punctate, finely granular, translucent.

Gelatin stab. Liquefaction saccate, with a yellowish film on the surface. Gelatin turbid, with a yellowish sediment.

Agar slant. An abundant opaque, smooth, moist, glistening layer, becoming bright orange.

Potato. Growth abundant, orange, or rather scanty.

Bouillon. Turbid, becoming clear, with abundant yellowish sediment.

Milk. Coagulated, acid. *Indol* negative. H_2S, slight.

Glucose bouillon. No gas; acid.

Pathogenesis. Virulence of cultures variable. *Rabbits:* Subcutaneous inoculations cause acute local inflammation and suppuration. The cocci multiply in the lymph spaces, and may be found within the leucocytes, and also invading the capillary walls; usually the cocci confined to the local centres of suppuration. Intraperitoneal injections cause suppurative peritonitis, either local or spreading, resulting in death. Subcutaneous inoculations of human subjects result in local suppuration.

Habitat. Widely distributed; in ulcerative endocarditis, osteomyelitis, and a variety of inflammatory and suppurative processes in the body. For etiological relations, see text-books.

VARIETIES.

Staphlococcus salivarius-pyogenes Biondi: Die path. Microorg. des Speichels, Zeitsch. f. Hygiene, II, 1887, 1094.

Micrococcus of Almquist: Zeitsch. f. Hygiene, X, 1891, 253.

In *pemphigus.* Organism possesses specific pathogenic power. Almquist, by inoculation of his own arm, caused bullæ like those of phemphigus.

Staphlococcus quadrigeminus Czaplewski: Centralblatt, Bakteriologie, XXV, 143.

In many points similar to *M. pyogenes-aureus.*

Diff. colonies on Löffler's blood serum show in a short time a transparency and liquefaction of the medium. Stain by Gram's method, but more easily decolorized. Grows more slowly in gelatin, and forms an air bubble in gelatin above the funnel of liquefaction. Pigment like *aureus,* but with a rose-colored tinge.

Habitat. Isolated from vaccine lymph.

65. M. aureus Dyar

M. cremoides-aureus, Dyar: l.c.

Morphology. Cocci 1.0–1.2 μ; associated irregularly.

Gelatin colonies. *Deep:* round, light yellow, opaque. *Surface:* cups of liquefied gelatin are quickly formed, and masses of opaque orange flocculi form
a ring about a clear central nucleus.

Gelatin stab. A liquefaction along the line of stab.

Milk. Coagulated, becoming peptonized.

Pepton-rosolic acid solution. Unchanged.

Lactose litmus. Reddened.

Agar slant. Growth limited, glistening, orange.

Glycerine agar. Growth scarcely chromogenic.

Nitrates. Not reduced.

According to Dyar, a variety of *M. cremoides* Zim., No. 54.

Habitat. Air.

66. M. flavus Flügge

M. flavus-liquefaciens Flügge: Die Mikroorganismen, 1886.

Morphology. . Cocci rather large, in twos, threes, and in groups.

Gelatin colonies. *Deep:* round, oval, often lobular. *Surface:* yellowish brown,
finely serrate, becoming surrounded by a zone of liquefied gelatin, containing the central colony with isolated radial processes from the latter;
colonies 4–6 mm.

Gelatin stab. Growth in depth beaded, yellowish; gelatin rapidly liquefied
with a yellowish flocculent sediment.

Potato. Growth irregular, intense yellow.

Habitat. From the air and water.

67. M. luteus Cohn

Beiträge Biologie, I, 1870.

Morphology. Cocci 1.0–1.2 μ, round, in twos and fours; form zoöglœa. Grow
at 37° C.

Gelatin colonies. *Deep:* round-elliptical, entire, granular, yellowish gray.
Surface: yellowish — yellowish white, raised, round-irregular. In 3 days
1–2 mm., becoming sunken. Microscopically, yellowish gray — gray-
brown; border undulate — lobed.

Gelatin stab. Growth in depth beaded, becoming slowly liquefied, cylindrical.
(According to Frankland, gelatin not liquefied.)

Agar slant. Growth citron-yellow.

Bouillon. Clear; yellowish sediment.

Milk. In 20 days partly coagulated; acid.

Potato. Growth thin, glistening, citron-yellow, becoming wrinkled (?).

Habitat. Air and water.

68. M. Uruguæ

M. No. 40 Conn: l.c., 1894, 78.

Morphology. Cocci 0.9 μ.

Gelatin colonies. A little pit, with a central granular yellowish nucleus, and an outer lobate rim.

Gelatin stab. In depth a narrow funnel or pit, which is quite dry. There is formed a thick syrup with a yellowish sediment and a yellowish scum. The gelatin never completely liquefies even after weeks of growth.

Agar slant. Growth Naples yellow, rough, dry.

Potato. Growth Naples yellow, dry, thick, mounded.

Milk. Coagulated in 5 days; acid.

Bouillon. Remains clear, with a slight flaky scum, and a slight sediment.

Habitat. Milk.

69. M. conglomeratus (Bumm) Flügge

Diplococcus citreus-conglomeratus Bumm: Der Mikroorg. der gon. Schleimhauterkrankungen Wiesbaden, 1885, 17.

M. citreus-conglomeratus Flügge: Die Mikroorganismen, 1886.

Morphology. Gonococcoid forms, frequently in tetrads, usually in conglomerate masses.

Gelatin colonies. Deep: lemon-yellow, with lobular projections. *Surface:* moist glistening, becoming cleft — scaly.

Gelatin stab. A crateriform liquefaction, with a yellow pellicle on surface.

Habitat. From gonorrheal pus, air, dust.

70. M. flavus Flügge

M. flavus-liquefaciens Flügge: l.c., 1886. (See No. 66.)

Morphology. Cocci large; in twos and in clumps.

Gelatin colonies. Deep: round-oval with sometimes lobular projections. *Surface:* yellowish brown, finely serrate; surrounded by a zone of liquefied gelatin, with radial extensions from the central nucleus.

Gelatin stab. In depth, yellowish colonies; on surface, yellowish colonies becoming confluent, and later depressed by liquefaction of the gelatin to a depth of 2 mm.

Potato. Growth deep yellow.

Habitat. Air, water.

71. M. rugosus

M. No. 2 Conn: l.c., 1893, 50.

Morphology. Cocci 1.0–1.2; never form chains. Grow at 37° C.

Gelatin colonies. A slight pit of liquefied gelatin becoming coarsely granular — fragmental.

Gelatin stab. Liquefaction shallow crateriform, only one-fourth of the gelatin liquefied after several weeks.

Agar slant. Growth dry, raised, limited, rugose, tenacious, sticky, Naples yellow.

Potato. Growth, thick, dry, rough — rugose, flesh-colored — salmon-brown.

Milk. Coagulated, alkaline; butyric acid and alcohol; orange masses on the surface.

Bouillon. Clear, no pellicle, a slight sediment.

Habitat. Milk.

72. M. orbicularis Ravenel

M. orbicularis-flavus Ravenel: l.c., p. 8.

Morphology. Cocci large, in irregular groups. Slight growth at 36° C.

Gelatin colonies. *Deep:* irregular, finely granular, yellow. *Surface:* in 3 days minute yellowish dots; in 5 days colonies 1 mm., round, entire, slightly elevated, becoming crateriform. Microscopically, a homogeneous centre, with granular entire margins.

Gelatin stab. On the surface a small yellowish button, becoming crateriform, and in 7 days a stratiform liquefaction, with a yellowish flocculent sediment.

Agar slant. Growth faint yellowish, glistening, limited, becoming canary-yellow.

Potato. Growth thin, moist, spreading, colorless, becoming thicker and yellow; then granular, moist, glistening.

Bouillon. Turbid, whitish sediment, becoming faint yellow.

Pepton–rosolic acid becomes slightly darker in 2 weeks.

Litmus milk. Unchanged.

Glucose bouillon. No gas. *Indol* slight.

Nitrites. Negative.

Habitat. Soil.

73. M. Tommasoli

Diplococcus citreus-liquefaciens Unna: Tommasoli, Monatsch. f. prakt. Dermatol., IX, 56.

Morphology. Cocci small oval, in twos, tetrads, clumps, and short chains.

Gelatin colonies. *Deep:* round-oval, entire, brownish yellow. *Surface:* in 4 ~~round~~, flat, grayish white; in 8 days 1–2 mm.; in 2 weeks lemon- crateriform liquefaction.

Gelatin stab. Surface growth thin, glistening, yellowish; in 3 weeks the gelatin is liquefied to a depth of 6 mm.; gelatin opaque, yellowish.

Agar slant. Growth yellowish brown.

Potato. Growth grayish yellow.

Habitat. Isolated from the skin of persons suffering from eczema suborrheicum.

74. M. versatilis Sternberg

M. tetragenus-versatilis Sternberg: Manual of Bacteriology, 1892.

Morphology. Cocci 0.5–1.5 μ; in tetrads and irregular groups. Stain by Gram's method.

Gelatin colonies. Round, opaque, light yellow — lemon-yellow. Liquefaction begins after several days, and progresses slowly.

Gelatin stab. In depth, slight growth; on surface, a crateriform liquefaction, with a yellowish viscid sediment.

Agar slant and *potato.* Growth thick, viscid, yellow, moist, glistening, spreading.

Pathogenesis. Rabbits and guinea pigs negative.

Habitat. Isolated from the excrement of mosquitoes which had sucked the blood of yellow fever patients, and from the air.

74 a. M. tetragenus-vividus Dyar

N. Y. Acad. of Sciences, VIII, 354.

Probably a variety of or identical with the preceding.

CLASS VIII. DISTINCTLY CHROMOGENIC. PIGMENT YELLOWISH ORANGE. GELATIN NOT LIQUEFIED

I. Cocci in tetragenous groups.
  75. *M. varians* (Dyar).
  76. *M. Vincenzii.*
II. Cocci in Staphlococcus groupings.
  A. Growth on potato yellowish.
    1. Pigment orange-yellow.
      77. *M. aurantiacus* (Schröter) Cohn.
    2. Pigment lemon-yellow.
      78. *M. cereus* Passet.
    3. Pigment brownish yellow.
      79. *M. orbiculatus* Wright.
  B. Growth on potato invisible.
    80. *M. aerius.*

75. M. varians (Dyar)

Merismopedia flava-varians Dyar, l.c., p. 346.

Morphology. Cocci 1.0 μ, in twos and tetrads.

Gelatin colonies. Deep : round-irregular, opaque, yellow. *Surface :* opaque, dull, light yellow ; edge slightly wavy.

Lactose litmus. Reddened.

Milk. Coagulated only on boiling.

Bouillon. Turbid ; yellow sediment.

Glucose agar. Growth bright yellow, opaque.

Potato. As before.

Nitrates. Reduced.

Habitat. Air.

76. M. Vincenzii

M. tetragenus-citreus Vincenzi : La Riforma Med., 1897. 758.

Morphology. Cocci in tetrads. Do not stain by Gram's method. Facultative anaerobic. Grow at 37° C.

Gelatin colonies round, yellowish ; gelatin liquefied in 6-12 days.

Gelatin stab. In depth, a beaded growth ; on surface, gelatin softened, not fluid.

Agar slant. Growth yellowish.

Bouillon. A citron-yellow sediment.

Milk. Not coagulated ; a citron-yellow sediment.

Pathogenesis. Rabbits, guinea pigs, and mice, negative.

Habitat. Isolated from the submaxillary lymphatic gland of a child.

77. M. aurantiacus (Schröter) Cohn

Bacteridium aurantiacus Schröter : Beitr. z. Biol., I, 1870, 119.
M. aurantiacus Cohn : Beiträge, I, 1870, 154.

Morphology. Cocci round — slightly oval, 1.3-1.5 μ ; occur singly, in twos and small clumps.

Gelatin colonies. Surface : round-oval, smooth, glistening, with orange-yellow centres. Microscopically, finely granular.

Gelatin stab. In depth, small yellow colonies ; on the surface, a similar development.

Agar slant. Growth orange-yellow, spreading.

Potato. Growth slimy, yellow ; pigment insoluble in alcohol and ether.

Habitat. Isolated from the air and from water.

78. M. cereus Passet

M. cereus-flavus Passet: Fortschritte d. Medezin, III, 1887.
Tils: Zeitsch. f. Hygiene, IX, 1890, 300.

Morphology. Cocci of variable sizes; in clumps and occasionally in chains.

Gelatin colonies. Lemon-yellow, becoming 1–2 mm.

Gelatin stab. In depth, growth scanty, yellow; on the surface, the growth resembles drops of stearine or wax, yellow, with elevated margins.

Potato. Growth citron-yellow.

Habitat. Isolated from a human abscess.

79. M. orbiculatus Wright

L.c., p. 432.

Morphology. Cocci rather large, in pairs, tetrads, and small clumps. Grow at 36° C.

Gelatin colonies. Deep: round, entire, opaque, yellowish. *Surface:* in 5 days rounded, slightly elevated, white, glistening, 1 mm., becoming 2–3 mm.; dark yellow, pale margins. Microscopically, brownish yellow — brown, dense, granular.

Gelatin slant. Growth brownish yellow, glistening, limited, rather thick.

Agar slant. Growth slight, composed of discrete and confluent colonies; grayish-yellowish.

Bouillon. Alkaline, with a yellowish stringy sediment.

Potato. Growth yellow, glistening, thick, spreading.

Litmus milk. Not coagulated; after several weeks slightly pink.

Pepton-rosolic acid solution. Slowly decolorized, alkaline. Indol, negative.

Habitat. Water.

80. M. aerius

M. No. 49 Conn: l.c. 1894, 81.

Morphology. Cocci 1.0. Slight growth at 35° C.

Gelatin colonies. Not characteristic.

Gelatin stab. In depth, slight growth; on the surface, a slight orange growth.

Agar slant. Growth thick, dark orange, quite transparent.

Potato. Growth invisible.

Milk. Unchanged.

Bouillon. Turbid, becoming in 4 weeks clear.

Habitat. Milk.

CLASS IX. DISTINCTLY CHROMOGENIC. PIGMENT REDDISH-
PINKISH — FLESH-COLORED

I. Gelatin liquefied.
    A. Gonococcoid forms.
        81. *M. roseus* (Bumm).
    B. Cocci in irregular groups.
        1. Grow only at body temperatures, 37° C.
        82. *M. rubescens.*
        2. Do not grow at body temperatures, 37° C.
        83. *M. coralinus* Centanni.
II. Gelatin not liquefied.
    A. Very large cocci, 3.0–5.0 μ in diameter.
        84. *M. Dantecii.*
    B. Cocci much smaller than in *A.*, and of average size.
        1. Colonies on gelatin remain very small.
        85. *M. lactericeus* v. Dobrzyniecki.
        86. *M. cinnabareus* Flügge.
        2. Colonies on gelatin larger than above.
        87. *M. Kefersteinii.*
        88. *M. carneus* Zimmerman.
    C. Cocci very small.
        89. *M. cerasinus* Eisenberg.

81. M. roseus (Bumm)

Diplococcus roseus Bumm : Der Mikroorg. der gon. Schleimhauterkrankungen,
1885, 25.

Morphology. Like gonococci, 1.0–1.5 μ.
Gelatin colonies. Surface: slightly elevated, pink ; microscopically granular,
irregular.
Gelatin stab. On the surface an abundant pink growth ; gelatin slowly lique-
fied after a long time.
Habitat. Air.

82. M. rubescens

M. roseus, Eisenberg : Bak. Diag., 1891, 408.

Morphology. Cocci 0.8–1.0 μ, singly and in irregular groups. At 37° C. no
pigment.
Gelatin colonies. In 3-4 days, minute, pink ; liquefaction progresses slowly.

Gelatin stab. In depth a good growth; on the surface growth colorless, becoming in 3–4 days round, pink, depressed. In 3 weeks one-half of the gelatin is liquefied, with a pink sediment.

Agar slant. Growth soft, dark pink.

Potato. In 3–4 days a cherry-red streak becoming spreading, and darker, like *B. prodigiosus.*

Habitat. Isolated from sputum of an influenza patient.

83. M. coralinus Centanni

Centralblatt f. Bakteriologie, XXIII, 1898, 308.

Morphology. From blood-agar cultures, rather large, round, in twos, threes, tetrads, and clumps. Stain by Gram's method. Aerobic. On media at room temperatures grow slowly; grow best on Vöges blood-agar. No growth at 37° C. Optimum temperature 20°–25° C.

Blood-agar colonies. In 48 hours small points, becoming in 6 days confluent, dark coral-red; agar not stained.

Agar slant. Growth slight; very small white colonies, becoming a delicate layer, becoming in 10 days rose-red.

Gelatin. Slight growth; coral-red colonies, with slow liquefaction after 20–25 days.

Gelatin stab. In depth but slight or no growth; on the surface a small coral-red colony; gelatin liquefied after many days. Grow well on Capaldi's egg-agar. On *milk agar* a soft white growth, becoming rose-red.

Bouillon. Slight growth; in 15–20 days medium a slight red—rose-red; clear. Grow better on addition of glucose.

Milk. Not coagulated, becoming in 15–20 days yellowish red; the cream layer remains white.

Potato. Slight growth; in 20–30 days rounded colonies of a dark carmine color. *Pigment:* Only slightly soluble in water and alcohol.

Pathogenesis. Rabbits and guinea pigs die in 4–5 days, with toxic symptoms and emaciation; at the point of inoculation a slight infiltration of a reddish color.

Habitat. Isolated from a contaminated plate culture.

84. M. Dantecii

Micrococcus   Danteci: Annales Pasteur Institut, 1891, 659.

Morphology. Cocci 3.0–5.0 μ, often with a fission line.

Gelatin colonies. Surface: Small, disk-shaped, red; grow slowly; ± 1 mm.

Gelatin stab. In depth, a slight yellowish growth; on the surface, light red, becoming later deep red.

Agar slant. Grows more rapidly than on gelatin.

Pathogenesis. Negative.

Habitat. Isolated from salted codfish which was covered with a red pigment and of an offensive odor.

85. M. lactericeus v. Dobrzyniecki

Centralblatt f. Bakteriologie, XXI, 1897, 834.

Morphology. Cocci under 1.0 μ, irregularly arranged. Stain by Gram's method.

Gelatin colonies. Surface: dust-like points, bright rose-colored; microscopically, round, entire, brownish, granular.

Agar colonies. As before.

Gelatin stab. In depth, a slight beaded growth; on the surface, in 2–3 days, a brick-red growth.

Agar slant. Growth moist, limited, brick-red.

Potato. Growth brick-red.

Pathogenesis. Negative for mice and rabbits.

Habitat. Isolated from the human mouth.

86. M. cinnabareus Flügge

Die Mikroorganismen, 1886.

Morphology. Cocci large, round, in twos, threes, and tetrads.

Gelatin colonies. Feeble growth; in 8 days, colonies 0.5–1.0 mm., raised, red, becoming vermilion; microscopically, light brown, round, transparent at their peripheries.

Gelatin stab. In depth, a beaded growth; on the surface, growth raised, pink-vermilion.

Potato. Scanty growth, vermilion.

Habitat. Air and water.

87. M. Kefersteinii

Micrococcus of red milk Keferstein: Centralblatt f. Bakteriologie, XXI, 1897, 177.

Morphology. Cocci in staphlococcus groups. Stain by Gram's method.

Gelatin colonies. In 4–6 days, scarcely visible, pale rose-colored, becoming larger, round, entire, concentric; microscopically, round, entire, granular.

Gelatin stab. In depth, a slight growth; no pigment; on the surface, as in gelatin colonies.

Bouillon. At 37° C. no growth; in 5–6 days, a slight sediment.

Pathogenesis. Mice negative.

Habitat. Isolated from red milk.

88. M. carneus Zimmerman

Bak. Nutz u. Trinkwässer, Chemnitz, 1890.

Morphology. Cocci medium-sized, ± 0.8 μ; in clumps. At 30°–33° C., but
slight growth.

Gelatin colonies. Deep: small, grayish white. *Surface:* round, slightly raised,
grayish — light red; microscopically, edges entire, homogeneous, grayish
red; later a darker centre, with lighter concentric zones.

Gelatin stab. Growth in depth fine, white, granular; on the surface, growth
round-irregular, thin, light red.

Agar slant. Growth deep flesh-red, with a play of violet.

Potato. Growth spreading, color of red lead, glistening, becoming dull.

Habitat. Water.

89. M. cerasinus Eisenberg

Micrococcus . . . List: Untersuch. gesunden Schafes vorkommenden Pilze, Inaug.
Diss. Leipzig, 1885, 17.

M. cerasinus-siccus Eisenberg: Bak. Diag., 1891, 34.

Morphology. Cocci very small, 0.2–0.3 μ; occur singly and in twos. Grow
best at 37° C.

Agar stab. No growth in depth; on the surface, growth dry, dull, spreading,
cherry-red.

Potato. Growth at 37° C., dry, spreading, red. Pigment insoluble in water,
alcohol, and ether; not affected by acids and alkalies.

Habitat. Water.

CLASS X. DISTINCTLY CHROMOGENIC. PIGMENT BLUISH BLACK

I. Gelatin liquefied.
   90. *M. fuscus* Adametz.
II. Gelatin not liquefied.
   91. *M. cyaneus* (Schröter) Cohn.

90. M. fuscus Adametz

Brauner Coccus Maschek: Bak. Untersuch. Leitmeritzer Trinkwässer, Jahresber. Kom-
nunal-Oberrealschule, Leitmeritz, 1887, 60.

M. fuscus Adametz: Die Bak. Nutz u. Trinkwässer, 1888.

Morphology. Cocci, elliptical forms.

Gelatin colonies. Round; microscopically, light brown — blackish, with fine
clefts.

Gelatin stab. Slight growth in depth; on the surface the liquefied gelatin is sepia-brown, with a pellicle on the surface; cultures have a foul odor.
Potato. Growth, slimy, brown-black.
Habitat. Water.

91. M. cyaneus (Schröter) Cohn

Bacteridium cyaneum Schröter : Cohn, Beiträge, I, 1870, 122.
M. cyaneus Cohn Beiträge, I, 1870, 156.

Morphology. Cocci elliptical; form zoöglœa.
Gelatin colonies. Small, round, well-defined; microscopically, bluish, surrounded by an irregular network.
Gelatin stab. Slight growth in depth; on the surface a slimy growth.
Potato. Slight growth, dark indigo blue. Pigment resembles litmus in color.
Habitat. Air and water.

SARCINA (Goodsir)

Single cocci spherical. Cells after division remain united. Division in three directions of space, resulting in eight-celled cubes, the packets from this increasing in geometric ratio. Non-motile; without flagella. Endospores not certainly present in any of the species.

SYNOPSIS OF THE GENUS

I. Without pigment on gelatin or agar.
  A. Potato cultures brownish yellow.
    1. *Sarc. pulmonum* Hauser.
  B. Potato cultures remain white — grayish white.
    1. Gelatin colonies microscopically very finely granular. Liquefaction of the gelatin slight.
    2. *Sarc. alba* Zimmerman.
    2. Gelatin colonies microscopically rather coarsely granular. Liquefaction of the gelatin rapid.
    3. *Sarc. canescens* Stubenrath.
II. On gelatin and agar, growth grayish yellow — greenish yellow — chrome-yellow.
  A. Gelatin colonies microscopically very finely granular.
    1. Growth on potato.
    4. *Sarc. flava* De Bary.
    5. *Sarc. lactis.*
    2. No growth on potato.
    6. *Sarc. subflava* Ravenel.

   B. Gelatin colonies microscopically rather coarsely granular; form beautiful regular packets.
- 1. Potato cultures at first dark gray, later yellowish brown.
  - 7. *Sarc. lutescens* Stubenrath.
- 2. Potato cultures remain grayish yellow.
  - 8. *Sarc. equi* Stubenrath.

   C. Gelatin colonies very coarsely granular. Form very beautiful, regular packets. Potato cultures from the start citron-yellow.
- 9. *Sarc. lutea* Flügge.

III. On agar and gelatin the growth is orange-yellow.
- 10. *Sarc. aurantiaca* Flügge.

IV. On agar and gelatin the growth is brownish — brownish yellow.
   A. Agar slant cultures smooth; pure brown.
- 11. *Sarc. cervina* Stubenrath.

   B. Agar slant cultures rugose, corrugated, yellowish brown, translucent.
- 12. *Sarc. fusca* Gruber.

V. On gelatin and agar the growth is red.
   A. Gelatin and agar stab cultures rose-red. Sarcina forms observed only in hay infusions.
- 13. *Sarc. rosea* Schröter.

   B. Gelatin and agar slant cultures, bright red. Sarcina forms observed only in hay infusions.
- 14. *Sarc. erythromyxa* (Overbeck).

1. Sarc. pulmonum (Hauser)

Deutsche Archiv klin. Med., XLII, 1887.

Morphology. Cocci on different media as small, and not very regular packets. Grow slowly, even at body temperatures.

Gelatin colonies. *Deep:* round, gray, opaque. *Surface:* small, round, punctiform, yellowish white; microscopically, round, entire, gray, opaque, becoming sunken in 2-3 weeks; central colony lacerate.

Gelatin stab. In depth, the growth is uniform, beaded, gray — yellowish gray; on the surface, in 20 days, growth 2-3 mm., gray, roundish, lacerate, soft, glistening, becoming sunken.

Agar colonies. *Deep:* round, dark, granular. *Surface:* like gelatin colonies, but whiter; microscopically, round, bright — dark gray, granular.

Agar slant. Growth limited, mealy, gray-white, translucent; border undulate.

Bouillon. Clear, granular sediment.

Milk. Not coagulated.

Potato. Slight growth; in 3–4 weeks, a limited brownish glistening streak.
Pathogenesis. Negative.
Habitat. Isolated from air passages of man.

2. Sarc. alba Zimmerman

Bak. Trink u. Nutzwässer, 1890, 90.

Growth on different media, white — grayish white, generally very thin.

3. Sarc. canescens Stubenrath

Lehmann-Neumann, Bak. Diag., 1896, 143.

Imperfectly described. Form beautiful, regular packets.

4. Sarc. flava De Bary

Morphology. Packets not regular in form.
Potato. Growth chrome yellow, glistening.
Gelatin. Liquefied or unchanged.
Habitat. Isolated from stomach contents.

5. Sarc. lactis

No. 45 Conn: l.c. 1894.

Morphology. Cocci 0.7 μ. Grow at 20°, and at 35° C.
Gelatin colonies. A large sunken colony, with a nucleus and a granular
    border.
Gelatin stab. In depth, a slight growth; on the surface, growth yellowish,
    dry.
Agar slant. Growth raised, slightly spreading, bright yellow.
Potato. Growth of a yellowish tinge.
Milk. Unchanged.
Bouillon. Clear, with sediment and flakes on the sides of the tube.
Habitat. Milk.

6. Sarc. subflava Ravenel

Memoirs National Acad. Sci., VIII, 1896.

Morphology. Packets square, and longer than broad, showing 4, 8, 16, 32,
    and more elements on each face.

Gelatin colonies. In 36–48 hours, minute yellowish dots, becoming yellowish
   granular entire disks, becoming in 4 days 1 mm., pale yellow and slightly
   sunken; liquefaction slow. Microscopically, pale yellow, entire, homo-
   geneous; later with irregular margins and slightly granular.
Agar slant. A yellowish band, 3 mm. wide, smooth, pale yellow, with irregular
   margins.
Gelatin stab. On the surface a white irregular button, and a slow crateriform
   liquefaction; in 10 days a stratiform liquefaction to a depth of 10 mm.
   Liquefied gelatin cloudy.
Potato. No growth.
Bouillon. Clear, with white flocculi at the bottom, and a pellicle on the surface. ·
Pepton-rosolic acid. Unchanged.
Litmus milk. Unchanged.
Glucose gelatin stab. A good growth in depth; no gas.   *Indol* negative:
   Growth at 36° C.
Habitat. Soil.

7. Sarc. lutescens Stubenrath

Sarc. livido-lutescens Stubenrath: Lehmann-Neumann, Bak. Diag., 1896.

Imperfectly described.   From stools in a case of enteritis.

8. Sarc. equi Stubenrath

Like *Sarc. lutea,* except as specified in the synopsis.
Habitat. Isolated from the urine of horses.

9. Sarc. lutea Flügge

Die Mikroorganismen, 1896, 182.

Morphology. On all media typical packets.
Gelatin colonies. *Deep:* round, dark yellow, entire, granular. *Surface:* round
   punctiform, sulphur-yellow; after 10–12 days, sunken. Microscopically,
   round, entire, yellowish, granular, becoming irregularly bordered.
Gelatin stab. In depth, growth filiform — slightly beaded; on the surface,
   growth round-irregular, moist, glistening, raised, citron-yellow, becoming
   in 10–12 days a liquefied funnel; in other cases no liquefaction.
Agar colonies. *Deep:* like gelatin colonies, but more coarsely granular.
   Surface: round, entire, raised, moist, glistening, sulphur-yellow. Micro-
   scopically, round, entire, granular; border more transparent.
Agar slant. A sulphur — chrome yellow layer of a buttery consistency.
Bouillon. Clear, with much sediment.
Milk. Coagulated.

Potato. Growth raised, glistening, limited; surface rough, sulphur — chrome yellow.

Glucose bouillon. Some acid. *Indol* slight. H_2S produced.

Habitat. Common in the air.

10. Sarc. aurantiaca Flügge

Lindner: Die Sarcina Organismen des Gärungsgewerbes, Berlin, 1887.

Morphology. Beautiful packets on all media.

Gelatin colonies. Small round punctiform, which soon sink in the liquefied gelatin; microscopically, round, entire, bright — dark yellow, amorphous or finely granular. Later a central lobate — lacerate colony, granular, within a funnel of liquefied gelatin.

Gelatin stab. In 36 hours colony on the surface sunken in the liquefied gelatin, becoming a funnel of liquefied gelatin, with a granular sediment.

Agar colonies. Roundish, entire, moist, glistening, slightly raised, orange. Microscopically, irregular — roundish; centre greenish brown; border brighter and more yellow.

Agar slant. Slightly raised, orange-yellow — orange-red, buttery consistency.

Bouillon. A flocculent turbidity with much sediment.

Milk. Coagulated and peptonized.

Potato. Growth yellow-orange, glistening, becoming raised, red-orange, dull, granular.

Glucose bouillon. Acid production slight. H_2S negative. *Indol* slight.

Habitat. Abundant in the air.

VARIETIES.

Indt. from the preceding. *Sarc. aurea* Macé. *Sarc. aurescens* Gruber.

11. Sarc. cervina Stubenrath

Lehmann-Neumann, Bak. Diag., 1896.

Gelatin colonies. Whitish, becoming bright brown, moist; slowly surrounded by a zone of liquefied gelatin; microscopically, coarsely granular, erose.

Gelatin stab. In depth, growth filiform — beaded; on the surface, colony small, bright brown, becoming slowly sunken in the liquefied gelatin.

Agar colonies. Like gelatin colonies.

Potato. Growth brownish white.

Habitat. From stomach contents in carcinoma.

1

12. Sarc. fusca Gruber

Arbeiten bak. Inst. Techn. Hochschule, Karlsruhe, I, 1895.

The morphology and cultural characters like *Sarc. pulmonum.*
Bouillon. Turbid, with granular sediment. Much acid produced in milk and
glucose bouillon cultures. H_2S produced.
Habitat. From stomach contents.

13. Sarc. rosea Schröter

Sarc. rosea Menge: Centralblatt f. Bakteriologie, VI, 596.

According to Lehmann and Neumann, identical with *M. roseus* Bumm,
No. 81.

14. Sarc. erythromyxa (Overbeck)

M. erythromyxa Overbeck: Nov. Act. d. Leop. Carol., LV, 1891.
Kral: Verzeichniss der abzugebende Bak.

Morphology. Usually only as cocci, diplococci, or tetrads; only once a Sarcina
form in hay infusion.
Gelatin colonies. Moist, grayish, becoming carmine-red. Microscopically,
not granular, border transparent, erose.
Agar slant and *potato.* Growth glistening, limited, intense red.
Milk. A red surface growth, not coagulated.
Bouillon. Turbid, with a granular sediment.
Glucose bouillon. Acid.

PLANOCOCCUS (Migula)

Engler and Prantl: Die Natürlichen Pflanzenfamilien, 1895.

Cocci spherical, occur singly, in twos, tetrads, or irregular groups. Division in
two directions of space. Cells freely motile. Flagella usually one,
attached to each cell.
A. Chromogenic; pigment yellow.
    1. *Planococcus citreus* (Menge) Migula.
B. Chromogenic; pigment pink — rose-colored.
    2. *Planococcus agilis* (Ali-Cohen).
C. Non-chromogenic.
    3. *Planococcus tetragenus* (Mendoza).

1. Planococcus citreus (Menge) Migula

M. agilis-citreus Menge : Centralblatt f. Bakteriologie, XII, 1892, 49.
Pl. citreus Migula, l.c.

Morphology. Cocci in pairs or sometimes short chains or irregular groups. Each coccus has a flagellum which is easily demonstrated by Löffler's method, and is about six times its diameter in length. Aerobic. Gelatin not liquefied. Form a yellow pigment.

Gelatin colonies. A diffuse cloudiness around the colonies, which extends over the plate.

Gelatin stab. A slight growth in depth ; on the surface, growth round, intense yellow.

Agar slant. Growth pale, thin, limited, becoming in three days yellow, more spreading, and viscid.

Bouillon. Turbid, a yellow viscid sediment; no scum.

Potato. Slight growth, becoming bright yellow.

Milk. Not coagulated. Optimum temperature 20°.

Habitat. Isolated from an infusion of peas ; probably from the air.

2. Planococcus agilis (Ali-Cohen)

M. agilis Ali-Cohen : Centralblatt f. Bakteriologie, VI, 1889, 33.

Morphology. Cocci round — slightly oval, 1.0 μ, as diplococci, short chains, or in tetrads. Stain by Gram's method. Flagella demonstrated. On media a rose-colored pigment.

Gelatin stab. For some time a dry hollow funnel; liquefaction does not commence until after 3-4 weeks.

Agar slant. Growth pinkish-red.

Potato. As before. No growth at 37° C.

According to Migula, l.c., this form shows in hay infusions Sarcina forms, and is classed by the latter as a Planosarcina.

3. Planococcus tetragenus (Mendoza)

M. tetragenus-mobilis-ventriculi Mendoza : Centralblatt f. Bakteriologie, VI, 1889, 566.

From morphological and cultural characters indt. from *M. tetragenus*.

PLANOSARCINA (Migula l.c.)

Cocci spherical, mostly in twos and tetrads. Division in three directions of space, but rarely do the cells remain united in packets, the latter usually observed only in sugar-free media ; cells freely motile ; flagella long or

short, usually one for each cell.   In artificial cultures, as a rule, an active
motility of the cells is not observed, and most of the individuals are non-
motile.

A. Chromogenic, pigment orange-reddish.
  1. *Planosarcina mobilis* (Maurea) Migula.
  2. *Planosarcina agilis* (Ali-Cohen).
B. Non-chromogenic. Growth on agar and gelatin grayish.
  3. *Planosarcina Samesii.*

1. Planosarcina mobilis (Maurea) Migula

Sarcina mobilis Maurea: Centralblatt f. Bakteriologie, XI, 1892.
Planosarcina mobilis Migula: l.c.

Morphology.   Cocci 1.4 μ, in typical packets, each cell with 1–2 flagella gener-
   ally about three times the length of the cell.
Gelatin.   Liquefied slowly, with formation of a brick-red pigment.
Agar slant.   Growth thin, orange- or brick-red.
Milk.   Not coagulated.
Potato.   No growth.

2. Planosarcina agilis (Ali-Cohen) Migula

See *Planococcus agilis* Ali-Cohen, No. 2.

3. Planosarcina Samesii

Eine bewegliche Sarcina Sames: Centralblatt f. Bakteriologie, IV, 1898, 664.

Morphology.   Cocci in packets of 8 elements, 3.0 μ square.   Stain by Gram's
   method.   Actively motile; flagella long, thick, spiral, 20–50 from a single
   packet; grow best in an alkaline medium.
Alkaline gelatin slant and *alkaline agar slant.*   A glistening grayish layer.
Alkaline gelatin stab.   In depth, growth abundant, beaded; in 10–14 days
   acicular outgrowths.   On the surface, growth grayish, spreading.
Alkaline gelatin colonies.   Deep, irregular, dark.   *Surface:* in 4–5 days round,
   gray, 5–10 mm.   On *Zettnow's Spirillum agar* (Centralblatt f. Bakteriol.
   XIX, 394), deep colonies, oval-irregular, dark — black.   Surface round,
   granular, grayish.
Potato.   No growth except when made alkaline with Na_2Co_3, then yellowish,
   becoming brown.

Glucose bouillon. No gas. *Litmus* reduced. *Indol* slight.

Pathogenesis. Subcutaneous and intraperitoneal injections into mice, white rats, and guinea pigs, negative.

Habitat. Isolated from manure and sewage.

BACTERIACEÆ

Cells short or long, cylindrical, straight. Without a sheath surrounding the chains of individuals; motile or non-motile; with or without flagella; endospores present or absent. No true branching.

A. Flagella absent, endospores present or absent.

BACTERIUM Ehrenberg, p. 117.

B. Flagella present.

    1. Flagella arising from any part of the body, peritrichic.

BACILLUS, Cohn, p. 199.

    2. Flagella attached to one or both poles.

PSEUDOMONAS Migula, p. 306.

BACTERIUM [1] Ehrenberg

Char. emend. by Migula.

Cells cylindrical, varying from short ovals to longer rods and filaments; without flagella; endospores present or absent, or, at least in a large number of the species, unknown.

SYNOPSIS OF THE GENUS

I. Without endospores, or at least their presence not reported.

    A. Aerobic and facultative anaerobic.

        1. Without pigment on gelatin or agar.

            a. Grow only at the body temperature. CLASS I, p. 118.

            b. Grow at room temperatures, 20°–22° C.

               * Gelatin not liquefied.

                   † Decolorized by Gram's method. CLASS II, p. 121.

                   †† Stained by Gram's method. CLASS III, p. 148.

               ** Gelatin liquefied. CLASS IV, p. 155.

[1] The writer has provisionally included in this genus all non-motile Bacteriaceæ, leaving it to future investigations to determine whether any of the forms which may have been wrongly placed in this genus do or do not possess flagella; it being assumed that a non-motile organism is devoid of flagella until the contrary is proven.

2. Produce pigment on gelatin or agar.
   a. Pigment yellowish on gelatin or agar.
       * Gelatin liquefied. CLASS V, p. 162.
       ** Gelatin not liquefied. CLASS VI, p. 168.
   b. Pigment reddish on gelatin or agar. CLASS VII, p. 173.
   c. Pigment of other colors than red or yellow. CLASS VIII, p. 179.
3. Fluorescent bacteria.
   a. Gelatin liquefied. CLASS IX, p. 180.
   b. Gelatin not liquefied. CLASS X, p. 180.
B. Obligate anaerobic. CLASS XI, p. 183.

II. With endospores.
A. Non-chromogenic; without pigment on gelatin or agar.
   1. Aerobic and facultative anaerobic.
      a. Rods not swollen at sporulation — *B. subtilis* type.
           * No growth at room temperatures, or below 22°-25° C.
             CLASS XII, p. 184.
           ** Growth at room temperatures, 20°-22° C.
             † Gelatin liquefied. CLASS XIII, p. 187.
             †† Gelatin not liquefied. CLASS XIV, p. 196.
      b. Rods swollen at one end at sporulation — *tetanus* type.
             CLASS XV, p. 198.
   2. Obligate anaerobic. CLASS XVI, p. 198.

CLASS I. WITHOUT ENDOSPORES. AEROBIC AND FACULTATIVE ANAEROBIC. WITHOUT PIGMENT. GROW ONLY AT BODY TEMPERATURES.

I. Grow best on blood, or on agar moistened with blood, or on especially prepared media.
  1. *Bact. influenzæ* (Pfeiffer) Lehm.-Neum.
  2. *Bact. pseudoinfluenzæ* (Kruse).
II. Grow well on blood-serum.
  3. *Bact. acuminatum* (Sternberg).
  4. *Bact. Lumnitzeri* (Sternberg).
  5. *Bact. conjunctivitidis.*
III. Grow only in the presence of pathological secretions, not on blood or blood serum.
  6. *Bact. vaginæ* Kruse.
IV. Cultural characters not known.
  7. *Bact. cancrosi* Kruse.

V. Grow very poorly, or scarcely at all at room temperatures.
   8. *Bact. Ægyptium* Trevisan.
VI. Occupying a rather nondescript place in Class I, is
   9. *Bact. abortivum.*

1. Bact. influenzæ (Pfeiffer) Lehmann-Neumann

Bacillus of influenza Pfeiffer: Deutsehe med. Wochensch., 1892, 28.
Bact. influenzæ Lehmann-Neumann: Bak. Diagnostik, 1896, 187.

Morphology. Bacilli small, slender, commonly in twos; stain with Löffler's alkaline blue and carbol-fuchsin. Decolorized by Gram's method.
On agar moistened with blood there develop in 24-48 hours small glassy drops; older colonies show a yellowish-brownish centre.
Nastiukow's solution.[1] In 24 hours at 37° small white flecks at bottom of tube, composed of chains of bacilli.
Nastiukow's agar.[1] Colonies small gray points; microscopically, round, yellow, translucent.
Pathogenesis. Intraperitoneal injections of one-third of an agar slant culture cause death. Intravenous injections into rabbits cause fever and muscular weakness; subcutaneous injections, knotty thickenings and suppuration.
Habitat. Nasal and bronchial secretions, urine, of man affected with influenza.

2. Bact. pseudoinfluenzæ (Kruse)

Pseudoinfluenza Bacillus R. Pfeiffer: Zeitsch. f. Hygiene, XIII, 1893, 383.
Bacillus pseudoinfluenzæ Kruse: Flügge, Die Mikroorganismen, 1896.

Like the preceding, but rods somewhat longer than those of true influenza.
Pathogenesis. Doubtful.
Habitat. From bronchopneumonia, otitis-media; also associated with influenza.

3. Bact. acuminatum (Sternberg)

Babes, no name: Sept. Prozesse d. Kindesalters, Leipzig, 1889.
C. septicus-acuminatus Sternberg: Manual of Bacteriology, 1892, 472.

Morphology. Bacilli with pointed ends, resembling those of mouse septicæmia, but rather thicker; stain unevenly.
Gelatin colonies. Small, round, flat, translucent; by coalescence a yellowish layer.

[1] Centralblatt f. Bakteriol., XVII, 492.

Pathogenesis. Non-pathogenic for mice and rabbits. Guineas die in 2 to 6 days of septicæmia.

Habitat. From blood and organs of a new-born infant with septicæmia.

4. Bact. Lumnitzeri (Sternberg)

Bacillus der putriden Bronchitis Lumnitzer: Centralblatt f. Bateriol., III, 1888, 621.
Bacillus of Lumnitzer Sternberg: Manual, 1892, 467.

Morphology. Bacilli 1.5–2.0 μ, with rounded ends and slightly curved.

Agar slant and *blood serum.* At 37°, small semi-spherical grayish white colonies, becoming coalescent, with odor of sputum in putrid bronchitis.

Pathogenesis. Intraperitoneal injections of mice cause, in 24 hours, a purulent peritonitis. Injections into lungs of rabbits, pneumonia, pleuritis.

Habitat. Sputum in putrid bronchitis.

5. Bact. conjunctivitidis

Diplobacillus de la conjonctivite subaigue Morax: Annales Pasteur Institut, 1896, 337.

Morphology. Bacilli 1.0–2.0 μ, ends rather squared, in twos and chains. Decolorized by Gram's method.

Blood serum. In 24 hours, moist points; in 2 days, small transparent liquefied depressions.

Pathogenesis. With smaller animals and man, inflammation of the conjunctival sac.

Habitat. Isolated from chronic inflammation of the conjunctiva.

6. Bact. vaginæ (Kruse)

Doderlein's Scheidenbacillus: Das Schidensekret u. seine Bedeutung f. d. Puerperalfieber, Leipzig, 1892.
Bacillus vaginæ Kruse: Flügge, Die Mikroorganismen.

Morphology. Bacilli middle-sized, rather slender. Grow in glucose bouillon containing one per cent of secretion. Can then be transferred to glycerin agar, where it produces dewy, drop-like colonies. Facultative anaerobic.

Pathogenesis. Doubtful.

Habitat. Isolated from vaginal secretions.

7. Bact. cancrosi (Kruse)

Bacillus des weichen Schankers Ducrey: Monat. f. Dermatol., 1889, Heft IX.
B. ulceris-cancrosi Kruse: Flügge, Die Mikroorganismen, 1896.

Morphology. Bacilli 0.5 : 1.5 μ; ends rounded, mostly contracted in the middle; in chains. Decolorized by Gram's method. Stain 15 minutes with Löffler's alkaline blue, and wash but a short time in alcohol.

Habitat. Isolated from secretions in soft shanker.

8. Bact. Ægyptium (Trevisan)

Conjunctivitis Bacillus Koch-Kartulis: Arbeiten Kaiserlichen Gesundheitsamte, III, 1887.
B. Ægyptius Trevisan: Gen. e. Spec. Batteriaceae, 1888.
B. conjunctivitidis Kruse: Flügge, Die Mikroorganismen, 1896.

Morphology. Bacteria 0.25 : 1.0 μ; in twos or chains in the pus cells. Decolorized by Gram's method.

Agar slant and *blood serum*. At 37° C., isolated colonies, becoming a confluent, glistening, elevated growth.

Slight growth on gelatin; no liquefaction.

Pathogenesis. Inoculations on the cornea of asses, dogs, guinea-pigs, and rabbits negative; on human conjunctiva positive in one out of six inoculations.

Habitat. Associated with conjunctival catarrh in Egypt.

9. Bact. abortivum

Bacillus of contagious abortion in cows, Bang: Zeitsch. f. Tiermedizin, Bd. I, 1897, Heft 1.

Morphology. Bacilli small, rods about the size of the tubercle bacillus; each rod contains 1–3 granules. Does not grow in ordinary culture media. In 5 per cent glycerin bouillon a scanty growth in 14 days. A scanty growth in liquid blood serum.

Pathogenesis. Inoculation experiments on cows positive.

Habitat. Associated with contagious abortion in cows.

THE SPECIES WHICH FOLLOW GROW ON ORDINARY NUTRIENT MEDIA AND AT ROOM TEMPERATURES

CLASS II. GELATIN NOT LIQUEFIED. DECOLORIZED BY GRAM'S METHOD

I. Obligate aerobic. ACETIC FERMENT GROUP.

    A. Grow well in Pasteur's fluid containing alcohol [1] with the formation of a membrane.

        10. *Bact. aceti* (Kütz.) Lanzi, emend. Beijerinck.

    B. Do not grow well in preceding fluid, and no membrane produced.

        1. Form long chains of more than four elements, also long involution forms.

---

[1] Tap water 100, alcohol 3.0, ammonium phosphate 0.05, chloride of calcium 0.01. See Beijerinck: Centralblatt f. Bakteriol., IV, 1898, 214.

a. On sterile beer or beerwort a membrane over entire surface.

    * Membrane stained blue with iodine solution.

        † Cultures in beer, yeast-water, or dextrose solution remain clear.

          11. *Bact. Pasteurianum* (Hansen) Zopf.

        †† Cultures in beer, yeast-water, or dextrose solution turbid below the membrane.

          12. *Bact. Kützingianum* Hansen.

    ** Membrane not stained blue with iodine solution.

        † Membrane on beer, yeast-water, and dextrose solutions rather thin. Fluid below the membrane more or less turbid.

          13. *Bact. Hansenianum.*

          14. *Bact. oxydans* Henneberg.

        †† Membrane on beer, yeast-water, and dextrose solution thick and gelatinous, not easily broken; fluid below the membrane clear.

          15. *Bact. acetosum* Henneberg.

        ††† Membrane on beer, yeast-water, and dextrose solution thick, tough, and coriaceous, showing a cellulose reaction with iodine and sulphuric acid.

          16. *Bact. xylinum* Brown.

b. On sterile beer or beer-wort an imperfect membrane as islands. Motile, with a polar flagellum. See *Pseudomonas flagellatum* Zeidler.

2. Do not form long chains of more than four elements, or elongated involution forms.

          17. *Bact. acetigenum* Henneberg.

          18. *Bact. aceticum* (Kruse).

II. Aerobic and facultative anaerobic.

  A. Gas generated in glucose bouillon.

    1. Gas generated in lactose bouillon abundantly.

      a. Milk rendered viscous — slimy.

          19. *Bact. viscosum* van Laer.

      b. Milk coagulated, but not rendered viscous — slimy.

        * Bouillon rendered turbid.

          20. *Bact. aerogenes* Escherich.

              BACT. AEROGENES GROUP.

          21. *Bact. capsulatum* (Sternberg).

          22. *Bact. chinense* (Hamilton).

** Bouillon not rendered turbid.
    23. *Bact. pallescens* Henrici.
c. Milk not coagulated.
    24. *Bact. fermentationis* Chester.

2. Little or no gas generated in lactose bouillon, bacilli, at least in the animal body surrounded by a capsule.
                    FRIEDLANDER BACILLUS GROUP.
a. Surface growth in gelatin stab cultures punctiform or convex, not flat or spreading.
    25. *Bact. pneumoniæ* Zopf.
b. Surface growth in gelatin stab cultures flat and spreading.
  * Milk not coagulated.
    26. *Bact. ozænæ* (Abel) Lehmann-Neumann.
  ** Milk slowly coagulated.
    27. *Bact. Wrightii.*
c. Surface growth in gelatin stab cultures indeterminate.
    28. *Bact. sputigenum.*

B. Very little or no gas generated in glucose bouillon.
           SEPTICÆMIA HEMORRHAGICA GROUP.
1. Milk coagulated.
a. Gelatin colonies of the aerogenes type.   Bacteria closely related to *Bact. aerogenes.*
    29. *Bact. limbatum* Marpmann.
    30. *Bact. nasalis.*
b. Gelatin colonies of the *coli* type.
  * Produce indol.
    31. *Bact. choleræ* (Zopf) Kitt.
    32. *Bact. gallinarum* (Kruse).
  ** Do not produce indol.
    33. *Bact. anaerogenes* (Lembke).

2. Milk not coagulated.
a. Litmus milk rendered acid or reddened.
  * Produce indol.
    34. *Bact. suicida* Migula.
    35. *Bact. bovisepticum* (Kruse).
  ** Do not produce indol.
    36. *Bact. pneumopecurium.*
b. Litmus milk blue, reaction amphoteric — alkaline.
  * Gelatin colonies smooth, not characterized as in **.

      † Produce indol.
         37. *Bact. sanguinarium* Moore.
         38. *Bact. avium.*
      †† Do not produce indol, or reaction doubtfully faint.
         39. *Bact. inocuum* (Kruse).
         40. *Bact. tiogense* (Wright).
    ** Gelatin colonies crimpled — scalloped — petaloid.
      † Colonies radially crimpled.
         41. *Bact. refractans* (Wright).
      †† Colonies petaloid.
         42. *Bact. rodonatum* (Ravenel).

   3. Milk coagulation not stated.
     a. Pepton-rosolic acid decolorized; cultures viscous.
        43. *Bact. zurnianum* List.
     b. Pepton-rosolic acid not decolorized; cultures not decidedly viscous.
        44. *Bact. Martizeni* (Sternberg).

   4. Milk coagulation not stated. Bacteria closely related or identical with either fowl cholera or swine plague.
     a. Strongly pathogenic to rabbits.
       * Distinctly pathogenic to guinea pigs.
         45. *Bact. cuniculicida* (Kruse).
      ** Slightly pathogenic to guinea pigs.
         46. *Bact. cuniculicida* var. *immobilis.*
     b. Pathogenic to rabbits only by intraperitoneal injections.
        47. *Bact. putidum.*

C. Gas production in glucose bouillon indeterminate.
   1. Colonies on gelatin of the *coli* type.
     a. Pathogenic bacteria.
       * Bacteria closely related to fowl cholera.
        † Decidedly pathogenic to rabbits, producing general septicæmia.
         § Associated with specific diseases of pigeons.
           48. *Bact. diphtheriæ* (Flügge).
           49. *Bact. columbarum* (Kruse).
         §§ Associated with a specific disease of rabbits.
           50. *Bact. cuniculi* (Kruse).
           51. *Bact. Beckii* (Kruse).
        §§§ Septic bacteria of mixed origin.

‡ Produce general septicæmia in guinea pigs, strongly pathogenic.

52. *Bact. dubium* (Kruse).

53. *Bact. felis* (Kruse).

54. *Bact. septicum* (Trevisan).

55. *Bact. vitulinum.*

‡‡ Slightly or negatively pathogenic to guinea pigs.

56. *Bact. purpurum.*

57. *Bact. Bienstockii* (Schröter).

†† Less strongly pathogenic to rabbits; associated with hemorrhagic infection of man.

58. *Bact. hæmorrhagicum* (Kruse).

59. *Bact. velenosum* (Kruse).

60. *Bact. nephritidis* (Vassale).

††† Not pathogenic to rabbits.

61. *Bact. aphthosum* (Kruse).

62. *Bact. dysenteriæ* (Kruse).

** Bacteria, slender, minute, like influenza bacillus.

63. *Bact. salivæ* (Kruse).

b. Non-pathogenic bacteria.

* Milk not coagulated.

† Milk rendered decidedly acid.

64. *Bact. acidum.*

†† Reaction of milk unchanged.

65. *Bact. Connii.*

** Action on milk not stated.

66. *Bact. nitrovorum* Jensen.

67. *Bact. filefaciens* Jensen.

2. Colonies on gelatin of the aerogenes type, non-pathogenic; mostly milk bacteria, probably of the aerogenes group.

a. Milk coagulated.

* Milk coagulated at room temperatures.

68. *Bact. punctatum.*

69. *Bact. Middletownii.*

** Milk not coagulated at room temperatures, and only at about 35° C.

70. *Bact. coccoideum.*

b. Milk rendered slimy.

71. *Bact. lactis* (Kramer).

c. Milk not coagulated.

72. *Bact. aromafaciens.*

ACETIC FERMENT GROUP

10. **Bact. aceti** (Kütz.) Lanzi, emend. Beijerinck

Essigmutter Kützing: Jour. prakt. Chemie, XI, 1837, 385.
Ulvina aceti Kützing: Phycologie generalis, 1843, 148.
Mycoderma aceti Pasteur: Étude sur le Vinaigre, 1868, 106.
Bact. aceti Lanzi: N. Giorn. bot. Ital., 1876, 257.
Bact. aceti Beijerinck: Centralblatt f. Bakteriol., 2 Abt., IV, 1898.

In beer gelatin containing 10 per cent of cane sugar, very voluminous colonies of a slimy consistency, and causing a strong turbidity of the gelatin. Grow poorly in beer gelatin without sugar ; cane sugar inverted. Beijerinck holds this species to be distinct from *B. aceti* of Hansen.

Habitat. The quick vinegar ferment of Pasteur, living on the surface of beech wood shavings in the vinegar vats.

11. **Bact. Pasteurianum** (Hansen) Zopf

Mycoderma Pasteurianum Hansen: Compt. Rendu Carlsberg Lab., Copenhagen, 1879.
Bact. Pasteurianum Zopf: Spaltpilze, 1885, 64.

Morphology. The membrane consists of chains of elements 0.4–0.8 : 1.0 μ.
Wort gelatin colonies. Entire, without any rosette form, but with brain-like corrugations of the surface.
Sterile beer. In 24 hours, at 24° C., a rather thick dry membrane, minutely corrugated ; fluid clear.
Habitat. Beer and beer-wort, seldom in wine.

12. **Bact. Kützingianum** (Hansen)

Compt. Rendu Carlsberg Lab., Copenhagen, III, 1894, 265.

Morphology. Short, thick bacilli ; not in chains as in No. 11.
On sterile beer. A moist, smooth, slimy membrane, easily broken, with a tendency to rise on the walls of the tube.

13. **Bact. Hansenianum**

Mycoderma aceti Hansen: l.c., 1879.
Bact. aceti Zopf: Spaltpilze, 1885, 62.
Bact. aceti Brown: Jour. Chem. Soc., London, XLIX, 1886.

Morphology. Bacilli, short rods, rather more slender than No. 11, in chains, placed parallel with often hour-glass forms — long slender involution forms — irregularly swollen and often branched.
Wort gelatin colonies. Rosette-like, radiate.

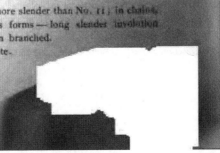

Beer or wort. A moist, slimy, smooth-veined membrane; fluid only slightly turbid. Minimum temperature, 4°-5° C.; optimum temperature, 34° C.
Habitat. Beer and beer-wort.

14. Bact. oxydans Henneberg

Centralblatt f. Bakteriol., 2d Abt., III, 1897, 223.

Morphology. Bacilli, 0.8–1.2 : 2.4 : 2.7 μ; in chains rather loosely jointed; show rotatory motion (Henneberg, l.c.). Involution forms as filaments with bud-like swellings; fluid turbid below the membrane. Optimum temperature, 18°–21° C. Minimum temperature, 8° C.; maximum, 30°-33° C.
Habitat. Beer, etc.

15. Bact. acetosum Henneberg

L.c., IV, 1898, 14.

Morphology. Bacilli, 0.4–0.8 : 1.0 μ, in chains of elements. Optimum temperature, 28° C.; minimum, 8° C.; maximum, ± 36° C. Involution forms like *B. Pasteurianum.*

16. Bact. xylinum Brown

Jour. Chem. Soc., London, XLIX, 1886, 432.

Characters as specified in the Synopsis.

17. Bact. acetigenum Henneberg

L.c., IV, 1898, 15.

Morphology. Bacilli, 0.8–1.2 : 1.2–1.4 μ; no chain formation of the elements; enlarged ellipsoidal — coccoid involution forms. On beer and yeast-water a thin, very tough membrane, which sinks in fragments to the bottom, a new membrane forming; fluid often turbid. With H_2SO_4 and iodine, often a cellulose reaction of the membrane. According to Henneberg, bacilli show a rotatory motion. Optimum temperature 38° C.

18. Bact. aceticum (Kruse)

Bacterium C of sour dough Peters: Bot. Zeitung, 1889.
Bacillus aceticus-Petersii Kruse: Flügge, Die Mikroorganismen, 1896, 355.

Morphology. Bacilli 0.8–1.6 μ, rounded — pointed, singly and in twos, rarely in chains of four. In old cultures, short, swollen involution forms.
Gelatin colonies. Round, convex, becoming flatter and more spreading.

Gelatin stab. No growth in depth; on the surface, growth flat, spreading, irregular.

Glucose yeast water. Turbid; on the surface a very thin pellicle, which is slimy and easily broken, with a tendency to rise up on the walls of the tube. In yeast water, with 5 per cent of alcohol, a strong acetic acid production.

Habitat. Isolated from sour dough.

19. Bact. viscosum van Laer

Bact. viscosum-cerevisæ van Laer: Acad. royale de Belgique, 1889, 36.

Morphology. Bacteria 0.8 : 1.6–2.4 μ.

Gelatin colonies. Entire — erose, brownish.

Potato. Growth of white, watery, doughy colonies, which smell like foul fish.

Beer-wort. Rendered viscous; CO_2 evolved.

Milk and *glucose bouillon.* Rendered slimy, with gas production.

Habitat. Isolated from beer, yeast, and bread; causes a viscous fermentation.

BACT. AEROGENES GROUP

20. Bact. aerogenes Escherich

Bact. lactis-aerogenes Escherich: Fortschritte Medizin, 1885, No. 16–17.
Bact. aceticum Babinsky: Zeitsch. f. phys. Chem., XII, 1888, 434.
Bacillus aerogenes Kruse: Flügge, Die Mikroorganismen, 1896, 340.

Morphology. Bacilli 0.5–1.0 : 1.0–2.0 μ.

Gelatin colonies. *Deep:* round, granular, grayish brown. *Surface:* porcelain-white, round, convex; microscopically, yellowish, granular, darker in the centre.

Gelatin stab. In depth, a good growth; on the surface, a round, convex growth.

Bouillon. Turbid; a slight membrane.

Agar slant. Growth rather opaque, porcelain-white.

Potato. Growth moist, yellowish white, with development of gas and a cheesy odor.

Milk. Coagulated; strongly acid. *Indol* negative.

Pathogenesis. Variable; for the smaller animals, usually pathogenic only in large doses (toxæmia).

Habitat. Milk, fæces, air, water, etc.

VARIETIES

Often difficult to differentiate from *B. coli*. Scheffer (Archiv f. Hygiene, XXX, 1897, 291) describes a variety of *B. aerogenes* which closely connects it with *B. coli*. Diff.. *B. coli*, when grown anaerobically, showed no change of morphology, while *B. aerogenes*, grown under the same conditions, developed abnormally long filaments. Guinea pigs immunized, on the one hand against *B. aerogenes*, and on the other against *B. coli*, gave sera which possessed diagnostic value; viz., aerogenes serum caused an agglutination of aerogenes bacilli, but not of *B. coli*; while coli serum caused an agglutination of *B. coli*, but not of *B. aerogenes*.

Bact. acidi-lactici Grotenfelt, Fortschritte Med., VII, 1889, 124. Indistinguishable from the preceding.

Habitat. Isolated from fæces, water, milk.

Bact. a and b Guillebeau: Ann. Micrograph., XI, 225. Indistinguishable from the preceding, except that bacilli show a slight motility.

Habitat. Isolated from milk.

21. Bact. capsulatum (Sternberg)

Capsule Bacillus of Pfeiffer: Zeitsch. f. Hygiene, VI, 1889, 145.
Bacillus capsulatus Sternberg: Manual Bacteriology, 1892, 431.

Morphology. Bacilli thick, with rounded ends, usually 2–3 times their breadth; often in chains of 2–3 elements, or in filaments.

Gelatin colonies. Deep: oval, granular. *Surface:* flat, glistening, porcelain-white.

Gelatin stab. In depth, a good growth; on the surface, growth round, glistening, flat, porcelain-white.

Agar slant. Growth thick, pure white, viscous.

Potato. Growth moist, glistening, yellowish white, viscid.

Pathogenesis. Subcutaneous inoculations of mice cause death by septicæmia in 2–3 days. Intraperitoneal inoculations of guinea pigs and pigeons, death in 30 hours; septicæmia. Rabbits, refractory.

Habitat. Isolated from the blood of guinea pigs which died spontaneously. According to Strong (Centralblatt f. Bakteriol., XXV, 49), this bacillus generates gas actively in lactose bouillon.

22. Bact. chinense (Hamilton)

Bacillus capsulatus-chinensis Hamilton: Centralblatt f. Bakteriol., 2d Abt., IV, 1898, 230.

Morphology. Bacilli 0.6:4–8 μ, with a capsule. Rods within the capsule small, 0.5–0.7 : 4.0–6.0 μ, with 2–3 elements within a single capsule. In old cultures the rods disintegrate. Decolorized by Gram's method.

K

Gelatin colonies. Deep: round — oval, often a bright ring between centre and border. *Surface:* in 48 hours colony the size of a pin's head; white, glistening, convex; microscopically, grayish brown, opaque.

Gelatin stab. In depth, a good growth, white; often gas produced; surface growth convex.

Agar colonies. Deep: fusiform. *Surface:* slimy glistening drops.

Agar slant. Growth thick, slimy, spreading; condensation water, slimy turbid.

In *glycerine agar*, gas production.

Bouillon. Turbid; a delicate membrane, which adheres to walls of tube, becoming thicker, and sinking; later a slimy — flocculent sediment, and clear medium above. Culture has the odor of walnuts.

Blood serum. Growth not so abundant as on agar; no liquefaction.

Potato. A thick, creamy layer, color of potato, with raised border; odor of trimethylamine and ammonia.

Agar stab. Gas production at 37° C.

Milk. Coagulated slowly; after 6 days at 37° C. acid, with a cheesy odor.

Do not grow on acid media, but best on one weakly alkaline. Maltose, glucose, and lactose fermented. Cane sugar not fermented. Litmus reddened.

Pathogenesis. Subcutaneous inoculation of mice causes death by general septicæmia in 24 hours. Guinea pigs, intraperitoneally, death in 36-48 hours, with peritonitis.

Habitat. Isolated from India ink.

23. Bact. pallescens Henrici

Bakterienflora des Käses, Baseler, Philos. Diss., 1894.

Indistinguishable from *Bact. aerogenes*, except that bouillon is not rendered turbid.

Habitat. Isolated from cheese.

24. Bact. fermentationis Chester

Report Del. College Ag. Expt. Sta., 1899.

Morphology. Bacilli 0.8 : 1.25-3.0 μ.

Gelatin colonies. Deep: round, entire, homogeneous, yellowish brown, 0.1-0.3 mm. *Surface:* macroscopically, moist, glistening, punctiform, 0.5-1.0 mm., later convex — flat, dull white, 1.0-1.5 mm. Colonies remain small after several weeks. Microscopically, round, entire, brownish in the centre, becoming gradually lighter toward the edge, or grayish yellow throughout, strongly refracting and amorphous.

Agar colonies. Deep : round — oval, brown, opaque, coarsely granular. *Surface:*
Round, flatly convex, moist, glistening, milky white — rather translucent ;
later round, thin, flat, slimy, translucent, 6 mm. Microscopically, light
yellowish brown, homogeneous, finely granular ; edge thin, indistinct :
later grayish brown, grumose — finely granular ; border thin, entire —
undulate.

Gelatin slant. A thin, opaque streak ; edges finely erose.

Agar slant. A flat, white, opaque, moist, glistening, slimy stripe, about 3
mm. wide.

Gelatin stab. Good growth in depth ; surface growth thin, flat.

Bouillon (neutral). Turbid ; in 3 weeks clear, with easily diffusible sediment.
No pellicle.

Potato. Growth thick, limited, moist, glistening, dirty white — light dirty
brown, becoming darker.

Milk. No change in consistency after 5 weeks, then coagulated on boiling.

Litmus milk. Rendered acid.

Blood serum. A narrow, moist, glistening, flat to raised, dirty white, slimy
stripe. *Indol* negative. Nitrates reduced to nitrites.

Habitat. Isolated from garden soil.

NOTE. — Under this head is placed : —

B. coli-immobilis (Kruse)

Unbeweglicher Fāces oder Kolonbacillus Germano-Maurea : Zeitsch. f. Hygiene, XII,
1892, 498.

B. coli-immobilis Kruse : Flügge, Die Mikroorganismen, 1896, 339.
Indistinguishable from *B. coli*, except as to motility.

Habitat. Isolated from fæces.

FRIEDLANDER BACILLUS GROUP
25. Bact. pneumoniæ Zopf

Pneumococcus Friedlander : Fortschritte Med., 1883, 715.
Bact. pneumoniæ-crouposæ Zopf : Spaltpilze, 1885, 66.
Bacillus pneumoniæ Weichselbaum : Centralblatt f. Bakteriologie, I, 1887, 589.

Morphology. Bacilli 0.5–0.8 : 0.6–3.5 μ, with rounded ends ; in animal body
with a capsule. Capsule only in milk cultures.

Gelatin colonies. Deep : round — oval, entire, brownish, opaque. *Surface :*
round, convex, white ; microscopically, round, entire, brownish — yellowish brown, opaque, with transparent borders.

Gelatin stab. In depth, growth beaded; on the surface, a convex growth.

Agar slant. Growth whitish yellow — gray, soft, glistening, spreading.

Bouillon. Turbid, with a slimy sediment.

Milk. Not coagulated (according to some authors coagulated).

Potato. Growth yellowish-grayish, slimy, glistening, thin, with gas bubbles.
   Indol slight. H₂S slight. Ethyl-alcohol, acetic, and formic acids produced (Frankland).

Pathogenesis. Variable. Subcutaneous inoculation of mice causes death by
   general septicæmia; also in guinea pigs and rabbits, by intraperitoneal
   and intravenous injections.

Habitat. Found in normal saliva. Associated with bronchopneumonia,
   bronchitis, and various inflammatory and purulent conditions.

VARIETIES

Probably varieties of the above: not sufficiently described to clearly differentiate from *Bact. pneumoniæ.*

B. capsulatus-mucosus Fasching: Centralblatt f. Bakteriol., XII, 1892, 304.

Habitat. Isolated from nasal secretions in influenza.

Capsule bacillus of Mandry: Fortschritte Med., VIII, 1890, No. 6.

Capsule bacillus of Kockel: Fortschritte Med., IX, 1891, No. 8.

Capsule bacillus of Dungern: Centralblatt f. Bakteriol., XIV, 1893, 546.

Capsulated canal water bacillus Mori: Zeitsch. f. Hygiene, IV, 1888, 53.

26. Bact. ozænæ (Abel) Lehmann-Neumann

Bacillus mucosus-ozænæ Abel: Zeitsch. f. Hygiene, XXI, 1896, 88.
Bact. ozænæ Lehmann-Neumann: Bact. Diagnostik, 1896, 204.

Morphology. Bacilli of variable length. Capsule in the body, occasionally
   in milk cultures.

Gelatin colonies. Transparent, watery, viscid.

Gelatin stab. Surface growth thin, spreading.

Agar slant. Growth watery, with a tendency to run down into the condensation water.

Milk. Not coagulated.

Potato. Growth watery, with gas bubbles only sparingly. *Indol* negative.
   Do not grow in acid gelatin.

Pathogenesis. Subcutaneous inoculation of mice causes death in 1–4 days, of
   septicæmia. Guinea pigs, subcutaneous inoculations negative; intraperitoneal injections cause peritonitis, etc. Rabbits refractory.

Habitat. Isolated from nasal mucus in coryza.

Bacillus of rhinitis-atrophicans Paulsen: Centralblatt f. Bakteriol., XIV, 1893, 249.

Not differentiated sufficiently to distinguish from the preceding.

27. Bact. Wrightii

Capsule bacillus of Malory and Wright: Zeitsch. f. Hygiene, XX, 1895, 220.

Morphology. Bacilli thick, with considerable variations in size; length usually 2–3 times their thickness; ends rounded. Slowly decolorized by Gram's method.

Gelatin stab. In depth, growth beaded; no gas; on the surface, growth translucent, gray, thin, not spreading.

Agar slant (with 1 per cent of glucose). Growth grayish, broad, delicate, translucent.

Bouillon. Turbid, with a delicate iridescent membrane.

Potato. Growth thin, delicate, colorless; no gas.

Milk. Slowly coagulated, acid; no odor.

Pathogenesis. Inoculations of mice cause death in 1–3 days with septicæmia. Subcutaneous inoculation of guinea pigs, a local suppuration; intraperitoneally, 0.2 cc., septicæmia and much colorless slimy exudate, death in 24 hours. Rabbits, 0.5 cc. into ear vein, death in 24 hours. Septicæmia and slimy exudate in body cavities.

Habitat. Isolated from a case of bronchopneumonia.

VARIETIES

Capsule bacillus of Nicolaier: Centralblatt f. Bakteriol., XVI, 1894, 601.

From purulent nephritis. Indt. from the preceding.

Keratomalacia infantum capsule Bacillus Loeb: Centralblatt f. Bakteriol., X, 1891, 369.

28. Bact. sputigenum

Bacillus aerogenes-sputigenus-capsulatus Herla: Archiv de Biol., XIV, 1895, 403.

Morphology. Bacilli thick, oval, rounded ends, often rather curved. Older cultures show filaments. Occur singly, now and then in twos. In blood preparations a capsule.

Gelatin colonies. Deep colonies remain very small. Surface colonies gray-white, with later a transparent border.

Gelatin stab. A nail-shaped growth, gray, not porcelain-white.

Agar slant. Growth grayish white, slimy.

Bouillon. Turbid, with a whitish sediment.

Milk. Not coagulated. In gelatin and agar much gas.

Pathogenesis. Mice die in 1–3 days of septicæmia; guinea pigs and rabbits refractory.

Habitat. Isolated from the blood of a mouse which had been inoculated with the sputum of a pneumonia patient.

29. Bact. limbatum Marpmann

Bact. limbatum-acidi-lactici Marpmann: Ergänzungshefte des Centralblatt f. allgemeine Gesundheitspflege, II, 122.

Morphology. Bacilli short, thick, with a capsule.

Milk serum gelatin colonies. 24 hours, punctiform, white, glistening; edges sharp.

Gelatin stab. Slight growth in depth; surface growth white, flat.

Litmus milk. 24 hours coagulated, slightly reddened. Grow at 37° C.

Habitat. Milk.

30. Bact. nasalis

Vorkommen von Frisch'schen Bacillen in der Nasenschleimhaut des Menschen u. der Thiere: Simoni, Centralblatt f. Bakteriol., XXV, 1899, 625.

Morphology. Bacilli rather large, oval, with a thick capsule, which commonly encloses two rods. In cultures smaller, more rod-like, and without a capsule. Stain readily.

Gelatin colonies. Round, much raised, homogeneous, opalescent, waxy; growth viscous.

Gelatin stab. Good growth in depth; no gas. Surface growth raised — convex, opalescent, becoming dirty white, never porcelain-white.

Agar slant. In 24 hours a moist, glistening, translucent, watery streak.

Glycerin agar colonies. 24 hours, 37°. *Deep:* dark, small, opaque. *Surface:* largest, the size of a pin's head, raised, translucent, whitish-grayish.

Bouillon. In 24 hours at 37°, a dense turbidity, with a delicate pellicle on the surface.

Milk. Not coagulated, not acid.

Potato. A raised, translucent, colorless, watery, glistening streak. No development in acid media. No gas in glucose bouillon.

Pathogenesis. Not pathogenic to guinea pigs and rabbits, except an infiltration at the point of injection.

Habitat. Isolated from nasal secretions in rhinoscleroma.

SEPTICÆMIA HEMORRHAGICA OR SWINE-PLAGUE GROUP

31. Bact. choleræ (Zopf) Kitt

Microbe du Cholera des Poles Pasteur: Compt. rend., LXC, 1880, 239, 952, 1030.
M. choleræ-gallinarum Zopf: Spaltpilze, 1885, 57.
Bact. choleræ-gallinarum Crookshank: Manual Bacteriology, 1887, 232.
Coccobacillus avicidus Gamalei: Centralblatt f. Bakteriol., IV, 1888, 161.
Bact. avicidum Kitt: Centralblatt f. Bakteriol., I, 1885, 305.
Bacillus des Kaninchensepticämia Koch: Aetiol-Wundinfectionskrank., Leipzig, 1878;
    Gaffky, Mitteilungen Kaiserl. Gesundheitsamte, I, 1881, 94.
B. cuniculicida Flügge: Die Mikroorganismen, 1886.

Morphology. Bacilli 0.4–0.6: 1.0 μ; show polar stain.
Gelatin colonies. Surface colonies like *B. coli*.
Agar slant. Aggregations of delicate colonies.
Potato. Growth waxy, translucent, gray-white, flat.
Bouillon. A slight turbidity.
Milk. Coagulated, acid.
Litmus milk reduced. Produce indol and phenol.
Pathogenesis. Subcutaneous injections of small doses produce septicæmia in chickens, pigeons, geese, ducks, etc.; also in rabbits and mice.
Habitat. Associated with chicken cholera, and septicæmia of rabbits.

32. Bact. gallinarum (Kruse)

Bacillus of infectious enteritis in fowls Klein: Centralblatt f. Bakteriol., V, 1889.
B. gallinarum Kruse: Flügge, Die Mikroorganismen, 1896, 416.

Morphology. Bacilli 2–3 times the length of fowl cholera. Cultural characters identical with the preceding.
Potato. 37°, no growth, later a brownish growth. Differs from the preceding mainly by its weaker pathogenic properties; only chickens affected by subcutaneous injections and by feeding. Stools loose, death in 7–9 days, with bacilli in the blood.
Habitat. Isolated from cases of enteritis in fowls.
VARIETY. *Bacillus of dysentery in turkeys and fowls* Lucet: Annales Pasteur Institut, 1891, 5. Not differentiated from the preceding.

33. Bact. anaerogenes (Lembke)

B. coli-anaerogenes Lembke: Archiv f. Hygiene, XXVII, 1896, 384.

Morphology. Bacilli 1.0: 2.0 μ. Morphology and cultural characters like *B. coli*.
Lactose bouillon. No gas; produces an amount of acid intermediate between *B. typhi* and *B. coli*.

Pathogenesis. Subcutaneous inoculations cause septicæmia in mice, guinea pigs, and rabbits; bacilli in the blood.

Habitat. Isolated from the fæces of a dog.

34. Bact. suicida (Migula)

Bacillus der deutschen Schweineseuche Löffler-Schütz: Arbeiten Kaiserl. Gesundheitsamte, I, 1886, 51, 376.
Bacillus of swine-plague Salmon: U. S. Dept. Ag., Bureau Animal Industry, 1886, 87.
Bact. suicida Migula: Die Natürlichen Pflanzenfam., 1895.
B. suisepticus Kruse: Flügge, Die Mikroorganismen, 1896, 419.

Morphology. Bacilli 0.5–0.6 : 1.0; others 0.7–0.8 : 1.8 μ, rounded ends; polar stain. Growth on gelatin rather variable; negative to very feeble.

Gelatin colonies. *Deep:* after some days round, entire, brownish, granular centres, and pale margins, 0.2–0.5 mm. *Surface:* like the former, but 4–5 times as large.

Agar colonies. *Deep:* in 24 hours 0.2 mm., round — lenticular, brownish, opaque, smooth or beset with knobs. *Surface:* round, entire, slightly convex, white — translucent; microscopically, centre brownish, granular toward margins, becoming homogeneous, translucent, with very delicate radial striations. Plates give off a disagreeable pungent odor.

Agar slant. Isolated colonies or a thin grayish translucent layer; growth in condensation water viscid.

Bouillon. Faintly turbid or granular in clumps — clear; sediment viscous; slightly acid.

Milk. Not coagulated, becoming slightly acid.

Potato. No appreciable growth; according to Karlinsky a very delicate, limited, straw-yellow growth.

Indol. Negative or only a trace. *Phenol:* present (Smith) or absent (Karlinsky).

Pathogenesis. Variable in bacilli from different outbreaks. Subcutaneous inoculations of rabbits, in virulent types, cause a rapid septicæmia; less virulent, death in 40 hours to 7 days. Peritonitis, hemorrhagic or diphtheritic, with bacilli in exudate; few bacilli in blood or organs. Attenuated forms cause death only after several weeks; local inflammatory reaction, circumscribed or spread over abdomen and thorax. *Guinea pigs:* as in rabbits, but slightly less susceptible. *Fowl* and *pigeons* refractory to subcutaneous inoculations; injections into pectoral muscle cause death in 36–48 hours.

Habitat. Associated with swine plague.

35. Bact. bovisepticum (Kruse)

Mikroparasiten bei einer Wild u. Rinderseuche Bollinger, 1878.
Bacterium der Wildseuche Kitt: Sitz. Ges. Morph. u. Physiol. in München, I, 1885.
Microbo del barbone dei bufali Oreste-Armanni: Atti d. R. Istit. d. incoragg. alle Scienz.
   Natur. Napoli Torn., 1886; ref. Centralblatt f. Bakteriol., II, 1887, 750.
B. bovisepticus Kruse: Flügge, Die Mikroorganismen, 1896, 421.

Morphologically and in cultures closely related to the preceding. See Canvena.
   Centralblatt f. Bakteriol., IX, 561. Phenol negative for buffelseuche;
   both indol and phenol produced by the other varieties.

36. Bact. pneumopecurium

Bacillus of sporadic pneumonia of cattle Smith: U. S. Dept. Ag.,
Bureau Animal Industry, 1895, 136.

Morphology. Bacilli like those of swine plague, except the presence of a
   capsule, 0.5–0.6:1.0 μ. Cultures become viscid with age. Growth on
   gelatin slight or invisible.
Agar colonies. Round, translucent, grayish, reaching 4 mm. in diameter.
Agar slant. Growth grayish, glistening, fleshy ; water of condensation viscid.
Potato. Growth not manifest.
Milk. Unchanged. *Phenol* produced. *Indol* absent or doubtful. Generally
   absent. No gas in glucose and saccharose bouillon.
Bouillon. A slight sediment, viscid when old.
Pathogenesis. Similar to the bacillus of swine plague.

37. Bact. sanguinarium Moore

U. S. Dept. Ag., Bureau of Animal Industry, 1895, 189.

Morphology. In the body, bacilli 1.0–1.3:1.4–1.8 μ, or coccoid forms, ends
   rounded ; involution forms common. In cultures, longer and more
   slender forms occur. Decolorized by Gram's method.
Gelatin colonies. Deep : grayish yellow, granular, 0.25 mm. *Surface* ·
   slightly spreading, granular, without markings.
Gelatin stab. Growth more abundant along the line of inoculation than on
   the surface.
Agar slant. Growth at 37°, grayish, glistening.
Potato. A delicate grayish, glistening growth ; often no development on
   acid potato.
Alkaline bouillon. In 24 hours a uniform turbidity, acid, becoming clear, with
   a granular sediment. In acid bouillon only a slight growth.
Milk. In 4 weeks, no change ; in 6 weeks, medium, opalescent, alkaline,
   saponified.

Glucose bouillon. No gas, acid. *Lactose* and *saccharose bouillon.* No gas, alkaline. *Indol* produced. *Phenol* negative.

Pathogenesis. Fowls inoculated into vein with 0.3 cc. of bouillon culture die in 3-13 days; pyrexia, crouching position, head drawn in, liver slightly enlarged, soft and fatty, spleen rarely enlarged, urates in tubules of kidneys, intestines show punctiform hemorrhagic spots, heart muscles pale with grayish points, only a few bacilli in blood and organs. *Rabbits:* injections of 0.2 cc. killed in 4-5 days; slight local reaction, necrotic areas in liver, enlarged and discolored spleen, infiltration of cells into follicles of cæcum and glands about ileocæcal valve.

Habitat. Associated with " infectious leukæmia " in fowls.

38. Bact. avium

Bacillus of roup in fowls Moore: U. S. Dept. Ag., Bureau of Animal Industry, Bulletin 8, 1895.

Morphology. Bacilli 0.8-1.5 : 0.8-1.2 μ; show polar stain. Decolorized by Gram's method; capsule (?).

Agar colonies. Minute grayish dots; odor pungent, similar to cultures of swine plague.

Agar slant. Growth gray, moist, glistening, slightly viscid.

Potato. No growth.

Alkaline gelatin. No growth.

Milk. Unchanged in 6 weeks.

Bouillon. In 24 hours at 36° clouded; in 2-3 days acid. No growth in acid bouillon.

Glucose bouillon. Acid; no gas.

Lactose bouillon. Remains alkaline; no gas. *Indol* produced. *Phenol* negative.

Pathogenesis. Injections of 0.1 cc. subcutaneously into rabbits caused death in 18-36 hours with lesions of virulent swine plague. Young fowls inoculated with 0.3 cc. subcutaneously died in 4 days; kidneys yellow with urates; cultures from blood and liver positive.

Habitat. Isolated from exudate of fowls in "roup" or diphtheria.

39. Bact. inocuum (Kruse)

B. lactis-inocuus Kruse: Flügge, Die Mikroorganismen, 1896, 352.

In its morphological and cultural characters similar to *Bact. aerogenes.*

Gelatin colonies. Porcelain-white, round — irregular, with characters approaching *B. coli.*

Potato. Growth brownish.

Habitat. Milk.

40. Bact. tiogense (Wright)

B. tiogensis Wright: l.c., 441.

Morphology. Bacilli medium-sized, plump; occur singly, in pairs, short chains, and in filaments.

Gelatin colonies. Round, milk-white, elevated, 2 mm.; microscopically, dark, opaque, with a greenish shimmer, becoming thinner, brownish, granular toward their margins. Grow in acid gelatin.

Agar slant. Growth grayish, glistening, rather limited.

Bouillon. Rendered turbid.

Potato. Growth gray-brownish, spreading.

Litmus milk. Decolorized; reaction amphoteric. *Indol* doubtful. No growth at 37° C.

Habitat. Water.

41. Bact. refractans (Wright)

B. refractans Wright: l.c., 442.

Morphology. Bacilli short, thick, medium-sized, in pairs or clumps.

Gelatin colonies. In 2 days round, white, slightly elevated, 1.0 mm.; microscopically, brownish, segmented — radially crimpled, with scalloped outlines.

Gelatin slant. Growth narrow, white, wrinkled.

Agar slant. Growth thin, narrow, composed of translucent colonies.

Pepton-rosolic acid. Unchanged or lighter in color.

Bouillon. Slightly turbid, with a slight pellicle, and a sediment.

Potato. Growth grayish — brownish gray, composed of minute colonies.

Litmus milk. Unchanged. *Indol* negative. Grow at 37° C.

Habitat. Water.

42. Bact. rodonatum (Ravenel)

B. rodonatus Ravenel: l.c., 40.

Morphology. Bacilli short ovals, rounded.

Gelatin colonies. Deep: yellowish brown with irregular edges, and a rosette structure. *Surface:* in 60 hours, 1.0 mm.; grow slowly with a rosette structure, becoming distinctly petaloid on their edges; reddish brown centres and yellowish gray edges.

Agar slant. Growth thin, white, translucent, limited.

Gelatin stab. On the surface, growth thin, irregular, leafy, 4–5 mm.

Potato. Growth moist, glistening, yellowish-brownish.

Bouillon. Turbid, a thin pellicle.

Pepton-rosolic acid. Decolorized.

Litmus milk. Pure blue, becoming decolorized in 10 days. *Indol* negative.
  Grow at 37°.
Habitat. Water.

43. Bact. zurnianum (List)

Adametz, Bakt. Nutz u. Trinkwässer, Vienna, 1888; Dyar, l.c., 362.

Morphology. Bacilli 0.5 : 0.6–1.0; occur singly and in short chains. Cultures on solid media translucent, white, very viscous. *Indol* negative.
  A slight reduction of nitrates to nitrites.
Habitat. Isolated from the air.

44. Bact. Martizeni (Sternberg)

B. Martizeni Sternberg: Manual of Bacteriology, 1892, 651.

Morphology. Bacilli 0.5 : 1.0–1.5 μ; occur singly and in short chains. Nitrates
  slightly reduced after 28 days (Dyar, l.c.).
Agar slant. Growth white, translucent.
Pepton-rosolic acid. Deepened in color (Dyar).
Habitat. Isolated from the liver of a yellow fever cadaver (Sternberg). From
  the air (Dyar (?)).

45. Bact. cuniculicida (Kruse)

B. cuniculicida-thermophilus Kruse: Flügge, Die Mikroorganismen, 1896.
Micrococcus sur une nouvelle septicémie du lapin Lucet: Annales Pasteur Institut,
  1889, 401.

Pathogenesis. Rabbits and guinea pigs inoculated subcutaneously and by
  feeding die in 1–3 days; septicæmia, spleen and liver enlarged, serous
  membranes inflamed, bacilli in organs.
Habitat. Associated with an epidemic of rabbits and guinea pigs.

46. Bact. cuniculicida var. immobile

Pathogenesis. Only slightly virulent to mice, guinea pigs, and pigeons; death
  only with large doses. Rabbits at autopsy show an inflammation of serous membranes.
Habitat. Associated with a spontaneous rabbit plague.

47. Bact. putidum

Bact. gracilis-cadaveris Sternberg: Manual of Bacteriology, 1892, 733.

Morphology. Bacilli 1.0–2.0 μ; in chains.
Gelatin stab. Beaded below, branched outgrowths above; on the surface,
  growth thick, white.
Potato. Growth creamy.

Bouillon. Turbid, with a bad odor.
Pathogenesis. Pathogenic for rabbits only by intraperitoneal injections.
Habitat. From human liver of a cadaver.

48. Bact. diphtheriæ (Flügge)

Bacillus der diphtheriæ bei der Taube Löffler: Mitteilungen Kaiserlichen Gesund-
heitsamte, 1884, 421.
Der Löffler Bacillus Babes-Puscariu: Zeitsch. f. Hygiene, VIII, 1890, 376.
B. diphtheriæ-columbarum Flügge: Die Mikroorganismen, 1886.

Morphology. Bacilli somewhat longer and more slender than the bacillus of
fowl cholera.
Gelatin colonies. Similar to *B. typhosus.*
Agar slant. Growth gray, translucent.
Bouillon. Turbid.
Potato. Growth white — grayish. *Indol* negative.
Pathogenesis. Inoculations of mice and rabbits cause death, with necrotic
spots in liver containing bacilli; spleen enlarged. Pigeons infected
through wounds of the mouth show diphtheritic deposits containing
bacilli, with death in 1-3 weeks; bacilli in the organs after death.
Habitat. Associated with diphtheria in pigeons.

49. Bact. columbarum (Kruse)

Microbe maladie des palombes Leclainche: Annales Pasteur Institut, 1894, 490.
B. choleræ-columbarum Kruse: Flügge, Die Mikroorganismen, 1896, 417.

Morphology. Bacilli somewhat larger than bacilli of fowl cholera. Cultures
similar to the latter.
Bouillon. Not turbid, but a flocculent sediment.
Potato. Growth grayish yellow.
Pathogenesis. Subcutaneous inoculations of rabbits and guinea pigs cause
death in about 8 days. By feeding cultures to wild pigeons death follows
in 3-6 days, with symptoms of chicken cholera. Differentiated from fowl
cholera and No. 48 by its effect upon guinea pigs, growth in bouillon, etc.
Habitat. Associated with a disease of wild pigeons.

50. Bact. cuniculi (Kruse)

Bacillus der Darmdiphtheriæ des Kaninchens Ribbert: Deutsche. medizinische
Wochenschrift, 1887.
B. diphtheriæ-cuniculi Kruse: Flügge, Die Mikroorganismen, 1896, 412.

Morphology. Bacilli 1.0-1.4 : 3.0-4.0 µ.
Gelatin colonies. Similar to *B. coli.*
Potato. Growth flat, whitish, slightly spreading. *Indol* negative.

Pathogenesis. Subcutaneous and intraperitoneal inoculations of rabbits cause death in 3–14 days; in liver and spleen necrotic spots containing bacilli. Infections *per os* cause a diphtheritic inflammation of the intestines, etc.

Habitat. Associated with the above disease.

51. Bact. Beckii

Bacillus der Brustseuche des Kaninchens Beck: Zeitsch. f. Hygiene, XV, 1893, 363.
B. cuniculi-pneumonicus Kruse: Flügge, Die Mikroorganismen, 1896, 418.

Morphology. Bacilli like those of fowl cholera; show polar stain.
Gelatin colonies. Small, entire, granular, clear, becoming brownish.
Gelatin stab. Slight growth in depth.
Potato. At 20° C. no growth.
Agar slant. Growth porcelain-white — brownish.
Pathogenesis. Inoculations of rabbits into the lung result in cough, fever, rapid respiration, and death in 3–5 days of pneumonia and pleuritis, with much exudate containing bacilli. Subcutaneous inoculations cause a spreading necrosis and death without general infection. Differentiated from No. 50 by growth on potato and pathogenesis.
Habitat. Associated with lung plague of rabbits.

52. Bact. dubium Kruse

Ein neuer für Thiere path. Mikroorg. aus dem Sputum eines Pneumoniekranken Bunzl — Federn: Archiv f. Hygiene, XIX, 1893, 326.
B. dubius-pneumoniæ Kruse: Flügge, Die Mikroorganismen, 1896, 419.

Morphology. Bacilli short rods, with polar staining; longer and more slender on agar.
Gelatin colonies. Slightly spreading.
Agar slant. Growth composed of transparent colonies.
Potato. No growth.
Pathogenesis. Subcutaneous or intraperitoneal inoculation of rabbits, guinea pigs, mice, and pigeons cause death in 1–4 days; septicæmia, local œdema — necrosis.
Habitat. Isolated from rusty sputum of a pneumonia patient.

53. Bact. felis (Kruse)

B. salivarius-septicus-felis Fiocca: Annali dell' Instituto d' igiene dell' Università di Roma, 1892, II.
B. felis-septicus Kruse: Flügge, Die Mikroorganismen, 1896, 423.

Morphology. Bacilli short rods. Cultures like those of fowl cholera.
Milk. Not coagulated.
Potato. Growth thin, invisible.

Pathogenesis. Inoculations of mice, guinea pigs, and rabbits cause a general septicæmia.

Habitat. Isolated from the sputum of a cat.

54. Bact. septicum (Trevisan)

B. septicus-agrigenus Flügge: Die Mikroorganismen, 1886.
Pasteurella agrigena Trevisan: Genera, 1889, 21.

Morphological and cultural characters similar to those of fowl cholera. Pathogenic to mice, guinea pigs, and rabbits.

Habitat. Isolated from the earth.

B. septicus-hominis Mironoff: Centralblatt f. Gynäkologie, 1892, 42. From a case of septic infection of the uterus; and

B. canalis-parvus Mori: Zeitsch. f. Hygiene, IV. From canal water; not differentiated from the preceding, No. 54.

55. Bact. vitulinum

Bacillus der Septikämie bei einem Seekalbe Bosso: Centralblatt f. Bakteriol., XXV, 1899, 52.

Morphology. In the blood of rabbits inoculated with the virus, bacilli 0.9: 2.7, or 0.5 : 1.7 μ. Grow poorly in ordinary media and quickly lose their vitality and virulence after 8–10 days.

Gelatin colonies. Small, round, raised; border distinct, granulose, yellowish.

Agar slant. 37°, isolated, ash-gray colonies, becoming confluent; condensation water turbid.

Bouillon. In 8 hours, at 37°, a uniform turbidity, becoming clear in several days. Grow well on glycerin agar.

Pathogenesis. Subcutaneous inoculations cause death in rabbits in 20–60 hours; slight enlargement of the spleen, diffuse coloration of the flesh, bacilli in the blood. Guinea pigs die in 32–40 hours; intense reddening of the flesh, kidneys strongly congested, bacilli in small numbers in the blood. Mice refractory to doses which kill the preceding.

Habitat. Isolated from a septicæmia of the sea-calf (*Phoca vitulina*).

56. Bact. purpurum

Bacillus of purpura-hæmorrhagica Babes: Septische Proz. Kindesalters, 1889.
B. hæmorrhagicus-septicus Kruse: Flügge, Die Mikroorganismen, 1896, 424.

Morphology. Short rods with a capsule. Grow rapidly in gelatin.

Agar slant. Small transparent drops, becoming whitish yellow.

Potato. Growth composed of whitish drops.

Bouillon. Rendered turbid.

Pathogenesis. Inoculations of mice cause death in a few days, of hemorrhagic septicæmia. Only slightly pathogenic to dogs and guinea pigs.

Habitat. Isolated from a case of septicæmia in man.

57. Bact. Bienstockii (Schröter)

Bacillus aus fæces, No. III, Bienstock : Zeitsch. klin. Med., VIII, Heft 1.

B. Bienstockii Schröter: Pilze Schlesien, 1886, 163.

B. coprogenes-parvus Flügge: Die Mikroorganismen, 1886.

Morphological and cultural characters like those of fowl cholera.

Pathogenesis. Subcutaneous inoculations of mice cause death in 56 hours; œdema, with few bacilli in the blood. Inoculations of rabbits into ear cause death in 8 days, with erysipelas and diarrhœa.

Habitat. Isolated from fæces.

58. Bact. hæmorrhagicum (Kruse)

Bacillus der idiopathischen Blutfleckenkrankheit Kolb: Arbeiten Kaiserlichen Gesund-heitsamte, VII, 1891, 60.

B. hæmorrhagicus Kruse: Flügge, Die Mikroorganismen, 1896, 424.

Morphology. Bacilli 0.8 : 1–2 μ; capsule present or absent.

Gelatin colonies. Flat, erose.

Potato. Growth thin, moist.

Pathogenesis. Inoculations of mice cause death in 2–3 days, of septicæmia. Guinea pigs affected only by large doses. Rabbits often die by intra-peritoneal injections of 0.5–1.0 cc.

Habitat. Isolated from corpses of persons dead of septicæmia.

59. Bact. velenosum (Kruse)

Bacillus der hämorrhagischen Infektion Tizzoni-Giovannini: Ziegler's Beiträge, VII, 1889, 300.

B. hæmorrhagicus-velenosus Kruse: Flügge, Die Mikroorganismen, 1896, 425.

Morphology. Bacilli 0.2–0.4 : 0.7–1.3 μ.

Gelatin colonies. Flat, irregular, with floccose borders.

Potato. Growth invisible.

Pathogenesis. Inoculations of dogs, rabbits, and guinea pigs cause only loca-œdema, with fever, hemorrhagic nephritis, vomiting, bloody diarrhœa; spleen normal; necrosis of liver, and epithelium of kidney.

Habitat. Isolated from a case of purpura-hæmorrhagica.

60. Bact. nephritidis (Vassalle)

B. hæmorrhagicus-nephritidis Vassalle: Tizzoni-Giovannini, Ziegler's Beiträge, VII, 1889.

Morphology. Bacilli similar to those of fowl cholera, but less pathogenic to rabbits; strongly so for guinea pigs; intraperitoneal inoculations cause hemorrhagic nephritis.

Habitat. Isolated from a case of hemorrhagic nephritis.

61. Bact. aphthosum (Kruse)

Bacillus der Mundseuche des Menschen (Stomatis epidemica) Siegel: Deutsche med. Wochensch., 1891, No. 49.

B. aphthosus Kruse: Flügge, Die Mikroorganismen, 1896, 427.

Morphology. Bacilli short 0.5–0.7 μ — filaments; show polar stain.
Gelatin colonies. Small, entire, bluish white — yellowish.
Gelatin stab. Growth in depth, beaded.
Pathogenesis. Non-pathogenic to rabbits, guinea pigs, mice, dogs, and cats. Local infection through the mouth to young pigs and calves.
Habitat. Isolated from the liver and kidneys of cattle affected with "Maul" and "Klauenseuche."

62. Bact. dysenteriæ (Kruse)

Bacillus der weissen Ruhr der Kälber Jensen: Monatshefte f. prakt. Thierheilk., III, 1892, 92.

B. dysenteriæ-vitulorum Kruse: Flügge, Die Mikroorganismen, 1896, 412.

Morphology. Bacilli somewhat larger than those of fowl cholera; show polar stain. Cultures similar to those of *B. coli.*
Potato. Growth slimy, brownish.
Pathogenesis. The feeding of 5 cc. of bouillon culture to young calves gives a fatal diarrhœa, with bacilli in intestines and organs.
Habitat. Associated with dysentery of calves.

63. Bact. salivæ (Kruse)

B. saliva-minutissimus Kruse: Flügge, Die Mikroorganismen, 1896, 440.

Morphology. Bacilli small like those of influenza. Decolorized by Gram's method.
Gelatin stab. Surface growth flat.
Potato. Growth brownish.
Habitat. Isolated from secretions of the mouth.

L

64. Bact. acidum

No. 56 of Conn : l.c., 83.

Morphology. Bacilli 0.8–1.2 μ ; in pairs and chains. Grow at 35° C.

Gelatin colonies. Large, white, thin, translucent, irregular — lobate ; surface irregular.

Gelatin stab. Good growth in depth ; on surface, growth thin, with an irregular border.

Agar slant. Growth thin, white, spreading.

Bouillon. Slightly turbid ; sediment white.

Habitat. Milk.

65. Bact. Connii

No. 55 Conn : l.c., 83.

Morphology. Bacilli 0.8 : 1.0 μ. Slight growth at 35° C.

Gelatin colonies. White, spreading, granular, entire.

Gelatin stab. Surface growth thin, translucent, spreading.

Agar slant. Growth white.

Potato. Growth elevated, yellowish, spreading.

Bouillon. Clear, with flaky sediment.

Habitat. Milk.

66. Bact. nitrovorum (Jensen)

B. nitrovorus Jensen : Centralblatt f. Bakteriol., 2d Abt., IV, 1898, 450.

Morphology. On agar and in bouillon bacilli 0.5 : 0.5–2.0 μ. No polar stain.

Gelatin colonies. Deep : in 4–5 days scarcely visible. *Surface :* white, moist, glistening, entire, 0.2–0.5 mm.

Agar colonies. In 3 days, at 30° C. *Deep :* small, yellowish, entire. *Surface :* dirty white, slimy, 2–4 mm.

Gelatin stab. In depth, growth fine, pearly. On the surface, growth 6–10 mm., white, often concentric.

Gelatin slant. Slight growth, moist, glistening, bluish — yellowish, knotty, very limited.

Agar slant. Growth moist, slimy, grayish — bluish white, limited.

Nitrate bouillon. In one day a weak turbidity, becoming clear ; a faint pellicle, and a granular sediment. Nitrates reduced to nitrites.

Habitat. Isolated from horse manure.

67. Bact. filefaciens (Jensen)

B. filefaciens Jensen : Centralblatt f. Bakteriol., 2d Abt., IV, 1898, 409.

Morphology. On agar bacilli 0.5–0.7 : 0.5–1.5 μ ; in nitrate bouillon, 0.5–0.7 : 1.0–2.5 μ.

Gelatin colonies. In 3–4 days. *Deep:* very small, white, entire. *Surface:* thin, moist, glistening, stringy, becoming thicker.

Agar colonies. In 2 days, at 30° C. *Deep:* small, yellowish white, entire. *Surface:* moist, glistening, dirty white, stringy, becoming white and spreading.

Agar slant. Growth limited, raised in the centre; whitish — grayish.

Gelatin slant. Growth milky, raised in the middle.

Gelatin stab. In depth, growth pearly; on the surface, growth spreading, milk-white.

Nitrate bouillon. Weakly turbid, with a pellicle, becoming clear with a stringy sediment; denitrification.

Habitat. As a contamination of an old culture of *B. Stutzeri.*

68. Bact. punctatum

Bacillus No. 19 Adametz: Landwirthsch. Jahrbücher, 1889.

Morphology. Bacilli 0.8 : 2.0–3.0 μ, and filaments.

Gelatin colonies. In 4–6 days, compact, scarcely visible points; opaque to dark brown, not spreading.

Gelatin stab. Slight growth.

Milk. Rendered acid.

Habitat. Milk.

69. Bact. Middletownii

No. 53 Conn: l.c., 1894, 82.

Morphology. Bacilli short. Grow at 35° C.

Gelatin colonies. As raised beads, 1 mm.

Gelatin stab. On the surface, slightly spreading, pearly white, waxy; gas in the gelatin.

Agar slant. Growth white, elevated.

Potato. Growth whitish — brownish.

Milk. Acid, not peptonized.

Bouillon. Slightly turbid, with a sediment.

Habitat. Milk.

70. Bact. coccoideum

No. 16 Conn: l.c., 1893, 51.

Morphology. Ovals — coccoid forms. Grow at 35° C.

Gelatin colonies. A white bead, becoming thin, spreading; sometimes a raised central nucleus.

Gelatin stab. Growth nail-shaped; gas in gelatin frequently.

Agar slant. Growth thin, white, glistening, spreading — elevated.
Milk. Coagulated at 37°, with gas, acid, and peptonization of the medium.
Habitat. Milk, Mansfield-Conn.

71. **Bact. lactis** (Kramer)

Bacillus schleimiger Milch Löffler : Berliner klin. Wochensch., 1887, 631.
B. lactis-pituitosi Kramer : Die Bakteriologie Landwirthschaft, 1892, 24.

Morphology. Bacilli rather thick, rods breaking into coccoid segments.
Gelatin colonies. Round, entire — erose, 0.2–0.5 mm.; microscopically,
   brownish and radially striped.
Potato. Growth gray white, rather dry
Agar slant. Dirty white colonies.
Milk. Acid, slimy.
Habitat. Milk.

72. **Bact. aromafaciens**

No. 41 Conn : l.c., 1894, 41.

Morphology. Bacilli $1.1 : 6.0 \mu$; in twos.
Gelatin colonies. Punctiform, 1 mm.
Gelatin stab. Slight growth in depth ; on the surface, growth moist, elevated,
   of the aerogenes type.
Agar slant. Growth white, glistening.
Potato. Growth elevated, white — yellow.
Milk. Slightly acid, and slowly peptonized.
Bouillon. Turbid, with pellicle.
Habitat. Milk.

CLASS III.   GELATIN NOT LIQUEFIED.   STAINED BY GRAM'S
METHOD.

I. Gas produced in glucose bouillon.   BACT. ACIDI-LACTICI GROUP.
   A. Milk coagulated.
             73. *Bact. acidi-lactici* Zopf.
             74. *Bact. acidiformans* Sternberg.
             75. *Bact. astheniæ* Dawson.
   B. Milk not coagulated.
             76. *Bact. endometriditis* (Kruse).
II. No gas produced in glucose bouillon.
                              BACT. RHINOSCLERMATIS GROUP.
   A. Milk coagulated.
             77. *Bact. lacticum* (Kruse).

B. Milk coagulated (28 days) only after boiling.

    78. *Bact. crassum* (Kreibohm).

C. Milk coagulation variable.

    79. *Bact. rhinosclermatis* (Trevisan) Migula.

D. Milk not coagulated.

  1. Bacilli with capsules.

    80. *Bact. Bordonii* (Trevisan).

  2. Bacilli not distinctly surrounded by capsules.

    82. *Bact. Bossonis.*

    83. *Bact. Czaplewskii.*

III. Gas production in glucose bouillon not stated.

A. Obligate aerobic. Grow very slowly.

    84. *Bact. ureæ* (Miquel) Leube.

B. Not strictly aerobic, as before.

  1. Grow best on blood serum at 37° C.

    85. *Bact. Grawitzii* (Trevisan).

  2. Grow well on ordinary media.

   a. Grow at ordinary room temperatures.

    * Milk coagulated, acid.

    86. *Bact. tenue* (Pansini).

    ** Milk becomes thick, viscous.

    87. *Bact. subviscorum* Migula.

   b. Growth only takes place at temperatures above 27° C.

    88. *Bact. sanguinis* (Sternberg).

BACT. ACIDI-LACTICI GROUP

73. **Bact. acidi-lacti** Zopf

Milchsäurebacillus Hueppe: Mitteilungen Kaiserl. Gesundheitsamte, II, 1882, 337.
Bact. acidi-lactici Zopf: Spaltpilze, 1883, 65.

Morphology. Bacilli 0.5–0.6 : 1.0–2.0 μ; often in twos. Optimum temperature 37° C.

Gelatin colonies. Coli-like.

Gelatin stab. A nail, flat-topped.

Agar slant. Growth white — yellow.

Potato. Growth yellowish brown.

Milk. Coagulated, acid, with production of CO_2 and alcohol.

Lactose bouillon. Gas. Non-pathogenic.

Habitat. Milk.

74. Bact. acidiformans Sternberg

Manual of Bacteriology, 1892, 449.

Morphology. Bacilli short, thick rods to longer forms and filaments.

Gelatin colonies. Deep: round, opaque, homogeneous. *Surface:* round, translucent — opaque, slightly irregular and somewhat iridescent.

Gelatin stab. Like *Bact. pneumoniæ*; gas in the gelatin.

Agar slant. Growth glistening, creamy. In glycerin agar, much gas; medium becomes intensely acid.

Potato. Growth thick, milky white, semi-fluid.

Glycerin bouillon. A milky opacity, and much gas. A slight reduction of nitrates to nitrites.

Litmus milk. Reddened.

Pathogenesis. Intraperitoneal inoculations of guinea pigs and rabbits, 1.0–2.0 cc., cause death in 24 hours; bacilli in the blood in small numbers; spleen enlarged; intestines hyperæmic.

Habitat. Isolated from the liver of yellow fever cadaver.

75. Bact. astheniæ Dawson

U. S. Dept. Ag., Bureau of Animal Industry, 1898, 330.

Morphology. Bacilli 0.5 : 1.0–1.3 μ; ends rounded; occur singly and in pairs. Do not stain in acid or alkaline methyl-blue or in carbol-fuchsin, but do stain well in aqueous solutions of the latter. Stain by Gram's method.

Gelatin colonies. Deep: yellowish, with irregular lobulate margins. *Surface:* in 24 hours, 1.5 mm., round, raised, papillate in centre, with yellowish borders.

Gelatin stab. In depth, yellowish, discrete, closely packed colonies; on the surface, a brownish, spreading, deeply dentate growth, with delicate periphery.

Bouillon (acid). A dense turbidity, with pellicle on the surface. Medium yellowish green, with a putrefactive odor. Alkaline bouillon rendered slightly acid.

Agar slant. A good, white, opaque growth along line of stroke.

Potato. In 3 days a yellowish, creamy growth, with gas blisters.

Glucose bouillon. Rendered acid; much gas. $\dfrac{H}{CO_2} = \dfrac{3}{2}$.

Lactose bouillon. Rendered acid; moderate gas production. $\dfrac{H}{CO_2} = \dfrac{2}{1}$.

Saccharose bouillon. Rendered acid; but little gas. $\dfrac{H}{CO_2} = \dfrac{2}{1}$.

Milk. Coagulated in 24 hours; whey clear; highly acid, odorless. Neither indol nor phenol produced. Pathogenic for guinea pigs and rabbits.
Habitat. Isolated from the duodenal contents of chickens in "asthenia."

76. Bact. endometriditis (Kruse)

Typhus ahnlichen Bacillus Germano-Maurea: Ziegler's Beiträge, XII, 1893, 494.
B. endometriditis Kruse: Flügge, Die Mikroorganismen, 1896, 432.

Morphology. Bacilli medium-sized, of variable length, with a capsule.
Gelatin colonies. Coli-like.
Potato. Growth yellowish.
Lactose bouillon. Gas production. No gas in saccharose bouillon.
Pathogenesis. Doubtful.
Habitat. Isolated from a liver abscess.

BACT. RHINOSCHLERMATIS GROUP

77. Bact. lacticum (Kruse)

Described by Günther-Thierfelder: Archiv f. Hygiene, XXV, 1895, 2.
B. lacticus Kruse: Flügge, Die Mikroorganismen, 1896, 356.

Morphology. Bacilli 0.5–0.6 : 1.0 μ; in twos and short chains.
Gelatin colonies. Small, prominent.
Potato. Growth scanty.
Glucose bouillon. No gas, acid.
Lactose bouillon. As before.
Milk. Coagulated, odor aromatic; production of alcohol and fatty acids.
Habitat. Milk.

78. Bact. crassum (Kreibohm)

B. crassus-sputigenus Kreibohm: Flügge, Die Mikroorganismen, 1886.
Klebsiella crassa Trevisan: Genera, 1889, 25.

Morphology. Bacilli short, thick; in body with a capsule.
Gelatin colonies. Aerogenes-like, grayish white, large, granular.
Gelatin stab. A nail-shaped growth, with a round head.
Potato. Growth moist, grayish white.
Milk. As above (Dyar, l.c.).
Pathogenesis. Subcutaneous inoculations of mice cause death in 2 days of septicæmia. Rabbits succumb to intravenous injections.
Habitat. Isolated from sputum.

79. Bact. rhinosclermatis (Trevisan) Migula

Bacillus der Rhinoscleroms Frisch: Wiener med. Wochenschrift, 1882, 970.
Klebsiella rhinosclermatis Trevisan: Sul micrococco della rabbia, 1887, 8.
Bact. rhinosclermatis Migula: Die Natürlichen Pflanzenfam., 1895.

Morphology. Bacilli short, with rounded ends, usually united in pairs and surrounded by a capsule; or longer rods and filaments.

Gelatin colonies. Round, yellowish white; microscopically granular.

Gelatin stab. In depth, densely crowded colonies; on the surface, growth white, glistening, pulvinate.

Potato. Growth creamy white — yellowish-brownish, in which gas bubbles may develop.

Milk. Coagulated (Paltauf); not coagulated (Abel).

Pathogenesis. Mice affected by small doses. Intraperitoneal inoculations of guinea pigs and rabbits, in large doses, may cause death. Bacilli only sparingly in blood and organs.

Habitat. Found in newly formed tubercles in rhinoscleroma.

80. Bact. Bordonii (Trevisan)

Proteus hominis-capsulatus Bordoni-Uffreduzzi: Centralblatt f. Bakteriol., I, 1887.
Klebsiella Bordonii Trevisan: Genera, 1889, 25.
B. capsulatus-septicus Kruse: Flügge, Die Mikroorganismen, 1896, 345.

Morphology. Bacilli variable in size, often irregularly swollen or curved; occur singly and in pairs or long filaments. Rods surrounded by a capsule.

Gelatin colonies. Like those of *Bact. pneumoniæ*, raised, but with irregular borders, or coli-like.

Gelatin stab. Growth like *Bact. pneumoniæ*.

Agar slant. Growth spreading, translucent.

Potato. Growth spreading, moist, glistening, colorless.

Blood serum. As on agar; medium not liquefied. Gas produced in glucose bouillon according to Kruse; not so according to Bordoni.

Bouillon. Turbid, with a pellicle on the surface; no putrefactive odor.

Pathogenesis. Subcutaneous inoculation of mice causes death in 1-4 days of septicæmia. Intraperitoneal inoculations of rabbits and guinea pigs cause septicæmia and death in 2-3 days.

Habitat. Isolated from cadavers in cases of *Hadernkrankheit*.

82. Bact. Bossonis

Bacillus über eine neue Infektionskrank. des Rindviehs Bosso: Centralblatt
f. Bakteriol., XXII, 1897, 537; XXIII, 1898, 318.

Morphology.   Bacilli 0.5–0.8 : 1.5–2.7 μ.

Gelatin colonies.   Round, raised.   Microscopically, round — irregular, light
yellow, marmorated ; often semilunar colonies.

Gelatin stab.   In depth, confluent yellowish colonies.   On surface, growth
round, raised.

Agar slant.   White — straw-yellow colonies, becoming confluent ; water of
condensation turbid.

Potato (alkaline).   37° C., an abundant growth of transparent drops, with
slight tendency to become confluent ; on acid potato, but slight growth.

Milk.   Not coagulated, acid.

Glucose bouillon.   37°, turbid, much odorless gas and abundant sediment ; at
room temperatures but slight growth.   *Indol* negative.

Pathogenesis.   Subcutaneous inoculations of guinea pigs cause death in 36
hours ; strong hyperæmia of the intestinal viscera, peritoneum, and kidneys ; spleen not enlarged ; lungs and pleura normal ; bacilli in spleen
and kidneys, also in the black clotted blood.

Habitat.   Associated with an infectious disease of cattle, septicæmia.

83. Bact. Czaplewskii

Bacillus bei Keuchhusten Czaplewski : Centralblatt f. Bakteriol., XXII, 1897, 641.

Morphology.   Bacilli very minute, like *Bact. influenzæ.*   Grow best at 37°.
Bacilli in young cultures show polar stain.

Gelatin slant.   A narrow, grayish stripe.

Gelatin stab.   In depth, growth beaded, whitish yellow ; on the surface, colony
small, grayish, rather dry.

Bouillon.   37° C., scarcely turbid ; sediment stringy.

Potato.   No growth observed.

Pathogenesis.   Inoculation experiments either negative or inconclusive.

Habitat.   Isolated from sputum in whooping cough.

84. Bact. ureæ (Miquel) Leube

B. ureæ Miquel : Bull. Soc. Chimiq d. Paris, XXXI, 1879, 391.
Bact. ureæ Leube : Virchow's Archiv, C, 1885, 540.

Morphology.   Bacilli 1.0 : 1.5–2.6 μ.   Grow very slowly.

Gelatin colonies.   Flat, spreading, irregular.   Ferment urine.

Habitat.   Isolated from urine.

85. Bact. Grawitzii (Trev.)

Bacillus der Acne Contagiosa des Pferdes Dieckerhoff-Grawitz: Virchow's Archiv,
   CII, 148.
B. Grawitzii Trevisan: Genera, 1889, 13.
B. acnes-contagiosæ Kruse: Flügge, Die Mikroorganismen, 1896, 445.

Morphology. Bacilli very small, long-oval rods; occur singly, or in short
   chains. Stain with difficulty.
Gelatin stab. In depth, a beaded growth.
Potato. Scarcely any growth.
Pathogenesis. Rubbing cultures into the skin of horses causes characteristic
   pustules. Subcutaneous inoculations of rabbits, toxic symptoms; bacilli
   not spreading. In guinea pigs, by rubbing into skin, death with hemorrhagic serous inflammation. Mice die by subcutaneous injections in 1-10
   days; bacilli in the organs.
Habitat. Isolated from *acne-contagiosa* in horses.

86. Bact. tenue (Pansini)

B. sputigenes-tenuis Pansini: Virchow's Archiv, CXXII, 1890, 453.

Morphology. Bacilli small, of variable length; in twos and chains; in the
   body with a capsule.
Gelatin colonies. Round, raised — slightly spreading, yellow, concentric, radially striped.
Gelatin stab. In depth, growth beaded; on the surface, yellowish.
Potato. Growth yellowish, flat.
Pathogenesis. Mice and guinea pigs refractory to ordinary quantities of the
   virus. Subcutaneous inoculations of 0.5 to 1.0 cc. into rabbits cause death
   by septicæmia; slight local effect, hemorrhage in peritoneum; spleen
   enlarged.
Habitat. Associated with advanced phthisis and catarrhal pneumonia.

B. pyogenes-minutissimus Kruse. From pus in man. From the descriptions,
   not differentiated from the preceding.

87. Bact. subviscorum Migula

Bacillus schleimiger Milch Adametz: Landw. Jahrbücher, 1891, 185.
B. lactis-viscosus Krammer: Die Bakteriologie Landwirthschaft, II, 1892, 26.
Bact. subviscorum Migula: System des Bakterien, 1900.

Morphology. Bacilli 1.1-1.3 : 1.2-1.7 μ — filaments; with a capsule.
Gelatin colonies White, slimy, erose.
Agar slant. Growth dirty white, slimy.

Milk. In 5-10 days becomes thick, viscous; later, peptonized and clear; no special odor.

Habitat. Milk.

88. Bact. sanguinis (Sternberg)

B. sanguinis-typhi Sternberg: Manual of Bacteriology, 1892.

Morphology. Bacilli typhi-like, 0.5-0.8 : 1.0-2.5 μ.

Agar colonies. Blue-gray, translucent, irregular, becoming dry.

Potato. Growth invisible.

Milk. Not coagulated.

Pathogenesis. Slight or doubtful.

Habitat. Isolated from blood of typhoid fever patients.

CLASS IV. WITHOUT ENDOSPORES. AEROBIC AND FACULTATIVE ANAEROBIC. WITHOUT PIGMENT. GROW AT ROOM TEMPERATURES. GELATIN LIQUEFIED.

I. Colonies on gelatin plates roundish, not amœboid or proteus-like.

BACT. AMBIGUUM GROUP.

  A. Gelatin liquefied very slowly, or merely softened.

    1. Stained by Gram's method.

        89. *B. Thuillieri* (Trev.). Swine erysipelas; see Mycobacterium.

        90. *B. insidiosus* Trevisan. Mouse septicæmia; see Mycobacterium.

    2. Decolorized by Gram's method.

      a. Pathogenic bacteria.

        91. *B. mallei* Zopf; see Mycobacterium.

        92. *Bact. salmonica* (Kruse).

      b. Non-pathogenic.

        93. *Bact. vermiculosum* (Zimmerman).

        94. *Bact. incannum* (Pohl).

        95. *Bact. truncatum.*

  B. Gelatin liquefied rather quickly.

    1. Stained by Gram's method.

        96. *Bact. orchiticum* (Kruse).

        97. *Bact. pneumonicum* Kruse.

    2. Not stained by Gram's method, or indeterminate.

      a. Potato growth white, or color of the medium.

        * Potato growth dry, with irregular surface. Grow best at 37°C.

        98. *Bact. varicosum* (Sternberg).

** Potato growth scarcely visible. Grow well at 20° C.
  † Gas generated in nutrient gelatin and glucose bouillon.
    99. *Bact. aromaticum* (Pammel).
  †† No gas in nutrient gelatin (?).
    100. *Bact. nubilum* (Frankland) Lehmann-Neumann.
    101. *Bact. methylicum* (Loew.).
*** On potato, raised dry grayish colonies. Bacilli show chromatic granules in the interior of the rods.
    102. *Bact. Trambusti* Kruse.
  b. Potato growth yellowish — brownish.
    * Gelatin stab cultures show a saccate liquefaction.
      103. *Bact. aquatilis* (Zimmerman).
    ** Gelatin stab cultures a crateriform liquefaction, becoming stratiform.   ·
      104. *Bact. flexuosum* (Wright).
  c. Potato growth pinkish — salmon-colored.
      105. *Bact. ambiguum* Chester.
  d. Color of potato growth not stated.
      106. *Bact. convolutum* (Wright).
II. Colonies on gelatin plates amœboid or proteus-like.

                              BACT. RADIATUM GROUP.

    107. *Bact. radiatum* Chester.

BACT. AMBIGUUM GROUP

92. **Bact. salmonica** (Kruse)

Bacillus der Forellenseuche Emmerich-Weibel: Archiv f. Hygiene, XXI, 1894, 1.
B. salmonica Kruse: Flügge, Die Mikroorganismen, 1896, 322.

Morphology. Bacilli short rods. Do not grow at 37° C.
Gelatin colonies. Whitish-gray, becoming brownish, glistening, cholera-like, with rosulate markings.
Gelatin stab. In depth, beaded colonies, then air bubbles, with slow liquefaction, and production of an open canal, mostly air.
Bouillon. Clear, with a sediment.
Agar slant. Growth thin, moist, glistening, greenish yellow — brownish.
Potato. No growth.
Pathogenic to trout by inoculation of cultures.
Habitat. Isolated from a disease of trout.

93. Bact. vermiculosum (Zimmerman)

B. vermiculosus Zimmerman: Die Bak. Nutz u. Trinkwässer, 1890.

Morphology. Bacilli 0.8 : 1.5 μ, with a capsule. Grow at 37° C.
Gelatin colonies. Deep: round, gray, granular. *Surface:* spreading, lobed, marmorated. Gelatin slowly liquefied.
Agar slant. Growth moist, opalescent.
Potato. Growth yellowish gray, glistening.
Habitat. Water.

94. Bact. incannum (Pohl)

B. incannus Pohl: Centralblatt f. Bakteriol., 1892, 142; Dyar, l.c.

Morphology. Bacilli 0.4–0.5 : 0.6–1.0 μ; in chains. Gelatin liquefied slowly.
Agar slant. Growth translucent, white, streaked.
Pepton-rosolic acid. Unchanged.
Milk. Not coagulated.
Litmus milk. No acid.
Habitat. Isolated from a leaf of *Sarracenia purpurea.*

95. Bact. truncatum

B. No. XII Adametz: Landw. Jahrbücher, XVIII, 1889.

Morphology. Bacilli, 0.7–0.9 : 1.4–2.0 μ, ends truncate, often in filaments.
Gelatin colonies. In 4 days small, dirty, white, punctiform, surrounded by a zone of liquefied gelatin, becoming thin, irregular disks, lying in basins of liquefied gelatin; microscopically, at first round, entire, brownish, becoming irregular.
Gelatin stab. Growth on the surface thin, yellowish white, under which a slow liquefaction occurs.
Milk. Coagulated and peptonized.
Habitat. Isolated from Emmenthaler cheese.

96. Bact. orchiticum (Kruse)

Bacillus zur Rotzdiagnose: Zeitsch. f. Hygiene, XXI, 1896, 156.
B. orchiticus Kruse: Flügge, Die Mikroorganismen, 1896, 455.

Morphology. Bacilli like those of glanders.
Gelatin colonies. Like old cholera colonies.
Agar slant. Growth dense, white.
Blood serum. An orange pigment.
Bouillon. Turbid; no sediment.

Pathogenesis. Intraperitoneal inoculations of guinea pigs cause death in 4–5 days; enlargement of the testicles, tubercles on the diaphragm, testicles, etc.; with larger doses the peritoneum is affected as above. Mice inoculated subcutaneously die in 4–7 days; abscess, hemorrhagic œdema; with intraperitoneal injections, yellowish tubercles on the peritoneum. Rabbits nearly immune.

Habitat. Isolated from nasal secretions of a glandered horse.

97. Bact. pneumonicum (Kruse)

Pneumobacillus liquefaciens-bovis Arlong: Compt. rend., 99, 109, 116.
B. pneumonicus-liquefaciens Kruse: Flügge, Die Mikroorganismen, 1896.

Morphology. Short rods — coccoid forms. Cultural characters not fully described.

Potato. Growth whitish, to brown.

Pathogenesis. Intraperitoneal inoculations of guinea pigs and rabbits cause death; dogs but little affected.

Habitat. Isolated from exudate in lung plague of cattle.

98. Bact. varicosum (Sternberg)

Gombert: Rech. expér. microbes conjonctives, Paris, 1889.
B. varicosus-conjunctivæ Sternberg: Manual of Bacteriology, 1892, 474.

Morphology. Bacilli 1.0 : 2.0–8.0 μ; short, often constricted in the middle. Slight growth at 22° C.; optimum temperature 37°.

Agar colonies. In 4 days colonies 4 mm., with minute thorny projections; an opaque nucleus, and a yellowish, granular, transparent zone, from which proceed twisted, bent, tapering offshoots.

Gelatin stab. In depth, but slight growth; on the surface, growth round, flat, grayish white; liquefaction extends gradually downward.

Agar slant. A thin, dry, white, adherent film.

Potato. Growth dry, white, spreading; surface irregular, and margins fringed; later, reddish brown.

Pathogenesis. An inflammation of the cornea of rabbits from local infection.

Habitat. Isolated from the healthy conjunctival sac of man.

99. Bact. aromaticum (Pammel)

B. aromaticus Pammel: Iowa Ag. Expt. Sta. Bul. 21, 1893, 792.

Morphology. Bacilli 0.3–0.4 : 0.9–1.2 μ; ends rounded. Grow slowly at 35°.

Gelatin stab. Liquefaction crateriform — saccate, with gas production.

Agar slant. Growth white, spreading.

Milk. Coagulated, acid, slowly peptonized.

Blood serum. Growth dirty white, spreading; medium liquefied.
Potato. Slight growth, invisible unless moist, when there form yellowish white colonies.
Habitat. Isolated from cheese.

100. Bact. nubilum (Frankland) Lehmann-Neumann.

B. nubilus Frankland: Zeitsch. f. Hygiene, VI, 1889, 386.
Bact. nubilum Lehmann-Neumann: Bak. Diagnostik, 1896, 255.

Morphology. Bacilli slender, 0.3 : 3.0 μ, or filaments more or less curved.
Gelatin colonies. Like those of *Bact. murisepticum.*
Agar slant. Growth thin, opalescent.
Bouillon. Turbid, with a pellicle.
Potato. A scarcely visible growth. A slight reduction of nitrates.
Habitat. Water.

101. Bact. methylicum (Loew)

B. methylicus Loew: Centralblatt f. Bakteriol., XII; 1892, 462.

Morphology. Bacilli short, thick rods, 1.0 : 2.0–2.5 μ.
Gelatin colonies. In 2 days, round — oval, yellowish, entire, sharp, becoming liquefied; edges ciliate. Colonies often like those of cholera.
Gelatin stab. Liquefaction crateriform, like cholera, with a whitish-yellowish sediment. No liquefaction in depth.
Glucose gelatin stab. In depth, little or no growth, a small liquefied bubble on top.
Agar stab. In depth, no growth; on the surface, growth spreading, grayish white.
Potato. Growth like *B. typhosus*, pure white, adherent.
Bouillon. Growth like anthrax; medium clear; on the surface and adherent to the glass, a white ring. Grow well in 0.5 per cent methyl alcohol, 0.05 per cent dicalcium phosphate, and 0.01 per cent magnesium sulphate, on which it forms a reddish membrane. Possesses the ability to decompose formic acid salts.
Habitat. A culture contamination.

102. Bact. Trambusti (Kruse)

Discovered by Trambusti-Galeotti, not named: Centralblatt f. Bakteriol., XI, 1892, 717.
B. Trambusti Kruse: Die Mikroorganismen, 1896, 319.

Morphology. Bacilli 3–5 μ, size variable. Chromatin granules in the interior of the rods. Grow at 37°.
Gelatin colonies. Border irregular, surrounded by a zone of liquefied gelatin.

Agar colonies. Star-shaped, with broad, radiating outgrowths.
Bouillon. Not turbid, with a membrane.
Agar slant. Growth grayish.
Potato. Raised dry colonies.
Habitat. Water.

103. Bact. aquatilis (Zimmerman)

B. radiatus-aquatilis Zimmerman: Bak. Nutz u. Trinkwässer, 1890.

Morphology. 0.6 : 1.2–2.5 μ.
Gelatin colonies. Surrounded by a delicate " Strahlenkranz."
Gelatin stab. A saccate liquefaction, with a pellicle on the surface; medium turbid.
Agar slant. Growth gray, translucent.
Potato. Growth yellowish.

104. Bact. flexuosum (Wright)

B. flexuosus Wright: l.c., 460.

Morphology. Bacilli medium-sized, thick, rounded; also chains and filaments.
Gelatin colonies. Deep: round — oval, brownish, granular. *Surface:* in 2 days, whitish irregular clumps. Microscopically, twisted strands, dense in centre, edge irregular.
Gelatin stab. A crateriform liquefaction, becoming stratiform; medium alkaline.
Agar slant. Growth grayish, translucent, limited; the agar becoming slightly greenish.
Bouillon. Turbid, becoming slightly greenish.
Potato. Growth brownish, viscid, glistening, uneven.
Litmus milk. Coagulated, acid. *Indol,* a slight or doubtful reaction. Grow at 36°.
Glucose bouillon. Gas not produced.
Habitat. Water.

105. Bact. ambiguum Chester

Del. College Ag. Expt. Sta. Report, 1899.

Morphology. On agar and potato, short rods and ovals to longer forms 3–4 times their breadth; 0.7 : 1.0–3.0 μ. In neutral bouillon long filamentous forms appear with often involution forms. Stain with ordinary analine colors; decolorized by Gram's method. Flagella absent. Aerobic, grow only in open end of fermentation tube. Grow at 37° C.

Gelatin colonies. Deep: in 2 days round — irregular, nearly amorphous, homogeneous or finely granular, grayish — light yellowish; border entire — finely filamentous. *Surface:* in 2 days round, thin, flat, veily disks, 1–1.5 mm. In 7–10 days round, entire, depressed, due to the liquefaction of the gelatin. Microscopically, in 2 days round, entire, with a small, dense, central nucleus, surrounded by a broad, thin, pale, homogeneous, finely granular portion. In 7–10 days grayish, darker in centre, granular, with concentric zoning; border tuberculate — ciliate; other colonies show an erose — lobed border. Microscopic appearances rather variable.

Gelatin stab. In depth, a filiform growth, becoming ciliate — plumose. On the surface, colony round — irregular, thin, flat, entire — lobate. In 7 days a thin, flat, surface growth, under which is a crateriform — shallow funnel-formed liquefaction, which proceeds slowly.

Agar slant. A thin, moist, glistening, translucent, spreading veil.

Bouillon (neutral). A light uniform turbidity; no pellicle. Remains turbid after 5 weeks.

Milk. After 7–10 days, rather thin, watery, with a slight brownish discoloration, which, upon heating to boiling, gives a finely flocculent curd.

Litmus milk. In 3–5 days, medium becomes a deep blue, with a strong alkaline reaction.

Potato. In 24 hours, a scarcely visible, dry, creamy streak, becoming, in 5 days, light pinkish, rather thick and raised, dry and dull. In 10 days, color becomes a salmon-pink, later chocolate-brown. *Indol* negative. Nitrates reduced to nitrites.

No growth in 1.5 per cent normal HCl bouillon. Cultures show no marked odor. No gas in glucose bouillon, and no acid production.

Habitat. Soil.

106. Bact. convolutum Wright

B. convolutus Wright: l.c., 461.

Morphology. Bacilli large, occur singly and in pairs; also twisted chains and filaments.

Gelatin colonies. Deep: round, dark, granular. *Surface:* round, grayish white, translucent, woolly-looking, slightly opalescent, 2–4 mm.; microscopically, segmented — fissured; edge irregular; darker centre.

Gelatin stab. Liquefaction crateriform; proceeds slowly. Medium alkaline.

Agar slant. Growth translucent, grayish, limited; agar becomes greenish brown.

Bouillon. Turbid, with a pellicle; medium becomes greenish.

Potato. Growth elevated.

M

Litmus milk. Alkaline; not coagulated. *Indol* negative.

Glucose bouillon. No gas. No growth at 36°.

Habitat. Water.

107. Bact. radiatum Chester l.c.

Morphology. On agar short ovals — longer forms; 0.7 : 1.0–3.0 μ; occur singly, with little tendency to form filaments or chains. Decolorized by Gram's method. Flagella absent. Aerobic, grow only in open end of fermentation tube. No growth at 37° C.

Gelatin colonies. Deep: round, entire, amorphous, homogeneous, yellow-brown. *Surface colonies:* in 2–3 days, thin, flat, slightly depressed, 1 mm.; in 4–5 days, colonies 3–5 mm., depressed, due to liquefaction, with a large central dirty white nucleus, irregular or ameboid in form, with irregular radial extensions. Microscopically, in 2 days, an indistinctly defined yellowish brown nucleus, becoming lighter and colorless exteriorly; refraction low; edge faintly defined, erose, embayed, lobed. Coli — proteus-like.

Gelatin stab. In depth, a scanty filiform growth; on the surface, a crateriform liquefaction, becoming stratiform.

Agar slant. A pure white, dense, opaque streak.

Bouillon (neutral). A slight, uniform turbidity, which persists for 5 weeks. No pellicle, and only slight sediment.

Potato. A thick, raised, slimy whitish streak, becoming yellowish — brownish.

Milk. Coagulated after 2–5 weeks; medium may become slightly brownish, and has a foetid odor.

Litmus milk. For the first 10 days, a slightly darker blue, becoming, in 10 days, slightly acid. Early cultures reduced litmus. *Indol* negative. No gas in glucose bouillon, and slight acid production. No reduction of nitrates to nitrites. No growth in 1.5 per cent normal HCl bouillon. Cultures show only a slight foetid odor.

Habitat. Soil.

CLASS V. WITHOUT ENDOSPORES. AEROBIC AND FACULTATIVE ANAEROBIC. PRODUCE PIGMENT ON GELATIN OR AGAR. PIGMENT YELLOWISH, GELATIN LIQUEFIED.

I. Growth on potato.

    A. Growth on potato yellowish.

        1. Gelatin colonies radiate filamentous, branched — arborescent.

            108. *Bact. arborescens* (Frankland).

2. Gelatin colonies not characterized as above.
   a. Milk coagulated.
      * Litmus milk reddened.
            109. *Bact. oxylacticum* (Dyar).
      ** Litmus milk rendered alkaline or unchanged in reaction.
        † Gelatin liquefied rapidly.
            110. *Bact. desidiosum* (Wright).
        †† Gelatin liquefied very slowly.
            111. *Bact. Fischeri* Dyar.

   b. Milk not coagulated.
      * Litmus milk reddened, rendered acid.
        † Nitrates reduced to nitrites.
            112. *Bact. eta* (Dyar).
        †† Nitrates not reduced to nitrites.
            113. *Bact. fulvum* (Zimmerman).
      ** Litmus milk rendered alkaline or unchanged.
        † Agar smear cultures reddish brown, shading to yellowish.
            114. *Bact. rubidum* (Eisenberg).
        †† Agar smear cultures yellowish — orange.
          § Gelatin liquefied quickly, *i.e.* within a few days.
            115. *Bact. caudatum* (Wright).
          §§ Gelatin liquefied very slowly, *i.e.* within 2–4 weeks.
            116. *Bact. Kralii.*
            117. *Bact. helvolum* (Zimmerman).

   c. Milk coagulation not stated. Growth on liquefied gelatin crumpled.
      * Gelatin colonies moruloid.
            118. *Bact. plicatum* (Zimmerman).
      ** Gelatin colonies punctiform.
            119. *Bact. citreum* (Strassman-Stecker).

B. Growth on potato brownish.
   1. Short, broad bacilli.
            120. *Bact. bucallis* (Sternberg).
   2. Bacilli minute, slender.
            121. *Bact. pseudoconjunctivitidis* (Kruse).

II. Do not grow on potato.
            122. *Bact. æris-minutissimum* (Kruse).
            123. *Bact. dormitor* (Wright).

108. Bact. arborescens (Frankland)

B. arborescens Frankland: Zeitsch. f. Hygiene, VI, 1889.

Morphology. Bacilli 0.5 : 2.5 μ — filaments.

Gelatin colonies. As above described, very characteristic; centre yellowish, border iridescent. Medium liquefied slowly.

Gelatin stab. In depth, a turbidity of the medium; on the surface, a thin, iridescent layer, later a yellowish sediment in the liquefied medium.

Agar slant and *potato.* Growth deep orange. *Pigment* insoluble in water, soluble in alcohol.

Bouillon. Turbid; no membrane, and a yellowish sediment. Nitrates not reduced.

Milk. Not coagulated (Dyar).

Habitat. Water.

109. Bact. oxylacticum (Dyar)

B. oxylacticus Dyar: l.c., 369.

Morphology. Bacilli 1.0–1.3 : 1.7–2.5 μ; occur singly and in chains. Gelatin liquefied quickly.

Potato. Growth glistening, transparent, watery, marked with opaque white spots.

Agar slant. White, with a slightly ochreous tint.

Pepton-rosolic acid. Unchanged. Nitrates not reduced.

Habitat. From Kral's laboratory.

110. Bact. desidiosum (Wright)

B. desidiosus Wright: l.c., 443.

Morphology. Bacilli small, short, rounded; in pairs and clumps.

Gelatin colonies. In 3 days yellowish brown irregular clumps, 1.0 mm., within the surrounding liquefied gelatin; microscopically, dense, granular, brownish yellow, often darker in centre, irregular, and broken.

Gelatin stab. Liquefaction fusiform, with a bubble at top.

Agar slant. Growth brownish yellow, glistening, translucent, slightly spreading.

Potato. Growth brownish yellow, moist, elevated, rough.

Litmus milk. Clot viscid, yellow, with a yellow-brown serum. Medium alkaline — amphoteric.

Glucose bouillon. No gas.

Indol. A slight or doubtful reaction. No growth at 36° C.

Habitat. Water.

111. Bact. Fischeri (Dyar)

B. fischeri Dyar: l.c., 370.

Morphology. Bacilli 0.5–0.6 : 0.6–1.2 μ; occur singly.
Gelatin. Liquefied slowly, *i.e.* after 28 days. On solid media growth yellow-ish, but pigment forms slowly.
Milk. Coagulated slowly, often not until 28 days. A slight reduction of nitrates.
Pepton-rosolic acid. Unchanged.
Litmus milk. Unchanged.
Habitat. From Kral's laboratory.
According to Dyar, identical with *Photobacterium Fischeri* of Beijerinck.

112. Bact. eta (Dyar)

B. eta Dyar: l.c., 374.

Morphology. Bacilli 0.5 : 0.7–1.0 μ; occur singly.
Gelatin. Liquefied slowly.
Milk. Not coagulated.
Pepton-rosolic acid. Unchanged. Growth on solid media yellow, viscous.
Habitat. Air.

113. Bact. fulvum (Zim.)

B. fulvus Zimmerman: Bakt. Nutz u. Trinkwässer, 1890. Dyar: l.c.

Morphology. Bacilli 0.8 : 0.9–1.3 μ; occur singly in pairs and short chains. Optimum temperature 30° C.
Gelatin colonies. *Deep*: round — irregular, yellowish gray, granular. *Surface*: in 8 days, convex, 1 mm., reddish yellow.
Gelatin stab. In depth, a good growth; on the surface, growth round, con-vex, yellowish.
Agar slant. Growth yellowish, glistening.
Potato. Growth abundant, yellowish, glistening.
Milk. Not coagulated. Nitrates not reduced to nitrites.
Litmus milk. Reddened (Dyar).
Habitat. Water, etc.

114. Bact. rubidum (Eisenberg)

B. rubidus Eisenberg. See No. 145.

115. Bact. caudatum (Wright)

B. caudatus Wright: l.c., 444.

Morphology. Bacilli small, slender, with conical ends ; in pairs and as filaments.
Gelatin colonies. In 3–4 days, colonies 1–2 mm., yellow, translucent, smooth
— wavy ; microscopically, with a yellowish centre and a somewhat radiate,
light periphery.
Gelatin stab. In depth, growth villous ; on the surface, liquefaction crateri-
form, with yellowish flocculi.
Agar slant. Growth, yellow, glistening, translucent, slightly spreading.
Bouillon. Turbid, with a yellowish sediment.
Potato. Growth dark yellow, elevated, spreading ; surface uneven.
Litmus milk. Slightly decolorized.
Glucose bouillon. No gas. *Indol,* reaction doubtful. No growth at 36° C.
Habitat. Water.

116. Bact. Kralii

B. fuscus-liquefaciens Dyar: l.c., 376.

Morphology. Bacilli 0.5–0.6 : 1.0–2.0 μ ; occur singly and in short chains.
Gelatin. Slowly liquefied, *i.e.* after 14–50 days.
Agar slant. Growth bright orange, forming a crusty membrane. Nitrates
slightly reduced.
Litmus milk. Blue.
Habitat. Air (from Kral's laboratory).

117. Bact. helvolum (Zimmerman)

B. helvolus Zimmerman: Bakt. Nutz u. Trinkwässer, 1890. Dyar: l.c.

Morphology. Bacilli 0.5 : 1.5–2.5–4.5 μ. Optimum temperature 25° C.
Gelatin stab. Slight growth in depth ; on the surface, growth convex, becom-
ing spreading, and Naples yellow in color ; later, a crateriform liquefaction.
Agar slant. Growth of a Naples yellow color.
Potato. Growth yellow, becoming slightly greenish.
(*Gelatin.* Liquefied in 30–40 days. Reduction of nitrates variable. *Milk.*
Not coagulated. *Litmus milk.* Blue.) Dyar.
Habitat. Water, air.

118. Bact. plicatum (Zimmerman)

B. plicatus Zimmerman: l.c.

Morphology. Bacilli small, slender ; in twos or short chains. Optimum temperature 20° C.
Gelatin colonies. Yellowish white, irregular ; microscopically, rough, moruloid.

Gelatin stab. In depth, growth beaded, yellowish white; a whitish yellow crumpled membrane on the surface of the liquefied gelatin.

Potato. Growth thin, dry, yellowish gray.

Habitat. Water.

119. Bact. citreum (Strassman-Stecker)

B. citreus-cadaveris Strassman-Strecker: Bakterien bei der Leichenfäulniss, 1888.
B. Streckeri Trevisan: Genera, 1889, 17.

Morphology. Bacilli 0.6–0.9 μ; often in chains.

Gelatin colonies. Punctiform.

Gelatin stab. An air bubble above, under this a yellowish layer, under this a clear canal of liquefied gelatin, containing a yellowish sediment. Odor of H_2S.

Potato. Growth citron-yellow, dry. Non-pathogenic.

Habitat. Isolated from a human corpse 50 hours after death.

120. Bact. bucallis

Bacillus g. Vignal: Archiv Phys., VIII, 1886.
B. bucallis-minutus Sternberg: Manual of Bacteriology, 1892, 643.

Morphology. Rods scarcely longer than broad, 0.5–1.0 μ.

Gelatin colonies. In 48 hours, round, elevated, slightly yellowish, surrounded by liquefied gelatin.

Gelatin stab. In 48 hours the growth in depth is yellowish white, beaded; on the surface, a cup-shaped liquefaction; in 6 days a small funnel.

Agar slant. Golden yellow colonies, easily removed with a needle.

Bouillon. Turbid, with an iridescent pellicle.

Potato. Growth thin, yellowish, becoming brownish.

Habitat. Isolated from the saliva of healthy persons.

121. Bact. pseudoconjunctivitidis (Kruse)

Bacillus ägyptischen katarrhalischen Conjunctivitis Kartulis: Centralblatt f. Bakteriol., I, 1887, 289.
B. pseudoconjunctivitidis Kruse: Flügge, Die Mikroorganismen, 1896, 441.

Morphology. Bacilli minute, 0.25 : 1.0 μ. Cultures produce a canary-yellow pigment.

Gelatin stab. A nail-shaped growth with a flat head, yellow; liquefaction slow.

Potato. Growth limited, bright brown.

Habitat. Isolated from conjunctival secretions.

122. Bact. aeris-minutissimum (Kruse)

Flügge, Die Mikroorganismen, 1896, 441.

From descriptions, indistinguishable from the preceding, except that it produces pigment a little less strongly.

Habitat. Isolated from the air.

123. Bact. dormitor (Wright)

B. dormitor Wright: l.c., 442.

Morphology. Bacilli of medium size, with conical ends; variable, long pairs and filaments.

Gelatin colonies. In 2 days, yellowish points, 1 mm.; microscopically, yellowish, slightly granular; rough, sharp, bulging outlines, surrounded by a zone of liquefied gelatin.

Gelatin stab. A funnel of liquefaction, turbid; sediment bright yellow.

Agar slant. Growth glistening, translucent, yellowish.

Bouillon. Turbid; yellowish sediment, and a slight pellicle.

Litmus milk. Decolorized, amphoteric.

Glucose bouillon. No gas. *Indol* negative. No growth at 36°.

Habitat. Water.

CLASS VI. WITHOUT ENDOSPORES. AEROBIC AND FACULTATIVE ANAEROBIC. PRODUCE PIGMENT ON GELATIN OR AGAR. PIGMENT YELLOWISH. GELATIN NOT LIQUEFIED.

I. Chromogenic function weak, pale or grayish yellow.
   A. Gelatin colonies beset with thorny outgrowths when old.
       124. *Bact. spiniferum* (Tommasoli).
   B. Gelatin colonies not characterized as above.
     1. Litmus milk reddened, or rendered acid.
       a. Nitrates reduced to nitrites.
         * Milk coagulated after a long time by boiling.
          125. *Bact. subochraceum* (Dyar).
        ** Milk not coagulated.
          126. *Bact. domesticum* (Dyar).
          127. *Bact. amabilis* (Dyar).
       b. Nitrates not reduced to nitrites.
         128. *Bact. lacunatum* (Wright).

2. Litmus milk blue or reaction unchanged.
   a. Nitrates reduced to nitrites.
       129. *Bact. javaniensis* (Dyar).
   b. Nitrates not reduced to nitrites.
      * Cultures on solid media rugose.
       130. *Bact. palidor* (Dyar).
      ** Cultures on solid media not rugose.
       131. *Bact. ovale* (Wright).
II. Pigment strongly developed, yellow, color decided.
  A. Gelatin colonies compound.
       132. *Bact. luteum* List.
  B. Gelatin colonies simple.
   1. Gelatin colonies dry, granular. Bacilli stain irregularly like diphtheria bacilli.
       133. *Bact. striatum* (v. Besser).
   2. Gelatin colonies not characterized as above.
    a. On agar a wrinkled layer.
       134. *Bact. fuscum* (Zimmerman).
    b. Agar cultures not characterized as above.
       135. *Bact. constrictum* (Zimmerman).
       136. *Bact. solare* Lehmann-Neumann.
       137. *Bact. breve* (Frankland).

124. Bact. spiniferum (Tommasoli)

B. spiniferus Tommasoli: Monatsch. f. prakt. Dermatol., IX, 57.

Morphology. Bacilli 0.8–1.0 : 2.0 μ, bent, often parallel in bundles.
Gelatin colonies. Old colonies with thorny outgrowths.
Agar slant. Growth grayish yellow.
Potato. Growth yellow.
Habitat. Isolated from surface of human body.

125. Bact. subochraceum (Dyar)

B. subochraceus Dyar: l.c., 358.

Morphology. Bacilli 0.7 : 1.5 μ; occur singly and in short chains.
Gelatin colonies. *Deep:* round, dusky, yellowish. *Surface:* clear, irregular, slightly veined.
Litmus milk. Reddened.

On solid media. Growth translucent, ochreous — light orange.

Bouillon. A slight pellicle on the surface.

Pepton-rosolic acid. Color deepened. A slight reduction of nitrates to nitrites.

Habitat. Air.

126. Bact. domesticum (Dyar)

B. domesticus Dyar: l.c., 358.

Morphology. Bacilli 0.5 : 1.0 μ.

Gelatin colonies. Large, translucent, yellowish.

On solid media. Growth white — light yellow, spreading slowly.

Litmus milk. Red, becoming in 50 days blue.

Habitat. Air.

127. Bact. amabilis (Dyar)

B. amabilis Dyar: l.c., 358.

Morphology. Bacilli 0.7 : 0.8–1.0 μ; occur singly, in chains, and masses.

Gelatin colonies. Large, translucent, yellowish.

Agar slant. Growth white, limited, with a yellowish tint.

Potato. Growth thin, bright yellow. *Indol* negative.

Pepton-rosolic acid. Unchanged.

Habitat. Air.

128. Bact. lacunatum (Wright)

B. lacunatus Wright: l.c., 435.

Morphology. Bacilli small, short, rounded; in pairs — small clumps.

Gelatin colonies. Deep: round, entire, slightly granular, gray-brown. *Surface:* in 24 hours, thin, translucent, with grayish centre; very irregular — deeply cleft; microscopically, areolate — grained; colonies become 4–5 mm., thin, translucent, with a yellow haziness about their centres.

Gelatin slant. Growth thin, translucent, grayish — yellowish in centre.

Bouillon. Turbid.

Potato. Growth thin, viscid, dirty brownish.

Litmus milk. Not coagulated, acid, becoming brownish.

Glucose bouillon. No gas. *Indol* positive. No growth at 36° C.

Habitat. Water.

129. Bact. javaniensis (Dyar)

B. javaniensis Dyar: l.c., 359.

Morphology. Bacilli short — coccoid, 1.0–1.2 μ; occur in masses and short chains.

Agar slant. Growth thick, white, with an indistinct yellowish tinge.

Milk. Not coagulated.
Pepton-rosolic acid. Unchanged.
Habitat. Air.

130. Bact. palidor (Dyar)

B. fuscus-palidor Dyar : l.c., 361.

Morphology. Bacilli 0.5–0.7 : 1.0–1.3 μ; occur singly and in chains.
On solid media. Growth pale, whitish, orange, almost pinkish, wrinkled, with lobed edges ; a crusty, brittle texture.
On liquid media. A surface membrane.
Litmus milk. Blue.
Habitat. Air.

131. Bact. ovale (Wright)

B. ovalis Wright : l.c., 435.

Morphology. Bacilli medium-sized, short, rounded ; in pairs and as filaments.
Gelatin colonies. *Deep :* opaque, granular, entire, brownish. *Surface :* in 5–6 days, 1 mm., round, entire, elevated, glistening, yellowish, translucent, becoming yellow — brownish yellow ; gelatin slightly brownish.
Gelatin slant. Growth elevated, brownish yellow, limited, smooth — rugged ; gelatin assumes a brownish tint.
Agar slant. Growth pale yellow, thick, glistening, limited.
Bouillon. Clear, with a sediment.
Potato. Growth brownish yellow, moist, spreading, viscid.
Litmus milk. Decolorized, not coagulated. *Indol.* A slight reaction. No growth at 36° C.
Glucose bouillon. No gas.
Habitat. Water.

132. Bact. luteum List

Adametz, Bakt. Nutz u. Trinkwässer, 1888.

Morphology. Bacilli 1.1 : 1.3 μ. Optimum temperature 30° C.
Gelatin colonies. Irregular, flat, consisting of many clavate coarse granular zoöglœa masses : orange-yellow.
Milk. Coagulated.
Habitat. Water.

133. Bact. striatum (v. Besser)

B. striatus-flavus v. Besser : Ziegler's Beiträge, VI.

Morphology. Bacilli small, often bent.
Gelatin colonies. Thick, dry, granular, yellow.
Agar slant and *potato.* Growth sulphur-yellow.
Habitat. Isolated from nasal mucus.

134. Bact. fuscum (Zimmerman)

B. fuscus Zimmerman: Bak. Trink u. Nutzwässer, 1890; Dyar, l.c., 361.

Morphology. Bacilli 0.6 μ wide and of variable length. Optimum temperature 36° C.

Gelatin colonies. Deep : round — irregular, granular, grayish yellow — brown.
Surface : punctiform, brownish yellow centre, lighter border.

Agar slant. Growth crumpled, thick, chrome yellow.

Potato. Growth chrome yellow, friable.

Litmus milk. Blue.

Nitrates not reduced. *Bouillon.* A membrane or pellicle.

Pepton-rosolic acid. Unchanged. Dyar.

Habitat. Water.

135. Bact. constrictum (Zimmerman)

B. constrictus Zimmerman: l.c.

Morphology. Bacilli 0.7 : 1.5–6.5 μ; in chains. Grow only at room temperatures.

Gelatin colonies. Small, glistening, Naples yellow, erose edges.

Agar slant and *potato.* Growth yellow.

Habitat. Water.

136. Bact. solare Lehmann-Neumann.

Bak. Diagnostik, 1896, 258.

Morphology. Bacilli 0.3–0.4 μ; broad, short, and long — filaments.

Gelatin colonies. Round, yellow, glistening, translucent; microscopically, yellow, radially fibrous, borders filamentous.

Gelatin stab. Long, delicate outgrowths from the line of stab, yellowish. Growth on agar straw-yellow.

Potato. Growth soft, white, becoming yellowish.

Bouillon. Remains clear. No gas in glucose bouillon.

Milk. Unaltered. *Indol* negative; H_2S negative.

Habitat. Water.

137. Bact. breve (Frankland)

No name Rinatoro-Mori: Zeitsch. f. Hygiene, IV, 1888, 53.
B. brevis Frankland: Microorganisms of Water, 1894, 429.

Morphology. Bacilli 0.8–1.0 : 2.5 μ; show polar stain. Not stained by Gram's method.

Gelatin colonies. Grow very slowly at room temperature; after 2–3 weeks, colonies very minute, pale yellow, compact.

Gelatin stab. After 3 weeks, a thin, yellowish expansion on the surface; in depth, small colonies.

Agar slant. Grows at 35°, yellowish.

Potato. No growth.

Bouillon. A white cloudy deposit.

Habitat. Isolated from Berlin drain water.

CLASS VII. CHROMOGENIC. WITHOUT ENDOSPORES. AEROBIC AND FACULTATIVE ANAEROBIC. PRODUCE PIGMENT ON GELATIN AND AGAR. PIGMENT REDDISH.

I. Gelatin liquefied.

    A. Bacteria stained by Gram's method.

        138. *Bact. pyocinnabareum* (Kruse).

    B. Not stained by Gram's method.

      1. Milk coagulated.

        a. Litmus milk blue or decolorized, reaction alkaline or neutral.

            * Nitrates reduced to nitrites ; *pepton-rosolic acid solution* unchanged.

            139. *Bact. erythrogenes* (Grotenfelt).

          ** Nitrates not reduced to nitrites.

            140. *Bact. exiguum* (Wright).

        b. Litmus milk rendered acid.

            * Growth salmon pink.

            141. *Bact. epsilon* (Dyar).

          ** Growth brownish red or dark orange-red.

            142. *Bact. zeta* (Dyar).

      2. Milk not coagulated.

        a. Litmus milk rendered acid.

            143. *Bact. delta* (Dyar).

        b. Litmus milk not reddened, reaction unchanged or alkaline.

            * Pigment bright red or pinkish.

            144. *Bact. hæmatoides* (Wright).

            144 *a.* *Bact. amylovorum* (Burrill).

          ** Pigment brownish red.

            145. *Bact. rubidum* (Eisenberg).

II. Gelatin not liquefied.

    A. Gelatin colonies filamentous to floccose.

        146. *Bact. ferrugineum* (Dyar).

B. Gelatin colonies not characterized as above, or not described.

  1. Pigment, brick-red, or carmine to blood-red.

    a. Grow on fresh (acid) potato.

      147. *Bact. finitimum* (Dyar).

      148. *Bact. ovatum* (Bruyning).

    b. Scarcely grow on fresh (acid) potato.

      149. *Bact. Havaniensis* (Sternberg).

  2. Pigment brick-red.

      150. *Bact. latericium* (Adametz).

  3. Pigment bright pink to salmon-pink.

      151. *Bact. rhodochroum* (Dyar).

      152. *Bact. salmoneum* (Dyar).

138. **Bact. pyocinnabareum** (Kruse)

B. über rote Eiterung Ferchmin: Wratsch, 1892, Nos. 25–26.
B. pyocinnabareus Kruse: Flügge, Die Mikroorganismen, 1896.

Morphology. Bacilli 0.8 : 2.5 μ, to threads. Optimum temperature 37°.
Gelatin colonies. Irregular, erose, granular.
Gelatin stab. Liquefaction funnel-formed; red sediment.
Agar slant. Growth reddish, moist.
Potato. Growth yellowish to reddish.
Bouillon. Turbid; red membrane; trimethylamine odor.
Pathogenesis. Not pathogenic; toxic in large doses.
Habitat. From green pus.

139. **Bact. erythrogenes** (Grotenfelt)

Bact. lactis-erythrogenes Grotenfelt: Fortschritte d. Medizin, 1889, No. 2.

Morphology. Bacilli 0.2–0.3 : 1.1–4.0 μ.
Gelatin colonies. Round, grayish yellow to yellow; rose color to the surrounding gelatin.
Gelatin stab. Along line of stab a slight growth; surface growth thin, whitish, becoming yellow; liquefied gelatin of a pinkish tint.
Agar slant. Growth yellowish to weak yellowish red.
Milk. Coagulated slowly, becoming peptonized, amphoteric to alkaline; a layer of blood-red serum above the precipitated casein; above this the yellowish white cream layer.
Bouillon. Turbid, with a yellowish tint, and a disgusting odor.
Glucose-bouillon. No gas. *Indol* present. H_2S present.

Pepton-rosolic acid solution. Unchanged.
Habitat. Red milk, water, fæces of a child.

VARIETY. *B. erythrogenes-rugatus* (Dyar). "Differs from the preceding, in that the growth on agar is thin, membranous, and wrinkled;" Dyar, l.c., 374.

140. **Bact. exiguum** (Wright)

B. exiguus Wright: l.c., 447.

Morphology. Bacilli small, rounded; occur singly and in pairs and clumps.
Gelatin colonies. Deep: round to oval, with greenish centres and granular margins. *Surface:* round, pinkish, translucent disks; microscopically, granular; centres pink, lighter toward margin, becoming liquid and salmon-pink.
Gelatin stab. Liquefaction crateriform; pinkish sediment.
Agar slant. Growth thin, moist, pinkish.
Bouillon. Turbid; white sediment.
Potato. Growth spreading, moist, glistening, reddish yellow.
Litmus milk. Coagulated, decolorized, slightly peptonized, amphoteric.
Glucose-bouillon. No gas. *Indol* slight. A strong development at 35°–36°.
Habitat. Water.

141. **Bact. epsilon** (Dyar)

B. epsilon Dyar: l.c., 369.

Morphology. Bacilli 0.5 : 0.7–1.0 μ.
Agar slant. Growth translucent, pink.
Milk. Coagulated, at least on boiling.
Pepton-rosolic acid solution. Unchanged.
Habitat. Air.

142. **Bact. zeta** (Dyar)

B. zeta Dyar: l.c., 369.

Morphology. Bacilli 0.5 : 0.7–1.0 μ.
Gelatin. A slight development; liquefaction does not begin before 10 days.
Milk. On the surface, a red cream. On solid media, as above. Nitrates not reduced to nitrites.
Habitat. Air.

143. **Bact. delta** (Dyar)

B. delta Dyar: l.c., 368.

Morphology. Bacilli 0.5 : 0.8–1.0 μ; occur singly or in short chains.
Gelatin. Liquefied, beginning in 21 days.
Agar slant. Growth thin, red.

Milk. A red growth on the surface.

Potato. Growth glistening, light red; slow development. Nitrates not reduced to nitrites.

Pepton-rosolic acid solution. Unchanged.

Habitat. Water.

144. Bact. hæmatoides (Wright)

B. hæmatoides Wright: l.c., 448.

Morphology. Bacilli medium-sized, blunt; stain irregularly.

Gelatin colonies. Deep : round, granular, yellowish red. *Surface :* small, slightly elevated to vermilion-colored disks. Gelatin liquefied after a long time.

Gelatin slant. Growth bright red, rough, glistening.

Acid gelatin. A strong development.

Agar slant. Growth pink, of confluent elevated colonies.

Potato. Growth bright red, wrinkled, granular, spreading.

Litmus milk. Not decolorized; growth reddish; reaction alkaline.

Glucose bouillon. No gas. *Indol* absent. No development at 36°.

Habitat. Water.

144 a. Bact. amylovorum (Burrill)

M. amylovorus Burrill: Am. Naturalist, VII, 1893, 319.

Morphology. On agar bacilli 0.8:1.25 μ; occur singly and in pairs. Stain uniformly with aqueous fuchsin. Non-motile, flagella absent.

Gelatin colonies. Deep : round, entire, amorphous — slightly granular, light grayish brown, 0.1 mm. In 5 days the deep colonies are round, entire, opaque, yellowish, and about 0.5 mm. in diameter. *Surface :* colonies minute, punctiform, white, not exceeding 1 mm. Microscopically, they are round, entire, yellowish, opaque in centre, and granular toward the border.

Gelatin stab. In depth, a thin filiform growth; on the surface the growth is thin, leafy, whitish, irregular, glistening, and dry.

Bouillon (neutral). In two days at room temperature a distinct opalescence, with no surface growth or sediment.

Potato. In 24–48 hours, a porcelain-white, moist, glistening streak, which becomes spreading and is of a slimy or watery consistency. Later the growth becomes pinkish or flesh-colored.

Milk. Unchanged.

Litmus milk. Unchanged.

Agar slant. A white, moist, glistening streak, which later becomes pinkish or flesh-colored.

Glucose bouillon. No gas.

Habitat. Associated with *fire-blight* of the pear and apple.

145. Bact. rubidum (Eisenberg)

B. rubidus Eisenberg: l.c.; Dyar: l.c., 568.

Morphology. Bacilli 0.5–0.7 : 0.6–1.0 μ; occur singly and in short chains.

Gelatin colonies. Round, slightly granular, reddish in centres; gelatin liquefied slowly to rather quickly.

Agar slant. Growth translucent, reddish brown, sometimes shading into yellowish orange.

Potato. Growth brownish red.

Habitat. Air.

146. Bact. ferrugineum (Dyar)

B. ferrugineus Dyar: l.c., 375.

Morphology. Bacilli 0.6: 1.0 μ; occur in pairs and chains. Growth red to brick red, crusty, granular, scarcely wrinkled; grow slowly.

Gelatin colonies. Like round tufts of cotton.

Milk. Not coagulated; reaction unchanged; a brick-red growth on the surface. Nitrates not reduced. Lactose litmus blue.

Habitat. Air.

147. Bact. finitimum (Dyar)

B. finitimus-ruber Dyar: l.c., 361.

Morphology. Bacilli 0.5 : 0.6–1.0 μ; occur singly and in chains of 3–4.

Gelatin colonies. Round, entire, smooth. Growth on solid media, pink to bright red.

Bouillon. No membrane. Nitrates not reduced. Lactose litmus blue.

Habitat. Air.

148. Bact. ovatum (Bruyning)

B. ruber-ovatus Bruyning: Archiv Neerland Sci. Exact. et Nat., Ser. II, 1898, 297.

Morphology. Bacilli oval, 0.7–0.8 : 0.9–1.2 μ; occur sometimes in pairs or threes. Not stained by Gram's method. Indications of a capsule. Optimum temperature, 20°.

Potato. Growth bright red to vermilion.

Bouillon. Slight growth, becoming opalescent, with a scanty colorless sediment.

Litmus gelatin. Unchanged.

Habitat. Isolated from sorghum.

N

149. Bact. Havaniensis (Sternberg)

B. Havaniensis Sternberg: l.c., 718.

Morphology. Bacilli short ovals, 0.4–0.5 μ; occur usually in pairs; almost a
   coccus. Slight growth at 20°.
Gelatin colonies. Round, small, translucent, blood-red.
Gelatin stab. A carmine layer on surface.
Agar slant. Growth thick, carmine-red, moist, glistening.
Habitat. Yellow fever cadavers.

150. Bact. latericeum (Adametz)

B. latericeus Adametz: l.c.; Wright: l.c., 436.

Morphology. Bacilli 3–5 times their breadth, to filaments.
Gelatin colonies. Deep: round, granular. *Surface:* in 7 days, small, 1 mm.,
   round, entire, red to reddish brown to brick-red; microscopically, round,
   entire, red to reddish brown, strongly refracting.
Gelatin slant. Growth limited, elevated, glistening, brownish red to dark
   vermilion.
Acid gelatin. A strong development.
Agar slant. Growth limited, moist, glistening, reddish brown to yellowish
   brown.
Bouillon. Clear; a stringy sediment, alkaline.
Potato. Growth thin, moist, reddish.
Litmus milk. Decolorized, coagulated, amphoteric.
Glucose gelatin. No growth.
Pepton-rosolic acid solution. Color somewhat deepened, slightly alkaline.
   Indol negative. No growth at 36°.
Habitat. Water.

151. Bact. rhodochroum (Dyar)

B. rhodochrous Dyar: l.c., 362.

Morphology. Bacilli 0.5–0.6–1.0 μ; occur singly and in chains. Pigment
   bright pink.
Gelatin colonies. Round, entire, smooth.
Bouillon. No membrane.
Lactose litmus. Blue. Nitrates not reduced.
Habitat. Air.

152. Bact. salmoneum (Dyar)

B. salmoneus Dyar: l.c., 362.

Identical with the preceding, except that the pigment is salmon-pink.
Habitat. Air.

CLASS VIII. WITHOUT ENDOSPORES; AEROBIC AND FACULTATIVE ANAEROBIC; PRODUCE PIGMENT ON GELATIN AND AGAR; PIGMENT OF OTHER COLORS THAN YELLOW AND RED.

    I. Pigment gray to brown.
       A. Gelatin liquefied.
          153. *Bact. glaucum* (Maschek).
          154. *Bact. fuscescens* (Migula).
    II. Pigment violet to blue on gelatin or agar.
       A. Gelatin liquefied.
          1. Growth on potato dirty white to olive-green.
          155. *Bact. amethystinum* (Eisenberg).
          156. *Bact. indigonaceum* Classen.
    III. Pigment greenish.
          157. *Bact. allii* (Griffiths).

153. Bact. glaucum (Maschek)

B. glaucus Maschek; Adametz: l.c., 1888.

Morphology. Bacilli as slender rods.
Gelatin colonies. Round, entire; centre gray, edge brown, and radially folded, becoming slowly liquefied.
Gelatin stab. Gray bacterial masses.
Agar slant and *potato.* Growth gray.
Habitat. Water.

154. Bact. fuscescens (Migula)

B. fuscus-limbatus Scheibenzucker: Allgemeine Wiener Med. Zeitung, 1889, 171.
B. fuscescens Migula: System der Bakterien, 1900.

Morphology. Bacilli short.
Gelatin colonies. As brownish clumps.
Gelatin stab. Along line of stab, growth uniform, with short outgrowths. Surface growth spreading slightly; gelatin near stab slightly brownish.
Agar slant. The medium around the growth stained brown.
Potato. Growth brownish.
Habitat. Isolated from decayed eggs.

155. Bact. amethystinum Eisenberg: l.c.

Morphology. Bacilli 0.5 : 1.0–1.5 μ. Strong development at 37°.
Gelatin colonies. Like typhoid, becoming dark violet, metallic, crumpled.

Potato. Growth dirty white to olive-green.

Bouillon. A membrane on the surface, fluid brownish.

Habitat. Water.

156. Bact. indigonaceum Classen

Centralblatt f. Bakteriol., VII, 13, 1890.

See *Pseudomonas indigofera* Voges. Differs from the latter in being non-
motile.

157. Bact. allii (Griffiths)

B. *allii* Griffiths: Proc. Roy. Soc., Edinburgh, XV.

Morphology. 0.5–0.7 : 2.5 μ; occur singly and in pairs.

Agar slant. Growth a thick green layer. Pigment soluble in alcohol.

Habitat. Isolated from decaying onions.

CLASS IX. FLUORESCENT BACTERIA. GELATIN LIQUEFIED.

I. Gelatin liquefied slowly and feebly.

158. *Bact. immobile* (Kruse).

II. Gelatin liquefied quickly.

159. *Bact. graveolens* (Bordoin).

158. Bact. immobile (Kruse)

B. *fluorescens-non-liquefaciens* Eisenberg: l.c.

B. *fluorescens-immobilis* Kruse: Flügge, Die Mikroorganismen, 1896, 294.

Morphology. Bacilli 0.3–0.5 : 2.0 μ, to filaments.

Gelatin colonies. B. *coli*-like; green fluorescence in the gelatin; slight liquefaction. Cultural characters like *Pseudomonas putidum.*

Habitat. Air and water.

159. Bact. graveolens (Bordoni)

B. *graveolens* Bordoni-Uffreduzzi: Fortschritte d. Medezin, 1886, 157.

Morphology. Bacilli short, 0.8 μ.

Gelatin colonies. Irregular; bad odor.

Potato. Growth stinking.

Habitat. Isolated from the skin of man from between the toes.

CLASS X. FLUORESCENT BACTERIA. GELATIN NOT LIQUEFIED.

I. Do not grow on ordinary (acid) potato.

A. Grow in milk.

160. *Bact. smaragdinum* (Katz).

B. No growth in milk.

    161. *Bact. phosphorescens* (Cohn) Fischer.

    162. *Bact. Giardii.*

II. Grow on ordinary (acid) potato.

  A. Grow in nutrient bouillon in the absence of NaCl.

    1. Bouillon fluorescent, potato becomes black.

      163. *Bact. Lepierrei.*

    2. Bouillon not colored. Potato light brown. Colonies of the aerogenes type.

      164. *Bact. iris* (Frick).

  B. Do not grow in nutrient bouillon in the absence of NaCl.

    165. *Bact. Pflugeri* (Ludwig).

160. **Bact. smargadinum** (Katz)

B. smargadino-phosphorescens Katz: Centralblatt f. Bakteriol., IX, 1891, 343.

Morphology. Bacilli 1.0 : 2.9 μ; ends somewhat pointed; solitary and in pairs.

Gelatin colonies. Deep: 18 hours, oval, entire, 0.15 mm., concentric. *Surface:* 18 hours, thin gray to yellowish, slightly granular; margin translucent, slightly dentate; 20 days, 2 mm., flat, irregular; centre yellowish, with a slate-colored zone.

Gelatin stab. Along line of stab growth thin; surface growth flat, becoming 5 mm., with a stearin lustre.

Agar slant. Growth slight.

Bouillon. No growth without NaCl.

Milk. A glistening. sticky layer on the surface.

Potato. No growth when acid; neutralized with sodic phosphate growth thin, brownish. Cultures added to sea water caused a decided phosphorescence.

Habitat. Isolated from herring (Sidney).

161. **Bact. phosphorescens** (Cohn) Fischer

M. phosphorescens Cohn: Vesamaling van stukken betr het genes Slaatsoez in Nederland, 1878, 126.

Bact. phosphorescens Fischer: Zeitsch. f. Hygiene, II, 1887, 54.

Photobact. phosphorescens Beijerinck: Centralblatt f. Bakteriol., VIII, 1890, 716.

According to Kruse (Flügge, Die Mikroorganismen, 1892, 332), closely related to, or identical with, the preceding.

162. Bact. Giardii

Photobact. of Giardi : Compt. rend. Soc. Biol., 1896.
Bact. phosphorescens-Giardi : Kruse, Flügge, Die Mikroorganismen, 1896.

Morphology. Bacilli and cultural characters like 161, but bacilli smaller and
more coccoid.
Habitat. Isolated from crustaceæ.

163. Bact. Lepierrei

B. fluorescens pathogène Lepierre.
B. fluorescens or Lepierre : Kruse, Flügge, Die Mikroorganismen, 1896.

Morphology. Bacilli 1.5 : 2.0–3.0 μ; not stained by Gram's method.
Gelatin colonies. Round to yellowish brown, becoming, in three days, fluores-
cent; in 5 days the colonies are green.
Potato. No fluorescence, becoming black.
Bouillon. A fluorescence of the medium.
Milk. Not coagulated, alkaline; no fluorescence. *Indol* absent.
Glucose-bouillon. No gas. Optimum temperature 20°–30°. No fluorescence
at 37°.
Pathogenesis. Inoculation of guinea pigs, death in 1–6 days, with abscess of
liver and peritonitis.

164. Bact. iris (Frick)

B. iris Frick: Virchow's Archiv, CXVI, 1889, 290.
B. fluorescens-crassus Kruse: Flügge, Die Mikroorganismen, 1896, 294.

Morphology. Bacilli very small, slender.
Gelatin colonies. Of the aerogenes type; a green color, slowly developed.
Gelatin stab. Along line of stab no growth; surface growth an aerogenes-
like bead; a green fluorescence of a yellowish brown, to a dark green
color.
Potato. Growth light brown.
Bouillon. No membrane; not colored.
Habitat. Air and water.

165. Bact. Pflugeri (Ludwig)

M. Pflugeri Ludwig: Hedwigia, 1884, No. 3.
Photobact. Pflugeri Beijerinck: Centralblatt f. Bakteriologie, VIII, 1890, 716.

Morphology and cultural characters like *Bact. phosphorescens*, but somewhat
larger, longer, and more slender. Grows on potato.
Habitat. Isolated from fish.

CLASS XI. WITHOUT ENDOSPORES. OBLIGATE ANAEROBIC. GELATIN
NOT LIQUEFIED.

I. Grow at room temperatures and in nutrient gelatin.
   A. Gas produced in ordinary nutrient gelatin.
      166. *Bact. Welchii* Migula.
   B. Gas not produced in ordinary nutrient gelatin.
      1. Large, thick rods, over 1.0 micron in diameter.
      167. *Bact. emphysematosum* (Kruse).
      2. Very slender rods, 0.3 micron in diameter.
      168. *Bact. infecundum.*
II. No growth at room temperatures or in nutrient gelatin.
      169. *Bact. cadaveris* Sternberg.
      170. *Bact. pyogenes* (Kruse).

166. Bact. Welchii Migula

Bact. aerogenes-capsulatum Welch-Nuttall: Bull. Johns Hopkins Hospital, July, 1892.
Bact. Welchii Migula: System der Bakterien, 1900, II.

Morphology. Bacilli somewhat thicker than anthrax; chains with a capsule.
Gelatin cultures give much gas.
Gelatin colonies. 1–2 mm., gray white, oval to irregular with a few out-
   growths.
Milk. Coagulated; has a faint odor of old cheese.
Potato. Growth gray white.
Pathogenesis negative.
Habitat. Isolated from blood in a case of aneurism of the aorta.

167. Bact. emphysematosum (Kruse)

B. der Gasphlegmon Fraenkel: Centralblatt f. Bakteriol., XIII, 1893, 13.
B. emphysematosus Kruse: Flügge, Die Mikroorganismen, 1896, 242.

Morphology. Bacilli somewhat thicker than anthrax; occur in threads.
   Stained by Gram's method.
Gelatin. With addition of glycerine and formate of soda, active growth with
   much bad-smelling gas.
Glucose bouillon. Much gas.
Bouillon. Turbid.
Pathogenesis. Guinea pigs, by subcutaneous inoculation, a local non-sup-
   purating inflammation; later, a necrosis sometimes spreading to perito-
   neum and pleura.
Habitat. Isolated from gaseous phlegmon.

168. Bact. infecundum

Bact. filiformis-Havaniensis Sternberg: Manual of Bacteriology, 1892, 650.

Morphology. Bacilli long, slender, 0.3 micron in diameter.
Gelatin stab. Slight growth in depth; no growth on the surface.
Agar slant. Growth scanty, milky-white, branched.
Potato. No growth.
Pathogenesis. Non-pathogenic to rabbits and guinea pigs.
Habitat. Isolated from liver of yellow fever cadaver.

169. Bact. cadaveris Sternberg

Manual of Bacteriology, 1892.

Morphology. Bacilli 1.2 : 1.5–4.0 μ — short filaments. Do not grow on gelatin.
Glycerin bouillon. Rendered acid; no gas.
Pathogenesis. Guinea pigs, death with œdema by inoculation with a section of the liver containing the bacilli.
Habitat. Isolated from cadavers.

170. Bact. pyogenes (Kruse)

Bact. pyogenes-anaerobium Kruse: Flügge, Die Mikroorganismen, 1896, 244.

Morphology. Bacilli large. Do not grow below 22°. Cause a stinking suppuration in rabbits.
Habitat. Isolated from stinking pus from a rabbit which died spontaneously.

CLASS XII. WITH ENDOSPORES, AEROBIC AND FACULTATIVE ANAEROBIC, RODS NOT SWOLLEN AT SPORULATION. DO NOT GROW AT ROOM TEMPERATURES OR BELOW 22°–25°.

I. Minimum temperature of growth 40°–49°; optimum temperature 60°–70°.

THERMOPHILIC BACTERIA

A. Spores placed at the ends of the rods, oval.
    1. Growth on potato white to gray.
        171. *Bact. thermophilum,* I (Rabinowitsch).
        172. *Bact. thermophilum,* VII (Rabinowitsch).
        173. *Bact. thermophilum,* VI (Rabinowitsch).
    2. Growth on potato brownish.
        174. *Bact. thermophilum,* III (Rabinowitsch).
        175. *Bact. thermophilum,* V (Rabinowitsch).

3. Growth on potato not stated.
    176. *Bact. Miquelii* (Kruse).
B. Spores placed at the middle of the rods.
    1. Potato cultures red.
        177. *Bact. thermophilum*, IV (Rabinowitsch).
    2. Potato cultures gray, yellow, or brown.
        178. *Bact. thermophilum*, II (Rabinowitsch).
        179. *Bact. thermophilum*, VIII (Rabinowitsch).
II. Grow at body temperatures and on media containing blood-serum.
        180. *Bact. erythematis* (Kruse).
        181. *Bact. Colomiatii.*
III. Do not grow below 22°–25°.
        182. *Bact. laxæ.*

171. Bact. thermophilum, I (Rabin.)

B. thermophilus, I, Rabinowitsch: Zeitsch. f. Hygiene, XX, 1895, 154.
Morphology. Rods to filaments.
Agar colonies. At 62°, coarsely granular, erose.
Potato. White colonies.
Habitat. Widely distributed.

172. Bact. thermophilum, VII (Rabin.) : l.c.

Agar colonies. Granular, erose.
Potato. Growth white to gray.
Habitat. Excrement.

173. Bact. thermophilum, VI (Rabin.) : l.c.

Agar colonies. Centres grayish green, border clear.
Potato. Growth moist gray.
Habitat. Excrement.

174. Bact. thermophilum, III (Rabin.) : l.c.

Morphology. Bacilli thick.
Agar colonies. Small, entire.
Potato. Growth brown.
Habitat. Soil, milk, excrement.

175. Bact. thermophilum, V (Rabin.)

Agar colonies. Granular, colorless.
Potato. Growth gray brown, scanty.
Habitat. Excrement.

176. Bact. Miquelii (Kruse)

Un Bacille vivant au delà de 70° C. Miquel: Ann. de Microgr., 88.
B. thermophilus Frankland: Microorganisms of water, 1894, 488.
B. thermophilus-Miquelii Kruse: Flügge, Die Mikroorganismen, 1896, 269.

Morphology. Bacilli 1.0 μ, thick, and of variable length, to filaments. Optimum temperature 65°-70°.
Agar slant. At 43°, colonies white, elevated disks.
Bouillon. At 50°, turbid with a fragile membrane.
Habitat. Water.

177. Bact. thermophilum, IV (Rabin.) : l.c.

Morphology. Bacilli rods to filaments.
Agar colonies. Colorless, with outgrowths.
Habitat. Soil and excrement.

178. Bact. thermophilum, II (Rabin.) : l.c.

Morphology. Bacilli bent.
Agar colonies. Granular, grayish.
Potato. Growth grayish yellow colonies.
Habitat. Widely distributed.

179. Bact. thermophilum, VIII (Rabin.) : l.c.

Agar colonies. Colorless, round, entire.
Potato. Growth, moist, gray brown.
Habitat. Excrement.

180. Bact. erythematis (Kruse)

B. of erythema-nodosum Demme: Fortschritte d. Medizin, 1888, No. 7.
B. erythematis Kruse: Flügge, Die Mikroorganismen, 1896, 426.

Morphology. Bacilli 0.5-0.7 : 2.5 μ.
Blood serum. A paraffin-like, glistening streak, with radiations like fish fins.
Pathogenesis. Inoculations of guinea pigs cause an appearance of erythema-nodosum, and death.
Habitat. Associated with erythema-nodosum.

181. Bact. Colomatii

Discovered by Colomati: Breslauer arztliche Zeitsch., 1883, No. 4.

Morphology. Bacilli like those of mouse-septicæmia; in irregular masses.
Blood serum. 37°, growth of rosette-like forms, dull, glistening, of a fatty lustre.
Habitat. Isolated from xerotic masses in conjunctivitis.

182. Bact. laxae

Ein thermophiler Bacillus aus Zukerfabriksproducten Laxa : Centralblatt f.
Bakteriol., 2te. Abt., IV, 1898, 362.

Morphology. Bacilli in cultures like *B. vulgaris.*
Glycerin agar colonies. Deep: With white centres and dense outgrowths.
   Surface: Round, yellowish, crumpled, with root-like outgrowths, becoming
   in 24 hours 20 cm. or more in diameter.
Bouillon. With a crumpled membrane.
Potato. A tough, yellowish, ragged layer. Growth begins and increases up
   to 58°. Grow in acid alkaline or neutral media. Grow best in saccharine
   media, with production of gas and lactic acid. Pepton solutions without
   sugar rendered alkaline. Nitrates reduced to nitrites.
Habitat. Associated with Schaumgarung in sugar factories.

CLASS XIII. WITH ENDOSPORES; AEROBIC AND FACULTA-
TIVE ANAEROBIC, RODS NOT SWOLLEN AT SPORULATION.
GROW AT ROOM TEMPERATURES. GELATIN LIQUEFIED.

ANTHRAX GROUP

I. Stab cultures in gelatin arborescent.

   A. On the surface of bouillon cultures a membrane is more or less strongly
      developed.
      1. Potato cultures dense, rough, crumpled, or felt-like, to mealy.
            183. *Bact. mycoides* (Flügge).
            184. *Bact. brassicæ* (Pommer).
            185. *Bact. granulatum.*
            186. *Bact. maritimum* (Russell).
      2. Potato cultures show scattered colonies.
            187. *Bact. sputi.*
      3. Cultures on solid media not described.
            188. *Bact. sessile* (Klein).

   B. No membrane forms on the surface of bouillon cultures.
      1. Fatally pathogenic to mice, guinea pigs, and rabbits.
            189. *Bact. anthracis* (Cohn) Migula.
      2. Not pathogenic, or only a local effect, on guinea pigs.
            190. *Bact. anthracoides* (Kruse).
            191. *Bact. crystaloides* (Dyar).

II. Stab cultures in gelatin not distinctly arborescent.
   A. Gelatin colonies round, punctiform, not flat or spreading.
      1. Gelatin stab cultures show a saccate-funnel-formed liquefaction.
         a. Growth on potato yellowish.
            192. *Bact. aerophilus* (Flügge).
         b. Growth on potato whitish.
            193. *Bact. Trichomii* (Trevisan).
      2. Gelatin stab cultures show a crateriform-napiform liquefaction.
            194. *Bact. Markusfeldii.*
   B. Gelatin colonies flat or spreading.
      1. Margins of surface colonies floccose-ciliate fringed.
         a. Gelatin liquefied quickly.
            * Surface colonies, dense, felt-like, crimpled.
               195. *Bact. crinatum* (Wright).
            ** Surface colonies not crimpled.
               † Agar and potato cultures smooth.
                  196. *Bact. verticillatum* (Ravenel).
               †† Agar and potato cultures crimpled.
                  197. *Bact. gangliforme* (Ravenel).
         b. Gelatin liquefied slowly.
            198. *Bact. vermiculare* (Frankland).
      2. Gelatin colonies with irregular erose borders of the *B. coli* type.
         a. Agar smear cultures crimpled.
            * Milk rendered strongly acid, slimy.
               199. *Bact. viscosum.*
            ** Milk rendered alkaline.
               200. *Bact. rugosum.*
         b. Agar smear cultures smooth, thin.
            201. *Bact. granulosum* (Russell).
         c. Agar smear cultures thick, wavy.
            202. *Bact. filliforme* (Tils.).
      3. Gelatin colonies streaming, of the *proteus* type.
         a. Milk rendered acid, gelatin rapidly liquefied.
            203. *Bact. proteum.*
         b. Milk rendered alkaline, gelatin slowly liquefied.
            204. *Bact. truncatum.*
      4. Gelatin colonies entire.
            205. *Bact. turgidum.*
      5. Gelatin colonies not definitely described.
            206. *Bact. geniculatum* (Duclaux).
            207. *Bact. panis.*

183. Bact. mycoides (Flügge)

Wurzel bacillus Eisenberg: Bak. Diag., 1886, Tab. 4.
B. mycoides Flügge: Die Mikroorganismen, 1886, 324.
B. figurans Crookshank: Manual, 1886, 324.
B. ramosus Eisenberg: Bak. Diag., 1891, 126.
B. implexus Zimmerman: Bak. Trink. u. Nutzwässer, 1890.

Morphology. Bacilli 0.8 : 1.6–3.6 μ; rods square-ended, or scarcely rounded, in chains. Stained by Gram's method. Aerobic.
Gelatin colonies. Small, round, ciliate, becoming large, felted, arborescent; densest in their centres; gelatin liquefied rapidly.
Gelatin stab. In depth, an arborescent growth; liquefaction crateriform, becoming saccate; a membrane on surface of the liquefied gelatin.
Agar colonies. White-gray, moist, with characteristic root-like branchings.
Agar slant. Growth shows rhizoid or root-like branchings.
Potato. Growth like *B. subtilis.*
Milk. Peptonized, alkaline.
Glucose bouillon. No gas. *Indol* negative. H_2S negative.
Habitat. Soil, water, etc.

184. Bact. brassicæ (Pommer)

B. brassicæ Pommer: Mitth. Bot. Inst. Gratz, 1886, Heft I : A. Koch, Bot. Zeit., 1888.

Imperfectly described, indt. from the preceding.
Habitat. Turnip infusion.

185. Bact. granulatum

Bacillus No. 3 Pansini: Virchow's Archiv, CXXII.

Morphology. Bacilli large, rods with granulated plasma.
Colonies. Anthrax-like.
Gelatin stab. Like anthrax.
Potato. Growth white to reddish yellow.
Bouillon. A thin membrane on the surface; odor of rotten cheese.
Habitat. Sputum.

186. Bact. maritimum (Russell)

B. maritimus Russell; Bot. Gazette, XVIII, 1893, 440.

Morphology. Bacilli 1.5 : 3.5–6.0 μ; ends rounded, chains of variable length; protoplasm granular.

Gelatin stab. Liquefaction shallow, crateriform ; lined with bacterial growth ; from the base of the latter filaments radiate a short distance into the gelatin ; a barely perceptible growth in depth ; liquefaction becoming stratiform.

Agar slant. Growth dense, white, smooth.

Potato. Growth thick, grayish white, dull, and mealy.

Habitat. Sea-mud, Woods Hole, Mass.

187. Bact. sputi

Bacillus No. 4 Pansini : Virchow's Archiv, CXXII, 1890.

Morphology. Bacilli large, in long filaments.

Gelatin colonies. Radiate.

Potato. Growth of moist, dew-like drops.

Bouillon. Turbid, with a dense membrane.

Habitat. Sputum.

188. Bact. sessile (Klein)

B. sessilis Klein : Centralblatt f. Bakteriologie, VI, 1889, 10.

Morphology. Bacilli like anthrax ; spores, in size and form, like *B. subtilis,* but show polar germination.

Bouillon. With a membrane.

Pathogenesis. Non-pathogenic to mice.

Habitat. Isolated from the blood of a cow supposed to have died of anthrax.

189. Bact. anthracis (Cohn) Migula

First discovered by Rayer : Memoirs de la Soc. de Biol., 1850, 141.

Les infusoires de la maladie charbonneuse Davaine : Compt. rend., LXIX, 1864, 393.

Bacteridie du charbon Pasteur-Joubert : Compt. rend., LXXXIV, 900.

B. anthracis Cohn : Beiträge Biol., I, 2 Heft, 1875, 177.

Bact. anthracis Migula : Engler-Prantl, Die Natürlichen Pflanzenfamilien, 1895.

Morphology. Bacilli 1.1–1.2 : 3–10 μ ; ends square, in chains ; in animal body, with a capsule. Stained by Gram's method.

Gelatin colonies. Round, white, becoming in 3–4 days liquefied ; contents turbid ; microscopically show grayish centres, with clearer floccose borders.

Gelatin stab. An arborescent growth, becoming a crateriform to saccate liquefaction ; no membrane on the surface.

Agar colonies. Very characteristic, floccose.

Agar slant. Growth gray white, moist, glistening, mealy ; older cultures show pellucid dots.

Bouillon. Clear, with a heavy, flocculent growth.

Milk. Coagulated, peptonized, slightly alkaline.

Potato. Growth grayish white, elevated, dense. Spore germination polar. *Indol* negative. H₂S, negative or slight.

Pathogenesis. Mice inoculated subcutaneously die in 24 hours, of general septicæmia. Guinea pigs and rabbits die in 48 hours; spleen greatly enlarged, bacilli in blood and organs.

Habitat. Blood, etc., of anthrax subjects.

190. Bact. anthracoides (Kruse)

Described but not named Hueppe-Wood: Berliner klin. Wochenschrift, 1889, No. 16.
B. anthracoides Kruse: Die Mikroorganismen, 1896, 232.

Morphology and cultural characters like anthrax.

Pathogenesis. Subcutaneous inoculation of large doses into guinea pigs give only a local reaction.

Habitat. Earth and water.

191. Bact. crystaloides (Dyar)

B. crystaloides Dyar: l.c., 371.

Morphology and cultural characters like the preceding.

Habitat. Isolated from a contaminated plate.

192. Bact. aerophilum (Flügge)

B. aerophilus Flügge: Die Mikroorganismen, 1886.

Morphology. Bacilli slender, rods of variable length.

Gelatin colonies. In 40 hours, small, punctiform, oval to pyriform; no great increase of size.

Gelatin stab. Liquefaction broadly saccate.

Potato. Growth dull yellow, smooth, paraffin-like, becoming dry to granular.

Habitat. Air and water.

193. Bact. Trichomii (Trevisan)

Bacillo della gangrena senile Trichomi: Riv. internaz. di Med. e. Chir., III, 1886, 73.
B. Trichomii Trevisan: Genera, 1889, 13.
Bacillus of Trichomi Sternberg: Manual, 1892, 473.

Morphology. Bacilli 1.0–3.0 μ; solitary — pairs; rods often show a club-shaped thickening. Stained by Gram's method. Grow at 37°.

Gelatin colonies. In 24 hours, round, slightly dirty yellow; liquefaction begins in 36 to 48 hours.

Gelatin stab. Liquefaction funnel-formed in 48 hours, with gas bubbles.

Agar slant. Growth thin, white, spreading, membranous.

Potato. At 37°, dirty white, milky colonies, becoming confluent.

Pathogenesis. Subcutaneous inoculations of rabbits and guinea pigs show a gangrenous process, with death in 2–3 days.

Habitat. Isolated from a case of senile gangrene.

194. Bact. Markusfeldii

B. der trichorrhexis-nodosa Markusfeld: Centralblatt f. Bakteriol., XX, 1897, 230.

Morphology. Bacilli 0.4–0.6 : 1.7–2.2 μ. Grow at 39°.

Gelatin colonies. Roundish clumps, which quickly liquefy the gelatin.

Gelatin stab. In 48 hours, a semi-spherical liquefaction of the gelatin.

Agar colonies. In 24 hours, at 37°. *Deep colonies:* round, oval, white, brownish yellow; contour irregular, with ringlet-like outgrowths and small colonies from border. *Surface colonies:* white, opaque in centre; border granular to floccose.

Agar slant. White, moist colonies.

Milk. Coagulated.

Potato. Growth moist, white.

Habitat. Associated with this disease.

195. Bact. crinatum (Wright)

B. crinatus Wright: l.c., 453.

Morphology. Bacilli large, chains to segmented threads.

Gelatin colonies. Deep: dark opaque, round to oval, entire, granular margins. *Surface:* in 2 days, 1–2 mm., round, glistening, translucent, entire; in 3 days, dense, felt-like margins, crimpled, slightly sunken; margins also fimbriate to frayed.

Gelatin stab. Liquefaction broadly funnel-formed to stratiform.

Agar slant. Growth gray white, frosted.

Bouillon. Turbid, with white floculi.

Potato. Growth thick, creamy-white, viscid, spreading, becoming yellowish, caseous.

Litmus milk. Decolorized, amphoteric, becoming viscid, yellowish, caseous.

Glucose bouillon. No gas. *Indol* slight or doubtful. Grow at 37°.

Habitat. Water.

196. Bact. verticillatum Ravenel

B. verticillatus Ravenel: l.c., 13.

Morphology. Bacilli thick, 3–5 times their breadth, rounded. Spores only in potato cultures after 2 weeks. Facultative anaerobic.

Gelatin colonies. *Deep:* in 14 hours, 0.2 mm., white-gray, filamentous.
   Surface: in 14 hours, 1 mm., crateriform, entire, circular; border ciliate,.
   becoming in 24 hours 3 mm. in diameter.
Agar slant. Growth thin, dirty white, frosted; agar stained brown.
Gelatin stab. In depth, slight growth, rarely ciliate; liquefaction crateri-
   form, becoming funnel-formed.
Potato. Growth dry, white, smooth, glistening, becoming in 2 weeks pinkish.
Bouillon. Turbid, with a crumpled pellicle.
Litmus milk. Coagulated, decolorized, alkaline.
Glucose bouillon. No gas. *Indol* negative. Grow at 36°.
Habitat. Soil.

197. Bact. gangliforme (Ravenel)

B. gangliformis Ravenel: l.c., 34.

Morphology. Bacilli straight rods, rounded; length 5–6 times their breadth ;.
   in chains.
Gelatin colonies. *Deep:* arborescent. *Surface:* margins well defined,
   bordered by a corona of fine spear points; in liquefied gelatin the filamentous habit resembles the potato bacillus; in 2 days, a thick mycoderma on the surface, with lacework of bars radially placed on margins.
Agar slant. Growth dirty white, dry, with fern-like edges, becoming
   wrinkled.
Gelatin stab. In 24 hours, surface growth 3 mm., liquefaction crateriform,
   becoming in 4 days a funnel, becoming stratiform; a pellicle on the-
   surface.
Potato. Growth dry, crumpled, becoming dirty white, moist, slimy.
Bouillon. Turbid, with a mycoderma.
Litmus milk. Rendered acid; coagulated in flocculi; in 2 weeks, alkaline.
Glucose bouillon. No gas. Grow at 36°.
Habitat. Soil.

198. Bact. vermiculare (Frankland)

B. vermicularis Frankland: Zeitsch. f. Hygiene, VI, 1889, 384.

Morphology. Bacilli 1.0 : 2–3 μ — filaments; spores in chains.
Gelatin colonies. *Deep:* irregular. *Surface:* thin, flat, irregular; margins:
   of wavy bundles; centre rough, wrinkled.
Gelatin stab. In depth, a slight growth; on the surface growth glistening,
   gray; after some time liquefaction commences below the surface of the-
   colony.
Agar slant. Growth smooth, glistening, gray.

o

Bouillon. Clear, with a white flocculent precipitate.

Potato. Growth thick, flesh-colored. Nitrates reduced to nitrites.

Habitat. Water.

199. Bact. viscosum

B. *No. 17* Adametz: Landw. Jahrbucher, XVIII, 1889.

Morphology. Bacilli 1.0–1.2 : 3–4 μ — filaments.

Gelatin colonies. White; edges irregular.

Agar slant. Growth slimy, crimpled.

Milk. Becomes plastic viscous; odor of butyric acid.

Habitat. Milk and cheese.

200. Bact. rugosum

B. *No. 14* Adametz: l.c.

Morphology. Bacilli 1.0–1.2 μ in diameter; rods to filaments, often interwoven.

Gelatin colonies. In 3 days, 3–4 mm., white, slimy, slightly convex; microscopically, gray, coarsely granular — grumose, erose.

Gelatin stab. On the surface, a white membrane with fatty lustre under which the gelatin is quickly liquefied.

Agar slant. Growth thick, white, slimy, crumpled, becoming reddish yellow.

Milk. A flocculent precipitate, alkaline, becoming peptonized.

Habitat. Milk, cheese.

201. Bact. granulosum (Russell)

B. *granulosus* Russell: Zeitschrift f. Hygiene, IX, 194.

Morphology. Bacilli large and as filaments, contents granulated.

Gelatin colonies. *Deep:* small, round, glistening, opaque masses. Deep, thin, leafy; microscopically, concentrically to reticulately marked.

Gelatin stab. Liquefaction shallow, crateriform.

Agar slant. Growth composed of thin, white-yellowish colonies, becoming confluent.

Bouillon. Turbid; sediment.

Potato. Growth in 24 hours, moist, white, limited, becoming thicker, dull, waxy, brownish.

Habitat. Sea-slime.

202. Bact. filiforme (Tils)

B. *filiformis* Tils: Zeitsch. f. Hygiene, IX, 1890, 317.

Morphology. Bacilli 1.0–4.0 μ — filaments or chains.

Gelatin colonies. *Deep:* white, finely granular, with irregular edges. *Surface:* 3 days whitish, striped; microscopically, irregular, serrate; centres uneven, granular, yellowish; liquefied slowly.

Gelatin stab. In depth, slight growth; on the surface, growth moist; edge
serrate; liquefaction in 3–4 days.
Agar slant. Growth thick; surface wavy.
Potato. Growth thick, irregular, dirty white, becoming darker.
Habitat. Water.

203. Bact. proteum

B. No. 16 Adametz: l.c.

Morphology. Bacilli 1.2 : 3.0–5.5 μ — filaments and chains, often interwoven
and reaching a length of 80–100 microns. Involution forms.
Gelatin colonies. Streaming, branched, with the gross appearance of a mould
colony.
Gelatin stab. Rapid liquefaction on the surface and along line of stab; a
membrane on the surface.
Agar slant. Growth crumpled, membraneous.
Agar stab. An arborescent growth along line of stab.
Milk becomes gelatinous, acid, with a cheesy odor.
Habitat. Milk and cheese.

204. Bact. truncatum

B. No. 51 Conn: l.c., 1894, 81.

Morphology. Bacilli short, square-ended rods, 0.2–0.8 : 1.5 μ. Grow at 35°.
Gelatin colonies. Curled, proteus-like; liquefied slowly.
Gelatin stab. Liquefaction slow, crateriform to conical.
Agar slant. Growth spreading, dry, granular.
Potato. Growth thick, white, dry, velvety.
Milk. Coagulated, alkaline, slowly peptonized.
Bouillon. Clear, with a pellicle.
Habitat. Milk.

205. Bact. turgidum

B. No. 15 Adametz: l.c.

Morphology. Bacilli 1.2–1.4 μ thick and about three times as long; filaments
rare; involution forms.
Gelatin colonies. 1 cm., dark gray in centres, with lighter coarsely granular
borders.
Gelatin stab. Rapid liquefaction, with a membrane, with later a butyric acid
odor.
Milk. Slimy, flocculent, slightly acid; butyric acid formed.
Habitat. Milk and cheese.

206. Bact. geniculatum (Duclaux)

Tyrothrix geniculata Duclaux: Le Lait, Paris, 1887, 331.
B. geniculatus Trevisan: Genera, 1889, 16.
B. gonatoides Trevisan: Saccardo, Syllog. Fungorum, VIII, 1889, 964.

Morphology. Bacilli, very thick rods — long interwoven filaments.
Milk. Slowly coagulated; leucin, tyrosin, acetic acid, and ammonia produced.
Habitat. Milk and cheese.

207. Bact. panis

B. mesentericus-panis-viscosus I Orth: Zeitsch. f. Hygiene, XXVI, 1897, 404.

Morphology. Bacilli plump, rounded, 3–5 μ long; no flagella. Spores oval, placed in the middle of the rods. Stained by Gram's method.
Gelatin colonies. Flat; liquefaction crateriform, with a thick grayish white nucleus, and a membrane on the surface.
Gelatin stab. Liquefaction slow, becoming saccate.
Agar colonies. Gray brown, granular, with delicate outgrowths.
Agar slant. Growth bluish gray, translucent.
Potato. Growth white, becoming gray, at first slimy, smooth, becoming silky and rugose.
Milk. Coagulated, peptonized.
Glucose bouillon. No gas.
Bouillon. Only a faint turbidity after many days.
Lactose bouillon. No change of color. Optimum temperature 35°–37°. No anaerobic growth. Grow in acid media, rendering them slowly alkaline.
Habitat. Isolated from stringy bread dough.

CLASS XIV. WITH ENDOSPORES. AEROBIC AND FACULTATIVE ANAEROBIC. RODS NOT SWOLLEN AT SPORULATION. GELATIN NOT LIQUEFIED.

I. Colonies on gelatin papillate, small (about 1 mm.).
  208. *Bact. Mansfieldii.*
II. Colonies on gelatin spreading.
  209. *Bact. Schottelii* (Trevisan).
  210. *Bact. subtiliforme* (Schröter).
  211. *Bact. simile* (Schröter).

208. Bact. Mansfieldii

B. No. 18 Conn: l.c., 1893, 51.

Morphology. Bacilli 1.4–2.0 μ, in twos and threes.
Gelatin colonies. Round, white, punctiform, 1 mm., spreading only slightly.
Gelatin stab. Surface growth white, spreading, rather dry.
Agar slant. Growth white, rather limited to spreading.
Potato. Growth thick, spreading, mottled; somewhat raised in mounds, becoming brown and more uniform.
Bouillon. · Turbid, with a sediment and slight pellicle.
Habitat. Milk from Mansfield, Conn.

209. Bact. Schottelii (Trevisan)

Darmbacillus Schottelius: Der Rotlauf der Schweine, Wiesbaden, 1885.
B. coprogenes-fœtidus: Flügge, Die Mikroorganismen, 1886.
B. Schottelii Trevisan: Genera, 1889, 17.

Morphology. Bacilli shorter than *B. subtilis.* Spore germination polar.
Gelatin colonies. Deep: compact, pale yellowish. *Surface:* gray, translucent, spreading; strong putrefactive odor.
Potato. Growth thick, gray. Non-pathogenic.
Habitat. Isolated from the intestinal contents of swine.

210. Bact. subtiliforme (Schröter)

Bacillus I Bienstock: Zeit. f. klin. Med., VIII, 1884, 1–2.
B. subtiliformis Schröter: Pilz Schles., 1886, 160.
B. mesentericus Trevisan: Genera, 1889, 115.
B. fœcalis I Kruse: Flügge, Die Mikroorganismen, 1896, 215.

Morphology. Like *B. subtilis.* Gelatin colonies have the form of a mesenterium.
Habitat. Fæces.

211. Bact. simile (Schröter)

Bacillus II Bienstock: l.c.
B. similis Schröter: l.c.
B. coprocinus Trevisan: Genera, 1889, 15.
B. fœcalis II Kruse: Flügge, Die Mikroorganismen, 1896, 215.

Morphology. Like the preceding.
Gelatin colonies. White, glistening, smooth, becoming uneven, with lobular outgrowths.
Habitat. Fæces.

CLASS XV. WITH ENDOSPORES. AEROBIC AND FACULTATIVE
ANAEROBIC. RODS SWOLLEN AT ONE END AT SPORULATION
OF THE TETANUS TYPE.

I. Gelatin liquefied.

212. **Bact. gracile** (Zimm.)

B. gracilis Zimmerman: Bact. Nutz. u. Trinkwässer, 1890.

Morphology. Bacilli 0.8 : 2.4–3.6 μ — threads.

Gelatin colonies. Deep: entire, becoming ameboid — reticulated. *Surface:*
thin and spreading.

Gelatin stab. In depth, growth beaded; on the surface, after some weeks a
liquefaction of the surface.

Agar slant. Growth thin, bluish white.

Potato. Growth scanty.

Habitat. Water.

II. Gelatin not liquefied.

213. **Bact. canis**

B. des Hundestaupe Bruno-Galli: Centralblatt f. Bakteriol., XIX, 1896, 694.

Morphology. Bacilli 0.3 : 1.2–2.5 μ, often dumb-bell-shaped. Stained by
Gram's method.

Gelatin stab. In 24 hours, gas bubbles in depth; on the surface, growth
punctiform, white, waxy, becoming large and sinking into the gelatin,
producing a shallow funnel without apparent liquefaction.

Agar slant. Growth of small, white points, becoming white disks with
undulate borders.

Potato. Growth whitish, translucent.

Milk. Not coagulated. *Indol* negative.

Lactose bouillon. No gas.

Habitat. Urine, exudate, etc., of dogs.

CLASS XVI. WITH ENDOSPORES. OBLIGATE ANAEROBIC.

I. Rods not swollen at sporulation, malignant œdema type.

214. **Bact. anaerobicum** (Sternberg)

B. anaerobicus-liquefaciens Sternberg: Manual, 1892, 693.

Morphology. Bacilli 0.6 : 2–3 μ; filaments.

Gelatin colonies. White, granular.

Pathogenesis. Doubtful.

Habitat. Intestines of yellow fever cadavers.

215. Bact. terræ (Ucke)

Streptobacillus terræ Ucke: Centralblatt f. Bakteriol., XXIII, 1898, 1001.

Morphology. Bacilli 2.0 : 6–20 μ, and longer chains. Polar oval spores. Stained by Gram's method.

Gelatin. No growth at 22°.

Bouillon. A flocculent white sediment; a stronger growth in glucose bouillon.

Agar slant. Growth white, very scanty, limited, with finely erose border.

Agar stab. In depth, growth scanty, filiform.

Milk. Not coagulated.

Potato. No visible growth. No gas. No odor. Litmus reduced. An acid production in saccharine media.

Blood serum. Small, flat, yellowish colonies.

Habitat. Isolated from the soil.

II. Rods becoming latterly swollen or spindle-shaped at sporulation.

216. Bact. parvum

B. liquefaciens-parvus Luderitz: Zeitsch. f. Hygiene, V, 1889, 149.

Morphology. Bacilli 0.5–0.7 : 2–5 μ; filaments, often bent. Spore formation not distinct, but small round refractive bodies in the greatly thickened rods.

Gelatin colonies. Entire, becoming tuberculate with delicate outgrowths. Liquefaction of gelatin slow; but little gas.

Habitat. Soil.

BACILLUS Cohn, char. emend by Migula

Cells cylindrical, varying from short ovals to longer rods and filaments. Motile, with flagella attached to any part of the rod, varying from a few to numerous, and surrounding the entire body of the bacillus (peritrichic). Endospores present or absent, or at least in a large number of the species unknown.

NOTE. — Our imperfect knowledge of the great majority of the described species of bacteria, especially as regards the nature of their flagella, makes it impossible to properly classify many of them.

Those species which are known to possess peritrichic flagella belong properly to this genus, and are designated by a large **B** in black-faced type before each specific name. Those doubtfully placed in the genus are all so-called motile forms whose flagella are not described, and are designated by a B in plain type.

The author has therefore made this group the great lumber room into which are thrown all indefinitely motile forms. It is likely that many of the species here included, although they have been described as motile, or slightly so, are in reality non-motile, or at least devoid of flagella, and are therefore members of the genus Bacterium.

Without definite knowledge on all questionable points of this kind the author has not presumed to any private interpretations, and has strictly adhered to the facts as set forth by the authors of the species in question.

SYNOPSIS OF THE GENUS

I. Without endospores, or at least their presence not reported.
   A. Aerobic and facultative anaerobic.
      1. Without pigment on gelatin or agar.
         a. Colonies on gelatin plates roundish, not ameboid or proteus-like.
            * Gelatin not liquefied.
               † Decolorized by Gram's method. CLASS I, p. 201.
               †† Stained by Gram's method. CLASS II, p. 227.
            ** Gelatin liquefied. CLASS III, p. 230.
         b. Colonies on gelatin plates becoming streaming, forked, ameboid, twisted, irregular, cochleate.
            * Gelatin liquefied.
               Stained by Gram's method. CLASS IV, p. 244.
               Not stained by Gram's method. CLASS V, p. 246.
            ** Gelatin not liquefied. CLASS VI, p. 248.
      2. Produce pigment on gelatin or agar, chromogenic bacilli.
         a. Pigment yellowish on gelatin.
            * Gelatin liquefied. CLASS VII, p. 250.
            ** Gelatin not liquefied. CLASS VIII, p. 254.
         b. Pigment reddish on gelatin or agar.
            * Gelatin liquefied. CLASS IX, p. 256.
            ** Gelatin not liquefied. CLASS X, p. 259.
         c. Pigment brownish, black-gray on gelatin. CLASS XI, p. 260.
         d. Pigment blue-violet on gelatin or agar. CLASS XII, p. 261.
      3. Colonies colorless, or colored slightly yellowish or greenish, but with a yellow-green or blue-green fluorescence. CLASS XIII, p. 262.
   B. Obligate anaerobic. CLASS XIV, p. 265.
II. Bacilli produce endospores.
   A. Non-chromogenic; without pigment on gelatin or agar.
      1. Aerobic and facultative anaerobic.
         a. Rods not swollen at sporulation — *B. subtilis* type.

           † Potato cultures never developing a red pigment.
               Gelatin liquefied. CLASS XV, p. 266.
               Gelatin not liquefied. CLASS XVI, p. 282.
           †† Potato cultures developing a red pigment. CLASS XVII,
               p. 285.
      b. Rods becoming spindle-shaped at sporulation, Clostridium type.
          CLASS XVIII, p. 287.
      c. Rods swollen at one end at sporulation, Tetanus type. CLASS
          XIX, p. 290.
   2. Obligate anaerobic.
      a. Rods not swollen at sporulation. CLASS XX, p. 292.
      b. Rods becoming laterally swollen or spindle-shaped at sporula-
         tion. CLASS XXI, 295.
      c. Rods swollen at one end at sporulation. CLASS XXII, p. 302.
  B. Chromogenic; produce pigment on gelatin or agar. CLASS XXIII,
     p. 304.

CLASS I. WITHOUT ENDOSPORES. AEROBIC AND FACULTATIVE
ANAEROBIC. COLONIES ON GELATIN PLATES ROUNDISH, NOT
AMEBOID OR PROTEUS-LIKE. GELATIN NOT LIQUEFIED. DE-
COLORIZED BY GRAM'S METHOD.

I. Gas generated in glucose bouillon.
  A. Milk coagulated. B. COLI GROUP.
    1. *Indol* produced.
      a. Phenol produced.
          1. *B. Marsiliensis* Kruse.
      b. No *phenol* produced.
          2. *B. coli* (Escherich).
          3. *B. Wardii.*
    2. No *indol* produced.
      a. Gelatin colonies of a distinctly colon type, indistinguishable
        from those of *B. coli.*
          4. *B. anindolicum* Lembke.
      b. Gelatin colonies of a character intermediate between the colon
        and *aerogenes* types (bacteria intermediate between *B. coli* and
        Bact. aerogenes).
          5. *B. enteritidis* Gartner.
          6. *B. chologenes* Kruse.
          7. *B. toxigenus.*

3. *Indol* production not stated.

        8. *B. brassicæ* Lehmann-Conrad.

B. Milk not coagulated.　Hog Cholera Group.

  1. *Indol* produced.

        9. *B. icterogenes* Kruse.

       10. *B. Poelsii.*

       11. *B. columbarum.*

  2. No *indol* produced.

    a. More or less gas produced in lactose bouillon.

       12. *B. Breslaviensis* Kruse.

    b. No gas in lactose bouillon.

       13. *B. Salmoni* (Trevisan).

       14. *B. levans* Lehmann-Wolffin.

    c. Gas production in lactose bouillon indeterminate.

       15. *B. loxiacida.*

  3. *Indol* production indeterminate.

       16. *B. morbificans* Basenau.

       17. *B. Silberschmidii.*

       18. *B. murium* Löffler.

II. No gas generated in glucose bouillon.　Typhoid Group.

  A. Milk coagulated.

       19. *B. intestinalis* Dyar-Keith.

  B. Milk not coagulated.

    1. Potato cultures whitish grayish or invisible.

     a. Bacteria from animal habitats, pathogenic; nearly identical in cultural characters with *B. typhosus.*

       * A pyogenic reaction at point of inoculation in guinea pigs.

         20. *B. meningitidis* Neumann-Schaeffer.

       ** No pyogenic reaction by inoculation into experimental animals.

         21. *B. typhosus* Zopf.

         22. *B. pseudo-typhosus* Kruse.

         23. *B. Billingsi.*

         24. *B. paradoxus* Kruse.

         25. *B. pestis* (Lehmann-Neumann).

     b. Soil and water bacteria, not so distinctly connected with *B. typhosus.*

       * Rosolic acid solution decolorized.

         26. *B. solitarius* Ravenel.

** Rosolic acid solution not decolorized.
    27. *B. geminus* Ravenel.
    28. *B. aquatilis-sulcatus-quartus* Weichselbaum.
    29. *B. primus-Fullesi* Dyar.
   Action on rosolic acid not stated.
    30. *B. tracheiphilus* Smith.
2. Potato cultures becoming yellowish-brownish.
  a. Grow well at the body temperature.
    * Produce *indol.*
      31. *B. pinatus* Ravenel.
    ** Do not produce *indol.*
     † Gelatin stab arborescent, *i.e.* with outgrowths.
       32. *B. Raveneli.*
     †† Gelatin stab not distinctly arborescent.
      § Milk rendered alkaline.
       33. *B. alcaligenes* Petruschky.
      §§ Milk reaction not stated.
       34. *B. Friedebergensis* Gaffky-Paak.
    *** *Indol* production not stated.
      35. *B. solanacearum* Smith.
  b. Do not grow at the body temperature.
    36. *B. Weichselbaumii.*

III. Gas development in glucose bouillon not stated. Bacteria of the Colon, Hog Cholera, and Typhoid Groups ; not classified.
 A. Milk not coagulated.
  1. Pathogenic to guinea pigs and rabbits.
    37. *B. Friedebergensis* Kruse.
  2. Pathogenic to pheasants, not so to guinea pigs and rabbits.
    38. *B. phasini.*
  3. Non-pathogenic.
    39. *B. Schafferi* v. Freudenreich.
    40. *B. rugosus.*
 B. Milk coagulation not stated.
  1. Gelatin colonies of the colon type.
   a. Pathogenic for birds (motile bacilli related to the bacillus of fowl cholera).
    * Not pathogenic to guinea pigs.
      41. *B. avisepticus.*
      42. *B. avium* Kruse.

** Only slightly pathogenic to guinea pigs.

    43. *B. meleagris.*

    44. *B. tetraonis.*

*** Pathogenic to guinea pigs.

    45. *B. cygneus.*

b. Not pathogenic to birds; scarcely pathogenic to other animals.

    46. *B. aerobius.*

    47. *B. pneumosepticus* Kruse.

c. Pathogenic to insects.

    48. *B. monachæ* Tubeuf.

2. Gelatin colonies of the aerogenes type.

  a. Pathogenic to the smaller animals.

    * Rabbits, general infection.

      49. *B. cuniculi.*

      50. *B. venenosus* Vaughan.

    ** Pyogenic to the smaller animals.

      51. *B. glischrogenus* Malerba.

  b. Non-pathogenic.

    * Do not grow at 37°, water bacteria.

      † Growth on agar smooth, not characteristic.

      52. *B. albus* Paglinni.

      53. *B. granulatus.*

    †† Growth on agar branched (rhizoid).

      54. *B. stolonatus* Adametz.

3. Colonies burr-like.

    55. *B. invisibilis* Vaughan.

    56. *B. venenosus* Vaughan.

4. Colonies show a coil-like (Knauelartig) structure.

    57. *B. murinus.*

5. Colonies not characterized as in 1–4.

B. DENITRIFICANS GROUP

a. Grow only with difficulty on the surface of gelatin plates.

    58. *B. denitrificans* Burri-Stutzer.

b. Grow on the surface in gelatin plates.

    * Produce only a faint turbidity in nitrate bouillon, becoming clear; a membrane on the surface.

      59. *B. Stutzeri* (Lehmann-Neumann).

    ** Produce a marked or strong turbidity in nitrate bouillon, with the formation of a membrane.

† Bacteria surrounded by a capsule.

    60. *B. centropunctatus* (Jensen).

†† Capsule formation at least not mentioned.

    61. *B. agilis* Ampola-Garino.

    62. *B. Hartlebii* (Jensen).

1. B. Marsiliensis Kruse

Bacillus of Marseilles swine plague Rietsch-Jobert: Compt. rend., CVI, 1888.

B. der Frettenseuche Ebert-Schimmelbusch: Virchow's Archiv, CXV, 1889, 282; Centralblatt f. Bakteriologie, XVI, 1894, 327.

B. spontanen Kaninschenseptikämie Ebert-Mandry: Fortschritte Med., VIII, 1890.

B. of swine plague Billings: Report Ag. Expt. Sta., Univ. of Nebraska, 1888.

B. of Texas fever Billings, l.c.

B. der Amerikanischer Rinderseuche Caneva: Centralblatt f. Bakteriologie, IX, 1891, 557.

B. Marsiliensis Kruse: Flügge, Die Mikroorganismen, 1896.

Morphology. Bacilli twice as long as broad, one-third smaller than *B. typhosus*; show the polar stain. Flagella peritrichic (4–5).

Gelatin colonies. Colon-like.

Potato. Growth yellowish gray.

Milk. Coagulated, acid.

Litmus milk. Reduced and reddened.

Indol and *phenol* produced. Cultures of Billings swine plague for old cultures as above, for new cultures approaching hog cholera. All cultural characters closely identical with those of the colon bacillus.

Pathogenesis. Variable for the different varieties of the species. Inoculations of rabbits give variable results, negative to slightly pathogenic; a general septicæmia often produced with the Ebert-Mandry bacillus. Inoculation of sparrows into the breast muscle causes death in 24–36 hours, with septicæmia, pleuritis, and pericarditis. Pathogenic to hens and ferrets; only slightly pathogenic to pigeons.

Habitat. Found in the blood and organs in ferret plague; associated with Marseilles swine plague, spontaneous septicæmia of rabbits, etc.

2. B. coli (Escherich)

Bact. coli-commune Escherich: Darmbak. des Säuglings, Stuttgart, 1886.

Neapeler Bacillus Emmerich: Deutsche med. Wochenschrift, 1884, No. 50.

B. Neapolitanus Fraenkel; Grundriss der Bakterienkunde, 1887.

Emmerich's Bacillus Eisenberg: Bak. Diag., 1886.

B. pyogenes-fœtidus Passet: Aetiol. eiterigen Phlegmon des Menschen, Berlin, 1885.

Morphology. Bacilli 0.4–0.7 : 1.3 μ; facultative anaerobic.

Gelatin colonies. Deep: round to lenticular, yellowish brown. *Surface:* flat, erose to lobate, marmorated.

Gelatin stab. A good growth in depth; surface growth flat, spreading.

Agar slant. Growth gray, white, moist, glistening, translucent.

Potato. Growth yellowish to yellowish brown.

Bouillon. A dense turbidity, with a heavy sediment.

Milk. Coagulated.

Litmus milk. Reduced, acid. Cultures have a fæcal odor. H_2S produced.

Lactose bouillon. Much gas.

Saccharose bouillon. Gas may or may not be produced; acetic, formic, and lactic acids produced. In bouillon, ammonia produced, and an alkaline reaction.

Pathogenesis. Variable; inoculation of mice with 0.1–1.0 cc. of a bouillon culture, intraperitoneally, causes death in 1–8 days; bacilli in the blood, and peritoneal exudate. One cc. of virulent varieties inoculated intraperitoneally into guinea pigs may cause death, with general peritonitis.

Habitat. In the intestines of man and animals, fæces, water, milk. Associated with a number of pathologic conditions — peritonitis, cystitis, cholera-nostras, etc.

VARIETIES

B. coli-dysentericum Ciechanowski: Centralblatt f. Bakteriol., XXIII, 445, 1898.

For varieties of *B. coli* see Pfaundler: Centralblatt f. Bakteriol., XXIII, 1.

Pottien: Zeitsch. f. Hygiene, XXII, Heft 1, describes a variety of *B. coli* which in the animal body shows a capsule, which gives *B. Zopfi*-like colonies, and which is strongly pathogenic to mice.

3. B. Wardii

Gas and taint producing Bacillus in cheese curd: Cornell Univ. Ag. Expt. Station, Bull. 158, 1899.

Morphology. Bacilli in bouillon 1.2–2 4 μ, with rounded ends; occur singly. Show a polar stain with carbol fuchsin, stain feebly but uniformly, with alkaline methyl-blue. Not stained by Gram's method. Flagella demonstrated. Optimum temperature 35°–38°. Facultative anaerobic.

Gelatin colonies. *Surface:* thin, spreading. 3–7 mm., wrinkled; border irregular. Microscopically, centre opalescent, border thin, translucent.

Gelatin stab. In depth, growth beaded; on the surface, growth thin, spreading.

Agar colonies. At 37°. *Deep colonies:* lenticular. 0.5–1.0 μ, grayish. *Surface colonies:* round, flat, entire, sharp, gray, moist, glistening, 2–4 mm.; not viscid. Have the odor of swine-plague cultures.

Agar slant. Growth thin, glistening, spreading; condensation water turbid.

Potato. Growth brownish yellow, becoming thicker and brown; not viscid.

Alkaline bouillon. Densely turbid; grayish sediment; acid, becoming alkaline. In old cultures, 3–6 weeks; a grayish pellicle on the surface.

Milk. Coagulated in about 3 days at 37°; serum clear; casein not digested; acid; odor sour.

Glucose bouillon. Gas, maximum in 2 days; growth in both arms; acid.

Lactose bouillon. Gas.

Saccharose bouillon. No gas; closed arm clear, remains alkaline. $H : CO_2 ::$ 2 : 1. *Indol* produced.

Habitat. Isolated from tainted, gassy, cheesy curd.

4. B. anindolicum Lembke

Archiv f. Hygiene, XXVII, 1896, 384.

Morphological and cultural characters like *B. coli*; differs in producing an amount of acid in milk intermediate between *B. coli* and *B. typhosus*.

Pathogenesis. Inoculations of 0.2 cc. of bouillon culture subcutaneously into mice cause death in 24 hours, with general septicæmia. Guinea pigs die by intraperitoneal injections of 0.7 cc. of bouillon culture.

Habitat. Isolated from the fæces of a dog.

5. B. enteritidis Gartner

Correspond. d. allg. Artzl. Vereins, Thuringen, 1888.

Morphology. Bacilli short, thick; stain unequally; capsule present or absent.

Gelatin colonies. Deep: brown. *Surface:* round, gray, translucent, granular.

Lactose bouillon. Gas.

Potato. Growth grayish white to grayish yellow, glistening.

Pathogenesis. Pathogenic to mice, guinea pigs, rabbits, pigeons, young sheep, and goats. Non-pathogenic to dogs, cats, rats, chickens, and sparrows. Mice and guinea pigs infected through the stomach, producing enteritis; bacilli found in the organs.

Habitat. Isolated from beef in meat poisoning.

6. B. chologenes Kruse

Discovered by Stern, no name: Deutsche med. Wochenschrift, 1893, No. 26.

B. chologenes Kruse: Flügge, Die Mikroorganismen, 1896, 374.

Morphology. Bacilli 0.5–1.3 μ.

Gelatin colonies. White; border erose; characters between the colon and aerogenes types.

Potato. Growth white to yellow; gas produced.

Milk. Coagulated in 1–2 days.

Lactose bouillon. Gas.

Saccharose bouillon. Gas.

Pathogenesis. Intraperitoneal inoculations of 0.5–1.0 cc. of bouillon cultures into mice cause death. With guinea pigs, subcutaneous inoculation causes abscess formation; intraperitoneal injections of larger doses cause death.

Habitat. Isolated from a case of angiocholitis and meningitis.

7. B. toxigenus

Bacillus of ice cream poisoning Vaughan-Perkins: Archiv f. Hygiene, XXVIII, 1896.

Identical with *B. coli*, but milk coagulated more quickly, with strong butyric acid odor. Grows on carrot; growth elevated, creamy; odor acid. (*B. coli* grows much less vigorously, and gives no odor.)

Pathogenesis. Pathogenic to rabbits, cats, dogs, mice, and rats.

Habitat. Isolated from poisonous ice cream.

8. B. brassicæ Lehmann-Conrad

Lehmann-Neumann, Bak. Diag., 1896, 232.

Morphological and cultural characters closely related to *B. coli*. Bacilli show 4–10 long slender flagella. Sometimes slightly colored by Gram's method. Ferments milk, sugar. Milk coagulated. Generates in sauer-kraut 80 per cent CO_2, 18 per cent H, and 2 per cent CH_4.

Habitat. Isolated from sauer-kraut.

9. B. icterogenes Kruse

B. of yellow atrophy of the liver Guarnieri: Acad. Med. Rom., XIV, 1887–88, fasc. 8.

B. icterogenes Kruse: Flügge, Die Mikroorganismen, 1896, 372.

Morphological and cultural characters like *B. coli*, but grow less vigorously.

Lactose bouillon. A small amount or no gas. No gas in saccharose bouillon.

Milk. Rendered slightly acid.

Pathogenesis. Intraperitoneal inoculations of guinea pigs cause septicæmia and degeneration of the liver.

Habitat. Isolated by Guarnieri from the liver and blood in acute yellow atrophy of the liver, and by Pasquale from typhoid stools.

10. B. Poelsii

Vleeschvergiftung te Rotterdam Poels-Dhont: Tweede rapport van de des Kundigen.

Morphology. Bacilli slowly motile. Grow on gelatin like *B. coli.* A very weak gas development in glucose bouillon. No gas in lactose and saccharose bouillon.

Bouillon. Rendered alkaline. *Indol* produced.

Milk. Not coagulated.

Pathogenesis. Intravenous inoculations of cows with 3 gelatin cultures resulted in death in 14 hours; bacilli in all organs and in the muscles.

Habitat. Isolated from beef in meat poisoning.

11. B. columbarum

Bacillus of pigeon cholera Moore: U. S. Dept. of Ag., Bureau of Animal Industry, Bull. No. 8, 1895.

Morphology. Bacilli 1.0 : 1.0–1.6 μ; size variable in different media; ends rounded; in the tissues, usually in pairs. Flagella not more than 8.

Gelatin colonies. Surface: small grayish dots; microscopically, yellowish, granular.

Agar slant. At 37°, growth grayish, glistening, not viscid.

Agar colonies. Convex, entire, 0.5–1.5 mm.

Potato. Growth thin, glistening, slightly yellowish. On acid potato no growth.

Alkaline bouillon. In 24 hours, turbid, slightly acid, with a thin, grayish membrane, becoming alkaline. In acid bouillon, only slight growth, reaction unchanged.

Milk. Not coagulated, strongly alkaline.

Glucose bouillon. Gas; $H : CO_2 :: 2 : 1$. No gas in lactose or saccharose bouillon; media rendered alkaline. *Indol* produced.

Pathogenesis. Intravenous inoculation of rabbits with 0.3 cc. of bouillon culture causes congestion of the internal organs. Intestinal mucosa reddened in patches. Subcutaneous inoculation causes death in 4–5 days, with purulent infiltration at the point of inoculation; in the liver, necrotic spots; spleen enlarged, dark-colored, and friable. Subcutaneous inoculation of guinea pigs, 0.1–0.2 cc., causes death in 8–10 days. Pathogenic to pigeons. Differs from hog cholera (1) bacilli larger; (2) in bouillon a delicate membrane, and in old cultures a deposit on the sides of the tube; (3) a marked *indol* reaction; (4) it is less rapidly fatal in small doses for experimental animals, and the lesions produced in rabbits are comparable to those following the inoculation of the more attenuated varieties. (Moore, l.c.)

P

12. B. Breslaviensis Kruse

B. Morseeler u. Breslauer Fleischvergiftung v. Ermenghem: Trav. Lab. d. Hygiène de Gand Bruxelles, 3, 1892.
B. Breslaviensis Kruse: Flügge, Die Mikroorganismen, 1896, 377.

Morphology. Bacilli 0.6–1.5 μ long, slender; 4–12 long flagella.
Gelatin colonies. Like *B. coli.*
Bouillon. Turbid, with a delicate membrane.
Potato. Growth yellowish, abundant.
Saccharose bouillon. Only a slight amount of gas.
Pathogenesis. The feeding and inoculation of mice and rabbits cause enteric symptoms; bacilli in the organs.
Habitat. Isolated from poisonous beef and veal which were the cause of meat poisoning.

13. B. Salmoni (Trevisan)

B. of swine plague or *swine fever* Klein: Report of the Local Gov. Board of England, 1877–78.
Hog-cholera bacillus Salmon-Smith: U. S. Dept. Ag., 1885.
B. der Schweinepest Bang-Selander: Centralblatt f. Bakteriol., III, 1886, 361.
Pasteurella Salmoni Trevisan: Genera, 1889, 21.
Amerikanische Schweineseuche Frosch: Zeitsch. f. Hygiene, 1890, 235.
Swine-fever Bacillus E. Kleine: Centralblatt f. Bakteriol., XVIII, 1895, 106.
B. suipestifer Kruse: Flügge, Die Mikroorganismen, 1896, 223.
Bact. cholera-suum Lehmann-Neumann: Bak. Diag., 1896, 233.

Morphology. Bacilli 0.6 : 1.2–1.5 μ. No characteristic polar stain. The central part of the rod frequently less stained than the periphery (Smith). According to Karliński, bacilli 0.6–0.8 : 1.2–2.0 μ, or longer rods; with alkaline methylene blue a polar stain. Flagella delicate, 3–4 times the length of the rod, peritrichic.
Gelatin colonies. Deep: round to oval, brown, homogeneous, or centre somewhat darker. *Surface:* colon-like, round, flat, irregular, grayish.
Gelatin stab. In depth, growth dense, gray, white, beaded; surface growth flat, rather small, white.
Agar slant. Growth grayish white, translucent to opaque, moist, glistening, and slimy.
Bouillon. A moderate or good growth, with much white sediment. Reaction not altered (Karliński).
Potato (alkaline). Growth straw-yellow to light brown, usually abundant. On acid potato the growth is scanty and white (Karliński).

Milk. Not coagulated, reaction unchanged or alkaline; medium rendered, opalescent.

Glucose bouillon. Gas; reaction acid. According to Karliński, gas is produced in glucose bouillon with bacilli fresh from the body, but is inconstant in cultures. No gas in saccharose bouillon.

Litmus milk. Unchanged, or a deeper blue. *Indol* and *phenol* not produced.

Pathogenesis. Pathogenic to mice and rabbits; death in 7-12 days; spleen enlarged, in liver necrotic spots, kidneys inflamed, bacilli in the organs. More attenuated varieties cause only an infiltration and ulceration of Peyer's patches and an infiltration of lymph glands.

Habitat. Associated with hog cholera. For variations of this species, see Smith: U. S. Dept. of Ag., Bureau of Animal Industry, Bull. 6, 1894, pp. 8-27.

14. **B. levans** Lehmann-Wolffin

Archiv f. Hygiene, XXI, 1894.

Morphology. Bacilli 0.6 : 1.8 μ, with numerous long flagella. Cultural characters like *B. coli*.

Glucose bouillon. Gas; H : CO_2 : : 1 : 3. No gas in saccharose bouillon. Lactic acetic and butyric acids in glucose bouillon.

Habitat. Isolated from sour dough.

15. **B. loxiacida** Tartakowsky

Archiv d. Veterinarwissenschaft, 1888.

Morphology. Bacilli 0.6-1.0 : 2.0-2.5 μ. Not stained by Gram's method.

Gelatin colonies. Surface : 2 mm., round to irregular, with entire borders; microscopically, gray-brown, radiately fibrous to granular on the border. Colonies become crumpled when dry.

Gelatin stab. A slight amount of gas in depth.

Agar slant. Growth abundant, moist, white.

Bouillon. Turbid, with an easily disturbed membrane.

Potato. A scanty growth. No growth on acid potato.

Blood serum. Growth moist, white.

Milk. Not coagulated.

Pathogenesis. Pathogenic to birds. Subcutaneous inoculation of guinea pigs cause a slight local swelling, with elevation of temperature. Intraperitoneal injections cause sero-fibrinous peritonitis and death in 1-2 days.

Habitat. Associated with an infectious disease of titmouse, crossbill, goldfinch, and canary birds.

16. **B. morbificans** Basenau

Archiv f. Hygiene, XX, 1894, 242.

Morphology. Bacilli 0.3–0.4 : 1.0–1.2 μ.

Gelatin colonies. Surface: papillate, yellowish, to flat and spreading; border
    erose. Microscopically, the colonies have a dark contour, within which
    is a clear zone and within this a yellowish granular to mottled centre.

Gelatin stab. Growth in depth filiform; surface growth thick, round, white,
    with an undulate border.

Potato. Growth moist, yellow, never brown.

Bouillon. Turbid, with a membrane.

Glucose bouillon. A small amount of gas. No gas in saccharose bouillon.

Litmus milk. Unchanged.

Pathogenesis. Pathogenic to mice, guinea pigs, and rabbits by subcutaneous
    and intraperitoneal inoculation, and by feeding. Calves and goats
    infected by feeding. Bacilli in the organs and in the muscles. Com-
    municated through infected meat.

Habitat. Isolated from the flesh of a cow with puerperal fever.

17. **B. Silberschmidii**

B. der Fleischvergiftung Silberschmidt: Correspondenz-Blatt für Schweizer
Aerzte, 1896, No. 8.

Morphology. Bacilli short rods. Flagella 4–8.

Milk. Not coagulated; no acid. Cultures have a faint sweetish odor.

Pathogenesis. Mice, guinea pigs, and rabbits were fed on infected meat with
    negative results. Intraperitoneal inoculation of guinea pigs caused death
    in 18–36 hours.

Habitat. Isolated from poisonous meat.

18. **B. murium** Löffler

B. typhi-murium Löffler: Centralblatt f. Bakteriol., XI, 1892, 129.

Morphology. Bacilli like *B. typhosus.*

Gelatin colonies. Deep: small, round, slightly granular, yellow-brown.
    Surface: like *B. typhosus.*

Gelatin stab. Surface growth flat.

Potato. Growth whitish-grayish. *Indol* and *phenol* production doubtful.

Milk. Rendered alkaline.

Pathogenesis. Subcutaneous inoculation of house mice and field mice cause death in 3 days. Bacilli in the organs; spleen enlarged. The latter also infected by feeding.

Habitat. Found by Löffler as the cause of an epidemic in mice.

19. B. intestinalis Dyas-Keith

Mass. Inst. of Technology Quarterly, VI, 3; ref. Centralblatt f. Bakteriol., XVI, 1894, 838.

Morphology. Bacilli 1.0 : 1.2 μ; somewhat thicker than *B. coli*. Grow at 37°. Cultural characters like *B. coli*.

Pathogenesis. Doubtful.

Habitat. Isolated from the excrement of the horse.

20. B. meningitidis Neumann-Schaeffer

Virchow's Archiv, CIX, 1887, 477.

Morphological and cultural characters like *B. typhosus*.

Potato. Growth gray white, viscid. No gas in lactose bouillon.

Pathogenesis. Subcutaneous inoculation of guinea pigs causes a pyogenic reaction.

Habitat. Isolated from a case of purulent meningitis.

21. B. typhosus Zopf

B. der Abdominaltyphus Eberth: Virchow's Archiv, LXXXI, 1880.
B. typhosus Zopf: Spaltpilze, 1885, 124.

Morphology. Bacilli 0.5–0.8 : 1–3 μ — filaments. Flagella peritrichic, 8–14, long, undulate.

Gelatin colonies. Deep : round, gray to yellowish brown, entire. *Surface :* at first small, punctiform, becoming flat, roundish, gray, glistening, with irregular borders ; microscopically, colorless, translucent, becoming grayish yellow, darker in the centre, marmorated ; border undulate to lobate ; strongly refracting.

Gelatin stab. Growth in depth filiform — beaded — tuberculate ; on the surface, growth thin, whitish, irregular.

Bouillon. Slightly turbid, less so than *B. coli*.

Milk. Not coagulated, only slightly acid.

Potato. Growth a pure white glistening streak, not very thick, or scarcely visible.

Agar slant. Growth thin, translucent, slimy, spreading.

Litmus milk. After some time a slight acid reaction; variable. *Indol* not produced. Nitrates reduced to nitrites. H₂S produced. Lactic acid produced in glucose bouillon.

Pathogenesis. Inoculation of experimental animals with moderate quantities usually negative; with large quantities death by toxæmia. Filtered cultures toxic to test animals.

Habitat. In the spleen in cases of typhoid fever; also in greater or less numbers in the intestinal lesions, mesentery glands, liver, bile, kidneys, etc.; also in the stools of typhoid patients, and in infected water.

22. B. pseudo-typhosus Kruse

Flügge, Die Mikroorganismen, 1896, 383.

Morphological and cultural characters identical with the preceding. Differentiated by the absence of the serum reaction. (See Zeitsch. f. Hygiene, XXI, 238.)

Habitat. Isolated by Pansini from a liver abscess, and by Lösener from the peritoneal fluid of a hog, water, etc.

23. B. Billingsi

Bacillus of corn-stalk disease of cattle Billings: Baumgarten's Jahresbericht, 1889, 184.

Morphology. Bacilli identical with *B. typhosus.* Cultural characters indistinguishable from *B. typhosus.*

Pathogenesis. Inoculations of mice, guinea pigs, and rabbits cause general septicæmia.

Habitat. Isolated by Billings from corn-stalk disease of cattle, and by Nocard from bronchopneumonia in oxen.

24. B. paradoxus Kruse

Typhus ähnlicher Bacillus Kruse-Pasquale: Zeitsch. f. Hygiene, XVI, 1894, 19.
B. paradoxus Kruse: Flügge, Die Mikroorganismen, 1896, 373.

Morphological and cultural characters like *B. typhosus.*

Potato. Growth spreading, invisible.

Lactose bouillon. No gas. *Indol* produced. Pathogenic for mice.

Habitat. Isolated from the liver from a case of dysentery in Alexandria.

25. B. pestis (Lehmann-Neumann)

Cocco-bacille de la peste Yersin: Ann. Pasteur Inst., 1894, 666.
Pest Bacillus Aoyama: Centralblatt f. Bakteriol., XIX, 1896; Zettnow, Zeitsch. f. Hygiene XXI, 1895-96, 165.
Bact. pestis Lehmann-Neumann: Bakt. Diag., 1896, 194.
B. pestis-bubonicæ Kruse: Flügge, Die Mikroorganismen, 1896, 429.

Morphology. Bacilli short, ovals to longer rods, 4-5 times their breadth; also chains of short elements. A uniform or polar stain. Flagella demonstrated by Gordon, one at the end, often one at the side, long, spiral. Grow better at room temperatures than at 37°.

Gelatin colonies. Deep: round, white to yellowish white. *Surface:* flat, with a granular border; do not grow larger than a pin's head. According to Klein, the colonies are small, gray, round to angular points, similar to young colonies of *B. vulgaris.*

Gelatin stab. In depth, growth often arborescent, like anthrax (Klein); surface growth flat.

Agar slant. Growth composed of confluent viscid colonies.

Potato. A scanty growth, white to gray.

Bouillon. Shows a turbidity of various grades, or with flocculent particles adhering to walls of the tube.

Milk. Not coagulated.

Litmus milk. Reddened in 24 hours. According to Klein, unaltered. *Indol* produced. Nitrates not reduced to nitrites.

Pathogenesis. Pathogenic to mice, rats, guinea pigs, and rabbits. An œdema at the point of inoculation; swelling of lymph spaces and congestion of the internal organs. Death in a few days; bacilli in the blood and organs.

Habitat. Isolated from suppurating glands, etc., in bubonic plague.

26. B. solitarius Ravenel

l.c., 29.

Morphology. Bacilli slender, straight, 3-7 times their breadth; ends rounded; occur singly. Aerobic. Bacilli rotatory, non-progressive; flagella not demonstrated.

Gelatin colonies. Deep: round, white, slightly granular; margins notched. Later the colonies show a zoned and marmorated structure. *Surface:* round, grayish, floccose to filamentous. In 70 hours, the colonies have a diameter of 1 mm.; show a gray-white central nucleus, with an irregular, indistinct, filamentous border, and an outer orange zone. In 7 days, the colonies are round, white, entire, and elevated.

Agar slant. Growth moist, glistening, porcelain-white, spreading.

Gelatin stab. In depth, growth filiform; surface growth elevated, umbilicate, 1 mm. in diameter.

Potato. Growth thin, whitish, becoming thick, pasty, and the color of putty.

Bouillon. Turbid; no pellicle.

Pepton rosolic acid. Decolorized in 7 days; alkaline.

Litmus milk. Becomes darker, and afterward is decolorized. *Indol* not produced. Grow at 35°-37°.

Habitat. Soil.

27. B. geminus Ravenel

B. geminus-minor Ravenel: l.c.

Morphology. Bacilli very short rods, with rounded ends, 2-4 times their breadth; occur singly. Motility slight; flagella not demonstrated.

Gelatin colonies. *Deep:* round, yellowish, granular, entire. *Surface:* yellowish, granular, entire, becoming in 1-2 days elevated, convex, dense, and pearly white. Colonies small, 1.5 mm.

Gelatin stab. A filiform growth in depth; on the surface, a bead, becoming larger and more spreading, with corrugated edges.

Potato. Growth thin, spreading, becoming, in 10 days, dirty white, moist, and glistening.

Bouillon. Turbid, with a slight pellicle.

Pepton rosolic acid. Becomes cherry-red in 10 days.

Litmus milk. Rendered alkaline. *Indol* produced. Nitrates reduced to nitrites. Grow at 35°-36°.

Habitat. Soil.

28. B. aquatilis-sulcatus-quartus Weichselbaum

Österreichische Sanitätswesen, 1889; Dyar, l.c., 359.

From descriptions (Dyar), indistinct from the preceding.

Habitat. Water.

29. B. primus-Fullesi Dyar

l.c., 360.

From descriptions, indistinct from *B. geminus*, except milk cultures emit a disagreeable odor.

Habitat. Water.

30. B. tracheiphilus Smith

Centralblatt f. Bakteriol., I, 2d Abt., 1895, 364.

Morphology. Bacilli 0.5–0.7 : 1.2–2.5 μ. Grow poorly in gelatin.

Agar slant. Growth thin, smooth, moist, glistening, milky-white, limited.

Agar stab. Lateral outgrowths from the line of puncture ; on the surface the growth is thin.

Potato. Growth thin, smooth, moist, glistening ; the color of the potato. Does not grow in alkaline media.

Bouillon. Turbid cloudy, but not turbid.

Glucose bouillon. Acid ; no gas. Cultures very viscid. No growth at 37°.

Habitat. Associated with a disease of melons and curcurbits.

31. B. pinatus Ravenel

l.c., 32.

Morphology. Bacilli slender, short, 3–5 times their breadth ; occur singly and in short chains. Motile ; flagella not demonstrated. Facultative anaerobic.

Gelatin colonies. Deep : round, entire, yellowish, granular. *Surface :* punctiform ; do not exceed 1 mm. ; centre yellowish brown, granular, border clear.

Gelatin stab. A good growth in depth ; on the surface, growth raised, 2 mm., porcelain-white, umbilicate.

Agar slant. Growth thin, glistening, watery.

Potato. Growth thin, colorless to light, dirty brown, smooth, moist, glistening.

Bouillon. Turbid, with flakes ; no pellicle on the surface.

Pepton rosolic acid. Decolorized, alkaline.

Litmus milk. Becomes a darker indigo blue. *Indol* produced. Nitrates reduced to nitrites.

Habitat. Soil.

32. B. Raveneli

B. geminus-major Ravenel: l.c., 27.

Morphology. Bacilli thick, ends rounded, of variable length ; occur singly and in short chains. Rods show deeply stained points, 2–3 in each rod. Motility slight ; flagella not demonstrated. Aerobic.

Gelatin colonies. Deep : brownish, granular, entire. *Surface :* like typhoid, but more granular and coarser.

Gelatin stab. Delicate offshoots from the line of stab ; on the surface, growth thin, spreading, with irregular borders.

Agar slant. Growth thin, translucent, spreading.

Potato. Growth honey yellow, moist, glistening, becoming chocolate-brown.

BACTERIOLOGY
Bouillon. Turbid, becoming clear.
Pepton rosolic acid. Slightly darker in 10 days.
Litmus milk. Amphoteric to slightly alkaline. *Indol* not produced.
Habitat. Soil.

33. B. alcaligenes Petruschky

Centralblatt f. Bakteriol., XIX, 1896, 187.

Morphology. Bacilli have perithrichic flagella. Indistinguishable from *B. typhosus*, except in the alkaline reaction in milk (not invariably constant). Differentiated also by the serum reaction.
Potato. Growth brown.
Habitat. Fæces.

34. B. Friedbergensis Gaffky-Paak

See No. 37.

35. B. solanacearum Smith

U. S. Dept. of Ag., Div. Veg. Path. Bull., XII, 1896.

Morphology. Bacilli 0.6–1.0 μ, variable; flagella several.
Gelatin colonies. Deep: round, yellowish to brownish, granular, entire. *Surface:* round, thin, white.
Gelatin stab. In depth, growth scanty; on the surface, growth thin, white.
Gelatin slant. Growth white, smooth, moist, glistening, with finger-like extensions into the gelatin.
Agar slant. Growth smooth, white, moist, glistening, becoming yellowish brown to brown; agar stained brown.
Bouillon. Zoöglœa in the upper layer; uniform turbidity on shaking.
Potato. Growth dirty white, becoming brownish to smoke-black.
Milk. Not coagulated, slowly saponified to a yellowish, translucent fluid.
Litmus milk. Rendered alkaline. Grow at 40°.
Glucose bouillon. No acid or gas production.
Bouillon. Rendered alkaline.
Habitat. Associated with a disease of tomato, egg-plant, and the Irish potato.

36. B. Weichselbaumii

B. aquatilis-sulcatus No. 5, Weichselbaum: Österreichische Sanitätswesen, 1889.
B. aquatilis-sulcatus Kruse: Flügge, Die Mikroorganismen, 1896, 410.

Morphological and cultural characters like *B. typhosus.* Diff. aerobic; in depth, in gelatin stab cultures, little or no growth. Do not reduce nitrates. Non-pathogenic.
Habitat. Water.

37. B. Friedbergensis Kruse

B. der Friedberger Fleichvergiftung Gaffky-Paak: Mitteilungen a. d. Kaiserl. Gesundheitsamte, 1890, 159.
B. Friedbergensis Kruse: Flügge, Die Mikroorganismen, 1896, 378.

Morphology. Bacilli about one-third smaller than *B. typhosus.*
Gelatin colonies. Deep : round, yellowish, homogeneous, often concentric.
    Surface : round, spreading; centre yellowish, border paler, marmorated;
    between aerogenes and colon types.
Gelatin stab. Good growth in depth; surface growth thin, spreading to the
    walls.
Agar slant. Growth grayish white, slimy.
Potato. Growth whitish to grayish yellow to reddish.
Pathogenesis. Pathogenic to mice, guinea pigs, and rabbits, by subcutaneous
    inoculation. Pathogenic to guinea pigs and mice by feeding; slightly so
    to dogs, cats, and rabbits.
Habitat. Isolated from poisonous sausage, in meat poisoning.

38. B. phasiani Kruse

Bacillus of an infectious disease of young pheasants Klein: Jour. of Path. and Bacteriol., 1893.
    B. phasiani-septicus Kruse: Flügge, Die Mikroorganismen, 1896, 410.

Morphology. Bacilli like *B. coli,* but smaller. Cultural characters like those
    of *B. coli.*
Pathogenesis. Inoculation of pheasants causes death in 24 hours of general
    septicæmia; fowls, pigeons, guinea pigs, and rabbits refractory.
Habitat. Associated with the above disease.

39. B. Schafferi v. Freudenreich

Ann. Micrographie, III, 1891.

Morphology. Bacilli 1.0 : 2–3 μ; threads.
Gelatin colonies. Deep : small, round, yellowish, granular. *Surface :* white
    (porcelain), spreading, slightly irregular.
Agar slant. Growth grayish to brownish.
Potato. Growth yellowish.
Habitat. Milk, cheese.

40. B. rugosus

B. No. 27 Conn: l.c., 1893, 54.

Morphology. Bacilli 0.8 : 1.3–2 μ. Grow at 35°.
Gelatin colonies. Transparent, elevated, spreading, wrinkled on edges.
Gelatin stab. In depth, growth beaded; on the surface, growth thin, transparent.
Agar slant. Growth white, elevated.
Potato. Growth thick, yellowish, spreading.
Milk. Alkaline, bad odor; peptonized to a brownish fluid.
Bouillon. Turbid, with a pellicle and sediment.
Habitat. Milk.

41. B. avisepticus

B. der Kanarienvögelseptikämie Rieck: Deutsche Zeitsch. f. Thier. Med., 1889.

Morphology. Bacilli 1.2–2.5 μ long; show a polar staining.
Potato. Growth grayish yellow, otherwise in cultures like *B. choleræ-gallinarum.*
Pathogenesis. Subcutaneous inoculations of mice cause septicæmia; of canary birds, sooty discoloration of the skin, liver necrosis, and septicæmia.

42. B. avium Kruse

B. de la diphthérie aviaire Loir-Duclaux: Ann. Pasteur Inst., 1894, 599.
B. diphtheriæ avium Kruse: Flügge, Die Mikroorganismen, 1896, 410.

Morphology. Bacilli like those of fowl cholera. Cultural characters not fully described.
Pathogenesis. Pathogenic for all kinds of birds. Subcutaneous inoculations of rabbits cause septicæmia, with death in 6–10 days; an exudate in pharynx.
Habitat. The cause of an epizootic among chickens, pigeons, turkeys, and canary birds in Tunis.

43. B. meleagris

B. of epizootic pneumo-pericarditis in the turkey MacFadyean: Jour. of Comp. Path. and Therap., VI, 1893, 334.

Morphology. Bacilli like those of fowl cholera.
Pathogenesis. Pathogenic to turkeys: nasal catarrh, "rattles" in the throat, pneumonia, and pericarditis; bacilli in the lungs and organs. Only slightly pathogenic to rabbits and guinea pigs.

44. B. tetraonis

B. of grouse disease E. Klein: Centralblatt f. Bakteriologie, VI, 1889, 593.

Morphology. Bacilli 0.4–0.6–1.0 μ, often coccoid.
Gelatin colonies. Deep: small, round. *Surface:* thin, spreading, irregular.
Gelatin stab. Surface growth flat.
Pathogenesis. By subcutaneous inoculations it is pathogenic to mice in 75 per cent and to guinea pigs in 50 per cent of the cases. Lungs hyperæmic, hepatized; spleen not enlarged; kidneys hyperæmic; bacilli in the blood and organs. In grouse: pneumonia, local hyperæmia of the intestines, enlargement of liver and kidneys. Bacilli in the blood and organs.

45. B. cygneus

Septikämie bacillus der Schwäne Fiorentini: Centralblatt f. Bakteriologie, XIX, 1896, 929.

Morphology. Bacilli like those of fowl cholera but larger, 0.5 : 1.5–2 μ threads.
Gelatin colonies. Granular, concentric; the border radiate ciliate.
Gelatin stab. In depth, a beaded growth; on the surface, growth white, lobed, and toothed.
Agar slant. Round, white colonies, becoming coalescent.
Potato. Growth of colonies, becoming coalescent, elevated, yellowish brown, with a bad odor.
Bouillon. Turbid, with a white sediment.
Pathogenesis. Pathogenic to rabbits, guinea pigs, hens, pigeons, geese; a comatose condition, followed by death. In affected swans, œdematous infiltration of the lungs; ecchymoses of serous membranes, hyperæmia of intestinal mucous membrane, and turbid degeneration of liver cells.

46. B. aerobius

Ein neuer B. des malig. Oedems Klein: Centralblatt f. Bakteriologie, X, 1891, 186.
B. pseudo-oedematis maligni Sanfelice: Zeitsch. f. Hygiene, XIV, 1893, 353.
B. oedematis-aerobius Kruse: Flügge, Die Mikroorganismen, 1896.

Morphology. Bacilli 0.7 : 1.6–2.4–24 μ.
Gelatin colonies. Deep: round, brownish. *Surface:* thin, spreading, typhoid-like, marmorated.
Gelatin stab. In depth, beaded; on the surface, growth thin, transparent, dentate.
Agar slant. Growth grayish white, smeary.
Bouillon. Turbid; no pellicle.

Potato. Growth viscid, yellowish.

Pathogenesis. Cultures quickly lose their virulence. Fresh, first generation cultures kill guinea pigs in 1 cc. bouillon culture doses. There is bloody œdema, with gas; a reddening of the muscles, and an enlargement of the liver and spleen.

Habitat. Isolated from guinea pigs which have been inoculated with fæces, earth, dust, etc.

47. B. pneumosepticus Kruse

Pneumonie bacillus Klein: Centralblatt f. Bakteriologie, V, 1889, 625.
B. pneumosepticus Kruse: Flügge, Die Mikroorganismen, 1896, 408.

Morphology. Bacilli 0.3–0.4 : 0.8–1.6 μ; in chains. Show the polar stain.

Gelatin colonies. *Deep:* small, round. *Surface:* thin, iridescent, spreading, erose.

Gelatin stab. Surface growth flat.

Agar slant. Growth whitish-brownish.

Potato. Growth thin, slimy, brownish.

Pathogenesis. Subcutaneous inoculation of mice is fatal in 60 per cent of the cases. Inflammation at the point of inoculation and in the lungs; spleen enlarged; hemorrhagic enteritis. Guinea pigs die in 25 per cent of the cases, with lobular pneumonia, pleuritis, etc.

Habitat. Isolated from rusty sputum of pneumonia patients.

48. B. monachæ v. Tubeuf

Forst. Naturwiss. Zeitsch. I, 1892, 34; ref. Centralblatt f. Bakteriologie, XII, 1892, 268.

Morphology. Bacilli 0.5 : 1.0 μ; occur singly, in twos, and in short chains.

Acid gelatin colonies. Transparent, opalescent, with mother-of-pearl lustre; microscopically, central portion ochre-yellow, sometimes zoned; edge erose, lobed.

Bouillon. Turbid.

Potato. Growth moist, gray.

Pathogenesis. Infection experiments positive.

Habitat. Found in the body fluids of diseased "nun-moth" larvæ (*Liparis monacha*).

49. B. cuniculi Lucet

B. septicus-cuniculi Lucet: Ann. Pasteur Inst., 1892, 558.

Morphology. Bacilli 1–3 μ.

Gelatin colonies. Smooth, very convex, slimy.

Bouillon. 39°–40°, growth in stringy masses. No growth on potato.

Pathogenesis. Subcutaneous inoculations of rabbits cause death in 24 hours; a local œdema; serous membrane inflamed; spleen enlarged; bacilli in all the organs. Subcutaneus inoculation of guinea pigs causes local abscess formation; intraperitoneal inoculation causes death. Chickens and pigeons refractory.

Habitat. Associated with a spontaneous epizootic of rabbits.

50. B. venenosus Vaughan

Am. Jour. Med. Sci., 1892, 107.

Morphology. Bacilli 2–4 times their breadth; ends rounded.
Gelatin colonies. Aerogenes-like.
Agar slant. Growth thin, white.
Potato. Growth moist, glistening, light brown. In Parietti's solution a good growth. In Uffelman's gelatin a feeble to a good growth.
Pathogenesis. Pathogenic to rats, mice, guinea pigs, and rabbits.
Habitat. Water.

51. B. glischrogenus Malerba

Malerba and Sanna-Salaris: Lavori Esequiti nell Institute fisiologico, Napoli, 1888.

Characters of *Bact. aerogenes.* Milk and urine rendered slimy.
Pathogenesis. Pyogenic to the smaller animals. Causes nephritis in dogs.
Habitat. Isolated from urine.

52. B. albus Paglinni

Giorn della Soc. Ital. d'igiene, IX, 1887, 587.
Weisser bacillus Eisenberg: Bak. Diag., 1888, 38.

Morphology. Bacilli short.
Gelatin colonies. Small, aerogenes-like.
Potato. Growth rugose, yellowish white, limited.
Habitat. Water.

53. B. granulatus

B. aquatilis-solidus Lustig: Diag. Bak. des Wassers, 1893.

Morphology. Bacilli 3 times their width.
Gelatin colonies. Aerogenes-like, granular.
Potato. Growth grayish white to yellowish. Nitrates reduced to nitrites.
Habitat. Water.

54. B. stolonatus Adametz

Mitth. Oest. Versuchstat. f. Braueri u. Malz. Wien, 1888, 844.

Morphology. Bacilli 2–3 times their breadth.
Gelatin colonies. Capitate, whitish-brownish.

Agar slant. Growth composed of large rhizoid colonies.
Potato. Growth dirty white.
Habitat. Water.

55. B. invisibilis Vaughan

Am. Jour. Med. Sci., 1892, 107.

Morphology. Bacilli large, ends rounded, 2–5 times their breadth. Grow at 37°.
Gelatin colonies. Pale yellow, burr-like, with irregular outlines.
Gelatin stab. A good growth in depth; a scanty growth on the surface.
Agar slant. Growth thick, white, limited.
Potato. Growth invisible. Grow in Parietti's solution and in Uffelmann's gelatin.
Habitat. Water.

56. B. venenosus Vaughan

B. venenosus-invisibilis Vaughan: Am. Jour. Med. Sci., 1892, 107.

Not clearly differentiated from the above.

57. B. murinus

B. of rat plague Issatschenko: Centralblatt f. Bakteriol., XXIII, 1898, 873.

Morphology. Bacilli variable in size. Flagella peritrichic.
Gelatin colonies. Round, brownish yellow, with coli-like structure.
Gelatin stab. A good growth in depth, with outgrowths; on the surface, a white layer.
Bouillon. A white scum and sediment.
Potato. Grows slowly, becoming in 6 days a bright yellow, scarcely visible layer.
Pathogenesis. Strongly pathogenic to rats and mice. By ingestion death of rats in 8–14 days; of mice, in 4–8 days. Non-pathogenic to rabbits and pigeons.
Habitat. Isolated from the spleen and liver of rats attacked in St. Petersburg by a plague.

58. B. denitrificans Burri-Stutzer

B. denitrificans, I, Burri-Stutzer: Centralblatt f. Bakteriol., I, ate Abt., 1895, 356.

Morphology. Bacilli; on agar small rods, 0.5 : 0.5–1.0 μ; in bouillon, 1.0–2.5 μ. Motility rarely progressive.
Gelatin colonies. Deep : small, white, entire. Do not generally appear at the surface; when they do they are soft, dry, with a translucent border, which is puckered and torn.

Gelatin stab. Growth filiform in depth; surface growth white, scarcely visible, covers the entire surface.

Gelatin slant. A limited yellowish white stripe, with a thin colorless border, and with outgrowths.

Agar colonies. In 1-2 days at 30°, the deep colonies are small, white, and entire; the surface colonies thin, soft, colorless, limited, with erose borders.

Nitrate bouillon. A membrane on the surface; reduction to nitrites.

Habitat. Straw, earth, and horse manure.

59. B. Stutzeri (Lehmann-Neumann)

B. denitrificans, II, Burri-Stutzer: Centralblatt f. Bakteriol., IV, 1898.
Bact. Stutzeri Lehmann-Neumann, Bak. Diag., 1896, 2te Abt., 1898, 408.

Morphology. Bacilli; on agar rods with rounded ends, often spindle-shaped, 0.5-1.0 : 1.5-2.5-4 μ; in bouillon, only 0.25-0.3 μ thick. Polar stain.

Gelatin colonies. In 4 days the deep colonies are small, white, entire; the surface colonies thin, whitish, translucent, with somewhat ragged edges.

Agar colonies. In 2 days at 30° the deep colonies are small, yellowish white, entire; the surface colonies thin, 5-10 mm.

Agar slant. Growth soft, flat, dry, grayish, becoming rather moist, glistening.

Gelatin slant. Growth thin, milky-white, with fine radial striping and undulations.

Gelatin stab. In depth growth pearly; on the surface growth milky-white, soft, dry, becoming somewhat slimy.

Nitrate bouillon. A weak turbidity; in 2 days a membrane. Nitrates reduced.

Habitat. Straw, earth, air.

60. B. centropunctatus (Jensen)

Bact. centropunctatum Jensen: Centralblatt f. Bakteriol., IV, 2te Abt., 1898, 410.

Morphology. Bacilli; on agar coccoid, 0.3-0.5 μ, with a capsule. In nitrate bouillon, morphology like *B. Stutzeri.* In anaerobic cultures bacilli large, ovoid, 0.5-1.0 μ.

Gelatin colonies. In 4 days the deep colonies are small, white, entire; the surface colonies thick, moist, glistening, grayish, becoming whitish.

Agar slant. Growth thin, soft, becoming in 2 days thick, glistening, grayish, slightly raised in the middle.

Gelatin slant. Growth milky-white, slimy, glistening, with long outgrowths.

Q

Gelatin stab. In depth, growth pearly ; on the surface growth milky-white.

Nitrate bouillon. In 2 days at 30° a strong turbidity, and a membrane on the surface ; nitrates reduced.

Habitat. Isolated from cow and guinea pig manure.

61. B. agilis Amp.-Gar.

B. denitrificans-agilis Ampola-Garino : Centralblatt f. Bakteriol., II, 2te Abt., 1896, 673 ; Jensen, l.c., IV, 1898, 408.

Morphology. Bacilli 0.3–0.5 : 1.0–2.5 μ, or smaller. No polar stain.

Gelatin colonies. In 5 days scarcely visible ; white, entire, homogeneous.

Agar colonies. Small, white, slimy, not spreading.

Agar slant. Growth grayish white, slimy, limited, or spreading where the medium is moist.

Gelatin slant. Growth in 6–8 days grayish white, slimy, limited, somewhat knobby.

Gelatin stab. In depth, growth pearly ; on the surface the growth in 8 days is grayish white, 1 mm. in diameter.

Nitrate bouillon. In 1–2 days turbid, with a membrane ; nitrates reduced.

Habitat. Isolated from cow manure.

62. B. Hartlebii (Jensen)

Bact. Hartlebii Jensen : Centralblatt f. Bakteriol., IV, 2te Abt., 1898, 449.

Morphology. Bacilli on agar and in nitrate bouillon 0.7 : 2–3–4 μ. Polar stain rare.

Gelatin colonies. In 4–5 days the deep colonies are small, white, entire ; the surface colonies, 1 mm., white, translucent, with entire or erose borders. In 2–3 weeks the colonies are watery, slimy, and 1–3 mm. in diameter.

Gelatin stab. In depth, a pearly growth ; on the surface the growth is white, slimy, and 3–6 mm. in diameter.

Gelatin slant. Growth white, moist, glistening, raised, limited.

Agar slant. In 2–3 days at 30° growth thick, grayish white, moist, glistening, watery.

Agar colonies. In 3 days at 30° the deep colonies are small, white, entire ; the surface colonies translucent, watery.

Nitrate bouillon. Rendered strongly turbid, with a membrane ; nitrates reduced.

Habitat. Soil.

CLASS II. WITHOUT ENDOSPORES. AEROBIC AND FACULTATIVE
  ANAEROBIC. COLONIES ON GELATIN PLATES NOT AMEBOID
  OR PROTEUS-LIKE. GELATIN NOT LIQUEFIED. STAINED BY
  GRAM'S METHOD.

I. Colonies on gelatin plates flat, spreading, *B. coli*-like
    A. Gelatin surface distinctly coli-like; rather thick, and yellowish brown
      by transmitted light, or granular.
        1. Gas generated in glucose bouillon.
          63. *B. muripestifer* Kruse.
          64. *B. aerogenes* (Schow).
        2. No gas generated in glucose bouillon.
          65. *B. Shigæ*.
        3. Action on glucose bouillon not stated.
          66. *B. colorabilis* Kruse.
    B. Gelatin surface colonies thin, translucent.
        67. *B. exanthematicus* Kruse.
        68. *B. accidentalis* Kruse.
II. Colonies on gelatin of the aerogenes type.
        69. *B. endocarditis* Weichselbaum.

63. B. muripestifer Kruse

B. der Mäuseseuche Laser: Centralblatt f. Bakteriologie, XI, 1892, 184.
B. muripestifer Kruse: Flügge, Die Mikroorganismen, 1896, 432.

Morphology. Bacilli short rods, with polar stain. Flagella peritrichic.
Gelatin colonies. Deep: round, brownish. *Surface:* spreading, coli-like.
Gelatin stab. Surface growth flat.
Bouillon. Turbid, with a slight membrane.
Potato. Growth brownish.
Litmus milk. Acid.
Pathogenesis. Subcutaneous inoculations of mice and field mice cause death
  in 2 days, and by feeding in 3-10 days; bacilli in all the organs. Patho-
  genic to rabbits, guinea pigs, and pigeons. Very similar to *B. murium*,
  but distinguished by the Gram reaction.
Habitat. Associated with a plague of field mice.

64. B. aerogenes Schow

Coccobacillus aerogenes-vesicæ Schow: Centralblatt f. Bakteriologie, XII, 1892, 749.
Morphology. Bacilli short coccoid; no threads.
Gelatin colonies. The deep colonies are small, round, and yellowish; surface
  colonies, flat, yellowish white, glistening; borders irregular to erose, becoming the size of millet seed and yellowish.

Gelatin stab. In depth the growth is beaded, with gas production; surface growth like a gelatin-plate colony.

Gelatin slant. Growth white, glistening, waxy, with folded edges, rather broad.

Potato. Growth thick, light yellow, with a granulated surface, raised, 3–4 mm.

Bouillon. Turbid, becoming clear, with a heavy white sediment. In urine a turbidity, alkaline reaction, and gas, with an aromatic odor.

Pathogenesis. Inoculations into dogs result in cystitis.

Habitat. Isolated from urine in cystitis.

65. B. Shigæ

B. of Japanese dysentery Shiga : Centralblatt f. BakterioL., XXXIII, 1898, 599.

Morphology. Bacilli short; ends rounded. Morphology quite like *B. typhosus*, showing involution forms.

Gelatin colonies. Sharp, yellowish, finely granulated, never very flat like *B. typhosus.*

Agar colonies. In 24 hours quite large, round, moist, glistening, brownish, translucent, becoming larger and irregular.

Potato. Growth scarcely visible, white, dry, becoming reddish brown after some weeks.

Milk. Not coagulated.

Glucose bouillon. No gas. *Indol* not produced. Bacilli show a distinct agglutination reaction with serum of persons ill with dysentery, but not with serum of sound men.

Pathogenesis. Subcutaneous inoculation of guinea pigs gives a strong infiltration at the point of inoculation, with subsequent suppuration. Intraperitoneal injections cause blood extravasation or peritoneal hemorrhage. Dogs fed on cultures show in 1–2 days slimy stools. Subcutaneous inoculations of man result in chills and fever, headache, etc. The point of inoculation is strongly infiltrated and painful. The serum of man so treated possesses agglutinating properties.

Habitat. Isolated from the dejecta of 34 cases of Japanese dysentery.

66. B. colorabilis Kruse

B. coli-colorabilis Kruse : Flügge, Die Mikroorganismen, 1896, 434.

B. der Gallenblase Naunyn : Deutsch. med. Wochenschrift, 1891, No. 5.

B. cuniculicida-Havaniensis Sternberg : Manual of Bacteriology, 1892.

Morphology. Bacilli short, thick, like *Bact. aerogenes*, often in twos and short filaments.

Potato. Growth grayish to yellowish brown.

Pathogenesis. Subcutaneous inoculations of mice cause death by septicæmia. Guinea pigs but slightly affected; rabbits negative (?).

Habitat. Isolated from the contents of the gall bladder and from fæces, and from yellow fever cadavers by Sternberg.

67. B. exanthematicus Kruse

B. der Typhus exanthematique Babes-Oprescu: Ann. Pasteur Inst., 1891, 273.
B. exanthematicus Kruse: Flügge, Die Mikroorganismen, 1896, 426.

Morphology. Bacilli 0.3–0.5 μ thick, often very short, and in 8-shaped forms.
Gelatin colonies. The deep colonies are round and yellowish brown; the surface whitish, translucent, spreading, irregular.
Agar slant. Growth glistening, gray, translucent.
Potato. Growth gray-brown, translucent.
Bouillon. Turbid, with a sediment and a membrane.
Pathogenesis. Pathogenic to mice, guinea pigs, rabbits, and pigeons; death in 2–4 days; local inflammation, enlargement of the spleen, and a brownish color of the organs; bacilli present.
Habitat. Isolated from a case of hemorrhagic infection in man.

68. B. accidentalis Kruse

Eine neue pathogene Bakteriumart im Tetanusmaterial Belfanti-Pescarolo: Centralblatt f. Bakteriologie, IV, 1888, 513.
B. accidentalis-tetani Kruse: Flügge, Die Mikroorganismen, 1896.

Morphology. Bacilli small, short, with polar stain.
Gelatin stab. In depth, growth beaded; surface growth thin, iridescent.
Potato. Growth yellowish, glistening.
Pathogenesis. Pathogenic to mice, guinea pigs, and rabbits. Death in a few days; bacilli in the blood; spleen swollen; often paralysis, with convulsions.
Habitat. Isolated from the wound pus of a person dead of tetanus.

69. B. endocarditis Weichselbaum

B. endocarditis-griseus Weichselbaum: Ziegler's Beiträge, IV, 119.

Morphology. Bacilli typhoid-like in size; diphtheria-like in form.
Gelatin colonies. Aerogenes-like; like Friedlander's bacillus, but of a grayer color.
Potato. Growth dry, greenish yellow — yellowish brown.

Pathogenesis. Subcutaneous inoculations of mice and rabbits cause local inflammation and suppuration.

Habitat. Isolated from a case of endocarditis.

CLASS III. WITHOUT ENDOSPORES; AEROBIC ·AND FACULTATIVE ANAEROBIC; COLONIES ON GELATIN PLATES NOT AMEBOID OR PROTEUS-LIKE; GELATIN LIQUEFIED.

I. Grow well on nutrient gelatin.

    A. Colonies on gelatin at all stages round, with no radiations from their edges.

        1. Gelatin liquefied rather quickly.

            a. Gas generated in glucose bouillon. B. CLOACA GROUP.

            Milk coagulated.

                70. *B. cloaca* Jordan.

            Milk not coagulated.

                71. *B. fermentationis.*

            b. No gas generated in glucose bouillon; milk not coagulated.

            Growth on potato smooth.

            Liquefaction of the gelatin in stab cultures crateriform-stratiform.

                72. *B. formosus* Ravenel.

            Liquefaction of the gelatin in stab cultures funnel-formed.

                73. *B. stoloniferus* Pohl.

            Growth on potato rough or folded.

                74. *B. antenniformis* Ravenel.

            c. Gas production in glucose bouillon not stated.

                * Potato cultures reddish, pinkish, or flesh-colored.

                    75. *B. bucalis* Sternberg.

                ** Potato cultures yellowish to brownish.

                  In gelatin stab cultures a funnel-formed liquefaction.

                    76. *B. hydrophilus* Sanarelli.

                    77. *B. pyogenes, var. liquefaciens* Lanz.

                    78. *B. liquefaciens* Frankland.

                  In gelatin stab culture a crateriform liquefaction.

                    79. *B. Matazooni.*

                *** Potato cultures whitish to grayish.

                  Milk coagulated.

                  Growth and liquefaction of the gelatin more rapid than the evaporation in stab cultures.

80. *B. delictatulus* Jordan.

Evaporation more rapid than the growth and liquefaction of
the gelatin in stab cultures; funnel partly empty.

81. *B. circulans* Jordan.

Milk not coagulated.

82. *B. putidus.*

83. *B. albus-putidus.*

d. Gas produced in ordinary gelatin or bouillon.

* Grow at 37° and more or less pathogenic.

84. *B. tachyctonus* (Fischer).

85. *B. dubius* Kruse.

** Do not grow at 37°, water bacteria.

86. *B. gasofòrmans* Eisenberg.

2. Gelatin liquefied very slowly.

a. Gas generated in glucose bouillon.

* Milk coagulated.

87. *B. Kralii* Dyar.

88. *B. lactis.*

89. *B. tartaricus* Grimbert-Fiquet.

** No growth in milk (?).

90. *B. halophilus* Russell.

b. No gas generated in glucose bouillon.

91. *B. nitrificans* Burri-Stutzer.

c. Gas production in glucose bouillon not stated.

* Grow on potato.

Rods scarcely longer than broad, thick ovals.

92. *B. guttatus* Zimmerman.

Rods several times longer than broad.

93. *B. inunctus.*

** Do not grow on potato.

Evaporation equals or excels the evaporation of the gelatin,
causing cavities in the latter.

94. *B. litoralis* Jordan.

Liquefaction equals or excels the evaporation of the gelatin.

95. *B. superficialis* Jordan.

B. Colonies on gelatin with filamentous borders or radiate.

B. Devorans Group.

1. Grow well upon potato.

a. Margins of gelatin colonies fibrillous-floccose.

* Growth on agar dry, dull, tough, becoming rough, warty.

96. *B. hyalinus* Jordan.
** Growth on agar thin, smooth, glistening.
    97. *B. pestifer* Frankland.
b. Gelatin colonies rosulate.
    98. *B. radiatus.*
    99. *B. reticularis* Jordan.
2. Little or no growth on potato.
    a. Gelatin stab cultures becoming an empty funnel from the evapora-
        tion of the slowly liquefying gelatin; rods short.
        100. *B. devorans* Zimmerman.
    b. Gelatin stab cultures not characterized as before; rods long
        and slender.
        101. *B. aquatilis* Frankland.
C. Colonies on gelatin erose-lobed, coli-like. B. DIFFUSUS GROUP.
    1. Grow on potato.
        a. Potato growth grayish to yellowish.
        102. *B. diffusus* Frankland.
        b. Potato growth yellowish brown; characters like *B. coli.*
        103. *B. sulcatus* Kruse.
    2. No growth on potato.
        104. *B. Havaniensis* Sternberg.
D. Colonies on gelatin irregular — fragmentary.
    105. *B. leporis.*
II. Grow poorly on ordinary nutrient gelatin unless urea is added.
    106. *B. Madoxi* (Miquel).

70. B. cloacæ Jordan

Report of the Mass. State Board of Health, 1890, 836.

Morphology. Bacilli 0.7–1.0 : 0.8–1.9 μ. Grow at 37°.
Gelatin colonies. The deep colonies are round and yellowish; the surface
    colonies are thin, bluish, entire — erose, with a dark centre and a clear
    outer zone; liquefaction crateriform.
Gelatin stab. Liquefaction napiform.
Bouillon. Turbid, with a slight membrane.
Agar slant. Growth porcelain-white.
Potato. Growth yellowish.
Milk. Acid. $\dfrac{H}{CO_2} = \dfrac{1}{3}$.

Habitat. Water, sewage. Moore (U. S. Dept. of Ag. Bureau of Animal Industry, Bull. 10, 1896, 45) holds *B. zeæ* Burrill (bacterial disease of corn) to be identical with *B. cloacæ.* The corn bacillus as described by Moore is characterized as follows: Bacilli 1.3–2 μ; occur singly, usually in short chains and clumps.

Agar slant. Growth grayish, viscid.

Gelatin stab. Gelatin slowly liquefied along needle track; on the surface the growth is grayish, beneath which the gelatin is softened; liquefaction occurs slowly; liquefied gelatin clear, with a viscid, grayish sediment; reaction alkaline.

Potato. Growth dull, grayish, not viscid.

Bouillon. Turbid in 24 hours.

Milk. Coagulated in 18 days.

Glucose bouillon. Gas and acid.

Saccharose bouillon. Gas.

Lactose bouillon. Gas developed more slowly.

Pathogenesis. Non-pathogenic to mice, guinea pigs, and rabbits.

71. B. fermentationis

B. fœtidus-liquefaciens Tavel: Ueber Aetiol. der Strumitis, Basel, 1892.

Gelatin stab. Liquefaction along the line of stab, with a bad odor.

Glucose bouillon. Gas.

Milk. Not coagulated.

Bouillon. Turbid, with a membrane.

72. B. formosus Ravenel

l.c., 12.

Morphology. Bacilli slender; ends rounded, 7–11 times their breadth. Motility slight.

Gelatin colonies. The deep colonies are round, entire, yellowish, and slightly granular; the surface colonies are round, entire, yellowish; centres gray, edges granular; later a concentric structure.

Agar slant. Growth white, moist, glistening, limited; edges notched.

Gelatin stab. Growth crateriform, becoming stratiform.

Potato. Growth moist, white, spreading to creamy.

Bouillon. Turbid, with a sediment.

Litmus milk. Alkaline, becoming decolorized in 10 days. *Indol* negative. Grow at 37°; optimum 20°.

Habitat. Water.

73. B. stoloniferus Pohl

Centralblatt f. Bakteriologie, XI, 1892, 142.

Morphology. Bacilli 0.8–1.2 μ.

Gelatin colonies. Round; borders sharp.

Gelatin stab. A funnel.

Agar slant. Growth white, thick, with streaming outgrowths.

Potato. Growth of small, pin-head colonies, which are spreading.

Lactose bouillon. No gas.

Litmus gelatin. Red.

Milk. Amphoteric.

Litmus milk. Unchanged in 24 days. *Indol* and *phenol* negative. Non-pathogenic.

Habitat. Swamp water.

74. B. antenniformis Ravenel

l.c., 25.

Morphology. Bacilli large straight rods with rounded ends, 8–10 times their breadth; occur singly. Actively motile.

Gelatin colonies. The deep colonies, oval, yellowish, granular; from the poles fine short projections like the antennæ of insects, disappearing in 36 hours. Surface colonies small, with orange-brown centres, with a fringe of wavy lines; border colorless, of parallel filaments, dentate; liquefaction crateriform, with a pellicle, becoming folded. In 7 days the colonies are 6 mm. in diameter, circular, with entire borders.

Gelatin stab. Crateriform, becoming stratiform.

Agar slant. Growth thin, smooth, grayish.

Potato. Growth invisible in 2 days; in 3–4 days the growth is spreading, and thrown into fine folds like herpetic vesicles, becoming putty-colored and dryer, and folds more numerous.

Bouillon. Scanty growth.

Litmus milk. Decolorized, becoming watery and slightly acid. *Indol* negative.

Habitat. Water.

75. B. bucalis Sternberg

B. j *Vignal:* Archiv Phys., VIII, 1886, 342.

B. *bucalis-fortuitus* Sternberg: Manual of Bacteriology, 1892, 685.

Morphology. Bacilli with square ends, 1.4–3.0 μ, often in pairs joined at an angle.

Gelatin colonies. In 48 hours small, round, white, becoming liquefied in 4–5 days.

Gelatin stab. Scanty growth in depth ; surface growth punctiform, becoming in 4–5 days spread over the entire surface ; liquefaction stratiform.
Agar slant. Growth of small, white, opaque colonies.
Bouillon. Turbid, with a membrane.
Potato. Growth thick, slightly spreading, pinkish.
Habitat. Isolated from saliva.

76. B. hydrophilus Sanarelli

B. hydrophilus-fuscus Sanarelli : Centralblatt f. Bakteriologie, IX, 1891, 222.

Morphology. Bacilli 0.6 : 1.3 μ — filaments.
Gelatin colonies. Round, translucent.
Agar slant. Growth bluish gray to brownish, thin.
Bouillon. Turbid, with a slight membrane.
Potato. Growth yellowish brown.
Pathogenesis. Subcutaneous inoculation causes hemorrhagic septicæmia in frogs, salamanders, fish ; also in mice, guinea pigs, and rabbits.
Habitat. Isolated from water, and from frogs dead of septicæmia.

77. B. pyogenes var. liquefaciens Lanz

Centralblatt f. Bakteriologie, XIV, 1893, 269.

Morphology. Bacilli 0.5–0.7 μ, of variable length. Cultures have a bad odor.
Gelatin colonies. Not described.
Gelatin stab. A funnel.
Agar slant. Growth thin, whitish, glassy.
Milk. Not coagulated.
Potato. Growth citron-yellow, with gas bubbles.
Pathogenesis. Intravenous inoculations of rabbits cause a multiple suppurative inflammation of the joints.
Habitat. Isolated from brain abscess after otitis-media.

78. B. liquefaciens Frankland

Microorganisms of Water, 1894. 461.

Morphology. Bacilli short, rather thick ; ends rounded.
Gelatin colonies. Round, entire, crateriform ; contents white, slimy.
Gelatin stab. Liquefaction napiform ; sediment whitish, granular.
Agar slant. A dirty white expansion.
Potato. Growth light yellow.
Habitat. Water.

79. B. Matazooni

B. No. 46 Conn: l.c., 1894, 80.

Morphology. Bacilli 0.4 : 0.8 μ, in chains.   Grow at 37°.
Gelatin colonies.   A central nucleus ; border crenate ; clear outer zone.
Agar slant.   Growth thin, whitish to yellowish.
Potato.   Growth yellowish brown.
Milk.   Slowly coagulated ; amphoteric to alkaline.
Bouillon.   Turbid, with a yellowish sediment.
Habitat.   Milk from Matazoon.

80. B. delictatulus Jordan

Report of the Mass. State Board of Health, 1890, 837.

Morphology. Bacilli 1.0 : 2.0 μ. Grow at 37°.
Gelatin colonies.   Whitish, homogeneous entire, with radiating edges ; in 2
    days a dark nucleus, with a clear zone of liquefied gelatin.
Gelatin stab.   In 2 days a funnel of liquefaction, with a surface membrane and
    a brownish sediment.
Agar slant.   Growth glistening, porcelain-white.
Potato.   Growth thin, gray.
Milk.   Acid.
Bouillon.   Turbid, with a sediment and a scum.   Nitrates reduced.
Habitat.   Water.

81. B. circulans Jordan

l.c.

Morphology. Bacilli 1.0 : 2-5 μ ; chains. Grow at 37°.
Gelatin colonies.   Round, brownish ; becoming depressions in the liquefied .
    gelatin.
Gelatin stab.   Slight growth along the line of stab ; a conical cavity, with a
    precipitate in the bottom, the liquefied gelatin drying out, leaving a partly
    empty cone.
Agar slant.   Growth thin, translucent.
Potato.   A scanty growth, the color of the potato.
Bouillon.   Turbid ; no membrane.
Milk.   Slowly coagulated and slightly acid.   Nitrates reduced to nitrites.
Habitat.   Water.

82. B. putidus Roger

B. septicus-putidus Roger: Revue de Med., 1893, 10.

Morphological and cultural characters like *B. vulgaris*.
Habitat. Isolated from a cholera corpse.

83. B. albus-putidus Maschek

Adametz, Die Bak. Nutz u. Trinkwässer, 1888.

From descriptions, not differentiated from the preceding.
Habitat. Water.

84. B. tachyctonus (Fischer)

Bact. tachyctonum Fischer: Deutsche med. Wochensch., 1894, Nos. 26–28.

Morphology. Bacilli medium-sized to filaments.
Gelatin colonies. Like cholera.
Gelatin stab. Liquefaction crateriform, becoming saccate; with a membrane.
Bouillon. Gas production, with a membrane.
Agar slant. Growth brownish.
Potato. Growth grayish brown, becoming reddish brown.
Pathogenesis. Subcutaneous and intraperitoneal inoculations of not too small
quantities cause septicæmia and bloody œdema in mice, guinea pigs, and
rabbits.
Habitat. Isolated from fæces in cholera nostras.

85. B. dubius Kruse

Flügge, Die Mikroorganismen, 1896.
Zur Bak. Differentialdiagnose der Cholera (Bacillus not named) Bleisch: Zeitsch.
f. Hygiene, XIII, 1893, 31.

Differs from the preceding in that the growth is pale yellow from the beginning.
On bouillon there is no membrane, and the cultures are less virulent.
Habitat. Isolated from fæces.

86. B. gasoformans Eisenberg

Bak. Diag., 1891, 107. Tils: Zeit. f. Hygiene, IX, 1890, 315.

Morphology. Bacilli small.
Gelatin colonies. Round, crateriform, spread rapidly.
Gelatin stab. Liquefaction saccate, turbid, with much gas.
Agar slant. Growth dirty white.
Potato. Growth light yellowish.
Habitat. Water.

87. B. Kralii Dyar

l.c., 376.

Morphology. Bacilli short, with rounded ends, 0.7 : 0.8 μ; occur singly.
Gelatin. Liquefied after 30 days. Nitrates reduced to nitrites.
Pepton rosolic acid solution. Unchanged.
Lactose litmus. Red, becoming blue.
Agar slant. Growth white, opaque.
Habitat. A culture from Kral's laboratory.

88. B. lactis

B. *b* Guillebeau : Ann. Micrographie, XI, 225.

Morphology. Bacilli 1.0 : 1.2 μ. Growth like *B. aerogenes.*
Milk. Quickly coagulated.
Habitat. Isolated from milk.

89. B. tartaricus Grimbert-Fiquet

Jour. Pharm. et de Chim., July 6, 1897.

Morphology. Bacilli 1.0-2.0 μ. Stain by Gram's method, facultative anaerobic.
Gelatin colonies. Resemble *B. coli;* liquefaction of the gelatin slow, beginning in 10–15 days.
Bouillon. Turbid, with a scum.
Milk. Coagulated in 8 days. Nitrates reduced to nitrites. *Indol* not produced. Fermentes glucose, lactose, saccharose, dextrin, mannit, with the
production of acetic and succinic acids, CO_2, H and alcohol. Decomposes
tartarates, with production of succinic and acetic acids, CO_2 and H.

90. B. halophilus Russell

· Zeitsch. f. Hygiene, XI, 1891, 200.

Morphology. Bacilli 0.7 : 1.5-3.5 μ. Grow only in gelatin; best in sea-water
gelatin.
Gelatin colonies. Round, grayish, white, translucent.
Gelatin stab. Liquefaction slow; evaporation causes an empty funnel. Cul-
tures rendered alkaline, with much gas.
Habitat. Sea water.

91. B. nitrificans Burri-Stutzer

Centralblatt f. Bakteriol., I, 1895, 2te Abt., 735.

Morphology. Bacilli 0.5 : 0.7–1.5 μ; involution forms; stain badly in aqueous analine colors.

Gelatin colonies. Deep colonies : round, gray. *Surface :* round, slimy, colorless, which after 8 days are sunken in the liquefied gelatin.

Gelatin stab. Surface growth spreading, colorless to bluish, 2–3 mm., which after a time begins to sink in the liquefied gelatin. In 3 weeks a crateriform to napiform liquefaction.

Bouillon. Slightly turbid, with a small amount of a whitish to reddish sediment. Oxidizes nitrites to nitrates.

Habitat. Isolated from the soil.

92. B. guttatus Zimmerman

Bak. Trink u. Nutzwässer, Chemnitz, 1890, 56.

Morphology. Bacilli 0.9 : 1.0 μ.

Gelatin colonies. Deep colonies have a brownish centre and bright borders. *Surface colonies* are small and round.

Agar slant. Growth gray, limited.

Potato. Growth slimy, yellowish gray.

Habitat. Water.

93. B. inunctus Pohl

Centralblatt f. Bakteriol., XI, 1892, 143.

Morphology. Bacilli 0.8–0.9 : 3.5 μ.

Gelatin colonies. Round, entire.

Gelatin stab. Radiating outgrowths from the line of stab; surface growth thick, glistening.

Agar slant. Growth whitish.

Potato. Growth slimy.

Habitat. Water.

94. B. litoralis Russell

Zeitsch. f. Hygiene, XI, 1891.

Morphology. Bacilli 2–4 times their breadth; grow slowly.

Gelatin colonies. The *deep colonies* in 3 days are small, brownish; the *surface colonies* are entire, shining to opalescent, and granular; liquefaction takes place in 5–8 days. The evaporation causes depressed colonies.

Gelatin stab. Growth in depth scanty; on the surface, the growth is thin, becoming depressed.

Agar slant. Growth slimy, white.

Bouillon. Turbid, with a membrane.

Habitat. Isolated from mud bottom of the Gulf of Naples.

95. B. superficialis Jordan

Mass. State Board of Health, 1890, 833.

Morphology. Bacilli 1.0 : 2.2 μ.

Gelatin colonies.. Deep colonies: round, segmented, cracked. *Surface colonies:* punctiform, translucent, slowly liquefied. Microscopically, round, homogeneous to finely granular; centres yellowish, brown edges, translucent.

Gelatin stab. In depth, none, or only a scanty growth; growth almost entirely on the surface.

Milk. Not coagulated in 20 days, slightly acid.

Bouillon. Slightly turbid; no membrane.

Habitat. Isolated from sewage.

96. B. hyalinus Jordan

Mass. State Board of Health, 1890, 836.

Morphology. Bacilli 1.0 : 5.4 μ to chains. Facultative anaerobic.

Gelatin colonies. In 24 hours, plainly visible to the naked eye; centre dark, translucent. Microscopically, centres coarsely fibrillous, with short fibrils radiating from the edges. In 2 days the colonies reach a diameter of 15 mm.

Gelatin stab. Liquefaction funnel-formed to saccate; in 8 days a highly tenacious scum.

Agar slant. Growth dry, dull, tough, becoming rough to warty.

Bouillon. Turbid, with a scum.

Potato. Growth as in agar slant.

Milk. Coagulated in 2 days, acid. Nitrates reduced to nitrites.

Habitat. Water.

97. B. pestifer Frankland

Phil. Trans. Roy. Soc., London, 1888, 277.

Morphology. Bacilli 1.0 : 2.3 μ; filaments. Grow slowly at room temperature.

Gelatin colonies. Deep colonies: irregular. *Surface colonies* show smooth centres, with margins of wavy bundles.

Agar slant. Growth glistening, translucent.
Potato. Growth thick, irregular, flesh-colored.
Habitat. Isolated from the air.
B. No. 9 Pansini, from sputum, and *B. pneumonicus-agilis* Flügge, indistinct from the preceding. Comp. *B. vermicularis* Frankland.

98. B. radiatus

B. aerogenes-meningitidis Centanni: Archiv per le Scienze Mediche, XVII, 1893, 1.

Morphology. Bacilli 0.3 : 2–2.5 μ; seldom in filaments.
Gelatin colonies. Daisy-shaped, rosulate.
Gelatin stab. Liquefaction slow, with gas.
Agar slant. Growth porcelain-white.
Bouillon. Turbid.
Potato. Growth grayish yellow, uneven, rough.
Pathogenesis. Subdural inoculations of rabbits cause death in a few hours to days or weeks, with progressive palsy, emaciations, and lung complications, hyperemia of meninges, etc.
Habitat. Isolated from two cases of meningitis.

99. B. recticularis Jordan

Report Mass. Board of Health, 1890, 834.

Morphology. Bacilli long, rather slender, ends slightly rounded, 1 : 5 μ; may occur in chains of 8–10 individuals. In many bacilli there are large vacuoles, with strongly refracting edges. Grow better at 37° than at 21°–23°.
Gelatin colonies. *Deep colonies* send out long spiral filaments, which under a low magnification look like so many jellyfish with streaming tentacles. The *surface colonies* at first form irregular expansions. The gelatin is liquefied so slowly that the liquid evaporates almost as soon as formed. The colonies then resemble slight hollows or cups in the gelatin, which show an irregular, reticulated structure.
Gelatin stab. The surface growth as in gelatin plates. Filamentous outgrowths from the line of stab.
Agar stab. Surface growth dry, dull, convex; a poor growth in depth.
Potato. In 2 days, growth white, dull, dry. In 5 days, growth of a characteristic woolly appearance.
Milk. Coagulated in 15–20 days at room temperatures, acid.

Bouillon. Becomes slowly turbid, with a slight stringy sediment. Nitrates reduced to nitrites.

Habitat. Water.

100. B. devorans Zimmerman.

Bak. Trink. u. Nutzwässer, Chemnitz, 1890.

Morphology. Bacilli 0.7 : 0.9–1.2 μ; occur singly and in pairs and chains.

Gelatin colonies. Deep colonies: small, white. *Surface colonies:* round, white, granular to filamentous; yellowish gray, margin fringed.

Gelatin stab. In depth, growth filiform, with a bubble above, under which there is a whitish growth. The funnel forms without visible liquefaction, with growth along the walls of the funnel.

Agar slant. Growth thin, gray, spreading.

Habitat. Water.

101. B. aquatilis Frankland

Zeitsch. f. Hygiene, VI, 1889, 381.

Morphology. Bacilli 2.5 μ; filaments 17.0 μ long. Grow very slowly in the usual culture media.

Gelatin colonies. Surface colonies show yellowish brown centres, from which twisted yellowish brown filaments are given off.

Gelatin stab. In depth, growth scarcely visible at first; liquefaction begins later, then progresses more rapidly; on the surface, a small yellowish colony.

Agar slant. Growth glistening, yellowish, limited.

Potato. Scarcely any growth, or a faint yellowish line only.

Habitat. Water.

102. B. diffusus Frankland

Zeitsch. f. Hygiene, VI, 1889, 396.

Morphology. Bacilli 0.5–1.7 μ; filaments.

Gelatin colonies. Deep colonies: round, granular, with erose edges. *Surface colonies:* thin, bluish gray, spreading. Microscopically, granular, erose-edged; a central nucleus, with a pale blue, irregular-edged, outer zone.

Gelatin stab. Growth only on the surface, which is thin, glistening, grayish yellow; the gelatin is slowly liquefied.

Agar slant. Growth thin, glistening, light yellow to creamy.

Bouillon. Turbid, sediment grayish yellow, with flocculi on the surface.

Potato. Growth thin, glistening, smooth, greenish yellow. Nitrates slowly reduced.

Habitat. Earth.

103. B. sulcatus Kruse

B. sulcatus-liquefaciens Kruse: Flügge, Die Mikroorganismen, 1896, 318.

Morphology. Bacilli medium-sized.
Gelatin colonies. Deep colonies: round, small, yellowish. *Surface colonies:*
   large, spreading, translucent, incised, marmorated; liquefaction slow.
Agar slant. Growth translucent, gray.
Potato. Growth yellowish brown.
Habitat. Water.

104. B. Havaniensis Sternberg

B. Havaniensis-liquefaciens Sternberg: Manual of Bacteriology, 1892, 686.

Morphology. Bacilli of variable length.
Gelatin colonies. Milky white, surrounded by erose, transparent borders;
   soon liquefied.
Gelatin stab. Liquefaction along the line of stab, turbid, becoming clear.
Agar slant. Growth brownish. No growth on potato. Not pathogenic to
   rabbits.
Habitat. Isolated from the surface of the human body.

105. B. leporis (Sternberg)

B. leporis-lethalis Sternberg: Manual, 1892, 453.

Morphology. Bacilli 0.5 : 1–3 μ; filaments.
Gelatin colonies. Deep: round, translucent, light yellow. *Surface:* trans-
   parent, resembling small fragments of broken glass; later liquefaction
   occurs.
Gelatin stab. Liquefaction infundibuliform.
Agar slant. Growth thin, white, translucent, glistening.
Potato. Growth thin, spreading, light yellow.
Pathogenesis. Intraperitoneal injections of 1–3 cc. of culture cause in 2–3
   hours a somnolent condition, with drooping head, and death from
   toxæmia.
Habitat. Isolated from the intestines of man ill of yellow fever.

106. B. Madoxi (Miquel)

Urobacillus Madoxi Miquel: Ann. d. Micrographie, 1889.

Morphology. Bacilli 0.5–1.0 μ; usually in pairs.
Gelatin colonies. Small, translucent, milky, with urea in the gelatin, colonies
   surrounded by a zone of crystals.

Gelatin stab. Liquefaction crateriform.
Agar slant. Growth whitish, with a slight greenish tint.
Bouillon. Turbid, with a membrane.
Habitat. Isolated from fermenting urine.

CLASS IV. WITH ENDOSPORES. AEROBIC AND FACULTATIVE ANA-
   EROBIC, COLONIES ON GELATIN PLATES BECOMING STREAMING,
   FORKED, AMEBOID, TWISTED, IRREGULAR, COCHLEATE. GELA-
   TIN LIQUEFIED. STAINED BY GRAM'S METHOD.

I. Gelatin colonies typical of the group, *i.e.* ameboid — cochleate.
   A. Agar smear cultures smooth.
      1. Potato cultures white, gray-yellowish, not distinctly brown.
         a. Milk coagulated.
            107. *B. vulgaris* (Hauser).
            108. *B. mirabilis* (Hauser) Trev.
            109. *B. No. VII* Pansini.
         b. Milk not coagulated.
            110. *B. sulphureus* (Holschewnikoff).
      2. Potato cultures brownish; cause septicæmia in mice.
            111. *B. septicus* (Babes).
   B. Agar smear cultures crumpled.
            112. *B. Strassmanni* Trev.
II. Gelatin colonies ciliate — radiate; related to *B. centrifugans.* Stained
      by Gram's method.
   A. Pathogenic to the smaller animals.
            113. *B. dysenteriæ* Kruse.
   B. Not pathogenic to the smaller animals.
            114. *B. Pansini.*

107. **B. vulgaris** (Hauser)

Proteus vulgaris Hauser: Ueber Faulnissbakterien, 1885.
B. proteus Trevisan: Genera, 1889.

Morphology. Bacilli o.6 : 1.2–4.0 μ — threads to chains, in floccose arrangement. Flagella numerous, peritrichíc.
Gelatin colonies. In 6–8 hours, small depressions, which contain liquefied
   gelatin and grayish white masses of bacteria; from the edge, ameboid
   processes.
Gelatin stab. Liquefaction saccate.
Agar slant. Growth slimy, moist, glistening, translucent.

Milk. Coagulated, acid, becoming yellowish.

Potato. Growth yellowish white, raised. Albuminous fluids give a putrefactive odor and an alkaline reaction.

Gas. In glucose and saccharose bouillon; no gas in lactose bouillon. $\frac{H}{CO_2} = \frac{2}{1}$. H_2S positive. *Indol* positive. Urea converted into ammonia.

Pathogenesis. Not properly pathogenic to the smaller animals. Injections of large quantities of filtered cultures cause toxæmia.

Habitat. Commonly found in putrefying fluids, water, etc.

108. B. mirabilis (Hauser) Trev.

Proteus mirabilis Hauser: l.c.

B. mirabilis Trevisan: Genera, 1889, 17.

A variety of *B. vulgaris*. Morphological and cultural characters as above; may liquefy gelatin a little more slowly. Deep colonies in gelatin cochleate.

Habitat. Isolated from putrefying fluids, etc.

109. B. No. VII Pansini

Virchow's Archiv, CXXII.

From descriptions, indistinguishable from *B. vulgaris*.

Habitat. Isolated from sputum.

110. B. sulphureus (Holschewnikoff)

Bact. sulphureum Holschewnikoff: Ann. de Microgr., 1889, 261.

In morphological and cultural characters, apparently identical with *B. vulgaris*.

Milk. Remains unaltered, but gradually becomes peptonized without coagulation, and has a yellowish color. H_2S produced on cooked egg.

Habitat. Water.

111. B. septicus (Babes)

Proteus septicus Babes: Septische Processe des Kindesalters, 1889.

B. proteus-septicus Kruse: Flügge, Die Mikroorganismen, 1896, 279.

Morphology. Bacilli middle-sized, 0.4 μ thick, length variable, comma forms to filaments.

Gelatin colonies. Like those of *B. vulgaris*.

Potato. Growth elevated, bright brown.

Agar slant. Growth a reticulated layer.

Pathogenesis. Subcutaneous inoculations of mice cause death by septicæmia.

Habitat. Isolated from the organs of a child dead with septicæmia symptoms.

112. B. Strassmanni Trev.

B. albus-cadaveris Strassmann: Zeitsch. f. Medicinalbeamte, 1888.
B. Strassmanni Trevisan: Genera, 1889, 103.

Morphology. Bacilli 0.7 : 2.5 μ; filaments.
Gelatin colonies. Ameboid, liquefied rapidly; have a bad odor.
Agar slant. Growth crumpled.
Potato. Growth thin, whitish yellow, granular.
Pathogenesis. Mice and guinea pigs die of toxic symptoms with comparatively small doses.
Habitat. Isolated from the blood of a four-days-old cadaver.

113. B. dysenteriæ Kruse

B. of Japanese dysentery Ogata: Centralblatt f. Bakteriol., XI, 1892, 264.
B. dysenteriæ-liquefaciens Kruse: Flügge, Die Mikroorganismen, 1896, 284.

Morphology. Bacilli slender rods, mostly in twos.
Gelatin colonies. With short radiations.
Gelatin stab. A funnel of liquefied gelatin, which is turbid and has a membrane on the surface.
Pathogenesis. Subcutaneous inoculations of mice cause local œdema. Subcutaneous inoculations of guinea pigs cause œdema; gray knots in the liver, spleen, and large intestines, with hemorrhagic infiltration of the large intestines.
Habitat. Isolated from a case of Japanese dysentery.

114. B. Pansini

B. IX Pansini : Virchow's Archiv, CXXII, 1890.

Morphology. Bacilli of variable length.
Gelatin colonies. Edged with radiating filaments.
Agar slant. Growth gray, transparent, stringy.
Potato. Growth stains the medium green; bad odor. Non-pathogenic.
Habitat. Isolated from the sputum of a consumptive.

CLASS V. WITH ENDOSPORES. AEROBIC AND FACULTATIVE ANA-
EROBIC. COLONIES ON GELATIN PLATES BECOMING AMEBOID,
ETC. GELATIN LIQUEFIED. NOT STAINED BY GRAM'S METHOD,
OR INDETERMINATE.

I. Potato growth whitish to grayish.
    A. Gelatin liquefied rather quickly; pathogenic.
        115. *B. murisepticus* Karliński.
        116. *B. Wesenbergii.*

B. Gelatin liquefied slowly; non-pathogenic.

   117. *B. larvicida* Dyar.

   118. *B. dendriticus* Lustig.

115. **B. murisepticus** Karliński

B. murisepticus-pleormorphus Karliński: Centralblatt f. Bakteriol., V, 1889, 193.

Morphology. Bacilli variable, coccoid-oval forms — spiraloid proteus forms.

Gelatin colonies. Like *B. vulgaris* and *mirabilis.* ,

Agar colonies. Like *B. vulgaris.*

Potato. Growth whitish gray, homogeneous, spreading.

Pathogenesis. Subcutaneous inoculations of mice cause death in 24 hours, of septicæmia.

Habitat. Isolated from pus.

116. **B. Wesenbergii**

B. der Fleischvergiftung Wesenberg: Zeitsch. f. Hygiene, XXVIII, 1898, 484.

Closely related to *B. vulgaris.*

Morphology. Bacilli 0.5–0.8 : 1.2–2.0 μ; in twos and, in old bouillon cultures, in chains. No involution forms. Not stained by Gram's method. Flagella peritrichic, mostly 8–12, rarely 20. Grow at 37°. Spores not observed.

Gelatin colonies. In 3 days, liquefied, with a central yellowish colony, from which outgrowths proceed; in 4 days, the whole plate is liquefied.

Gelatin stab. A good growth in depth; on the surface, liquefaction begins under the growth and extends downward.

Agar colonies. In 36–48 hours, at 37°, nearly the entire surface of the plate is covered with a moist, glistening, slimy layer, with often a brownish shimmer.

Bouillon. In 24 hours, turbid and alkaline, with a scum.

Milk. Coagulated, peptonized, alkaline.

Potato. Growth grayish white, slimy.

Glucose bouillon. Gas and acid production. *Indol* not produced. Cultures have a foul odor.

Pathogenesis. Subcutaneous inoculations of mice cause death.

Habitat. Isolated from poisonous meat in meat poisoning.

117. **B. larvicida**

Dyar: l.c.

Morphology. Bacilli 0.8 : 1.0 μ. Gelatin liquefied in 14–22 days.

Gelatin colonies. Proteus-like.

Milk. Coagulated, Nitrates reduced.

Agar slant. Growth thin, translucent. Lactose litmus red, becoming blue.

Habitat. Isolated from silk larvæ of *Clisiocampa fragilis.*

118. B. dendriticus Lustig

Diag. Bak. des Wassers, 1893, 99.

Morphology. Bacilli 0.5–0.8 : 0.8–2.0 µ.

Gelatin colonies. Large, raised, white, moist, glistening, with 8–10 branches; liquefied very slowly.

Gelatin stab. In depth, growth white, beaded; on the surface, a semispherical bead.

Agar slant. Growth a thin, iridescent layer.

Bouillon. A membrane adhering strongly to the walls of the tube.

Potato. Growth white, moist.

Habitat. Water.

CLASS VI. WITH ENDOSPORES. AEROBIC AND FACULTATIVE ANAE-ROBIC. COLONIES ON GELATIN PLATES BECOMING AMEBOID, ETC. GELATIN NOT LIQUEFIED.

I. Little tendency to grow in the depth of the gelatin, especially in gelatin stab cultures.

    119. *B. Zopfi* (Kurth).

    120. *B. Zenkeri* (Hauser).

II. Decided growth along the entire length of the needle in gelatin stab cultures.

    A. Growth on potato very slow and scanty, light, becoming yellowish, never brownish.

        121. *B. arborescens* Ravenel.

    B. Potato growth brownish, abundant; strongly pathogenic for mice and rabbits.

        122. *B. letalis* (Babes).

119. B. Zopfi (Kurth)

Bact. Zopfi Kurth: Ber. d. Deutsch. Bot. Gesellschaft, February, 1883.

Morphology. Bacilli like *B. mirabilis.*

Gelatin colonies. Colonies generally grow just beneath the surface, as branching zoöglœa, radiate and filamentous forms.

Gelatin stab. But little growth in depth, but in the upper portion of the line of puncture a radiately filamentous growth, as in plate cultures. But slight growth at 37°. *Indol* not produced, or doubtful.

Habitat. Isolated from the intestines of fowls.

B. alantoides Klein: Centralblatt f. Bakteriol., VI.

Proteus Zenkeri Kuhn: Archiv f. Hygiene, XIII.

According to Kruse, identical with the preceding.

120. B. Zenkeri (Hauser)

Proteus Zenkeri Hauser: Ueber Faulnissbakterien, 1885.

Indistinguishable from, and probably identical with, the preceding.

121. B. arborescens Ravenel

B. arborescens non-liquefaciens Ravenel: l.c., 39.

Morphology. Bacilli slender rods, 7–13 times their breadth; occur singly and in chains of several elements.

Gelatin colonies. In 48 hours bluish indistinct cloudy dots, easily over-looked, resembling colonies of *B. ramosus*, but less distinct and finer, *i.e.* radiate, filamentous, branched.

Gelatin stab. In depth, fine outgrowths, becoming beaded below; on the surface, growth irregular, white, concentric, thicker in the centre.

Agar slant. Growth a faint, colorless line, with lines of wavy colonies on each side.

Bouillon. Slightly turbid, becoming clear.

Pepton rosolic acid solution. Unchanged.

Litmus milk. Coagulated in 10 days, decolorized, acid.

Glucose bouillon. No gas. *Indol* negative. Optimum temperature 36°.

Habitat. Soil.

122. B. lęthalis Babes

Proteus lethalis Babes: Progrès Médical Roumain, 1889.

Morphology. Bacilli 0.8 : 1.5 μ; thick, short, flask-shaped rods to filaments.

Gelatin colonies. Raised, whitish, translucent; later, outgrowths, which branch on the surface.

Agar slant. Growth a thick, opaque, yellowish layer.

Potato. Growth brownish.

Pathogenesis. Subcutaneous inoculations of mice and rabbits cause death in 4 days, with local œdema, septicæmia, enteritis, peritonitis.

Habitat. Isolated from lung gangrene in man.

CLASS VII. WITHOUT ENDOSPORES. AEROBIC AND FACULTATIVE
ANAEROBIC. CHROMOGENIC; PIGMENT YELLOWISH. GELATIN
LIQUEFIED.

I. Gelatin colonies conglomerate — warty.
      123. *B. citreus* (Unna-Tommasoli) Kruse.
II. Gelatin colonies simple, not conglomerate or warty.
    A. Liquefaction of the gelatin near the surface in gelatin stab cultures, *i.e.*
        crateriform — stratiform.
        123 *a.* *B. Rheni.*
        124. *B. aurescens* Ravenel.
        125. *B. cuticularis* Tils.
    B. Liquefaction of the gelatin along the entire length of the needle in stab
        cultures, *i.e.* funnel-formed — saccate.
        1. Gelatin liquefied very slowly.
           126. *B. tremelloides* Tils.
        2. Gelatin liquefied rapidly.
           127. *B. Kornii.*
           128. *B. Schirokikhi.*
III. Gelatin colonies ameboid, proteus-like to zoöglœa masses.
      129. *B. dianthi* Arthur-Bolley.
IV. Gelatin colonies not described.
    A. Milk coagulated, at least on boiling.
        130. *B. Hudsoni* Dyar.
    B. Milk not coagulated.
        1. Nitrates reduced to nitrites.
           131. *B. theta* Dyar.
        2. Nitrates not reduced to nitrites.
           132. *B. gamma* Dyar.

123. B. citreus (Unna-Tommasoli) Kruse

Ascobacillus citreus Unna-Tommasoli: Monatschrift f. prakt. Dermatol., IX, 60.
B. citreus Kruse: Flügge, Die Mikroorganismen, 1896, 489.

Morphology. Bacilli 0.3 : 1.3 μ; occur singly or in bundles.
Gelatin colonies. Conglomerations of small spheres.
Gelatin stab. Small flakes in the liquefied funnel; on the surface a slimy
    citron-yellow layer.
Agar slant. Growth abundant, with honey-drop-like protuberances.
Potato. Growth slimy, citron-yellow, becoming in 2 weeks greenish yellow.
Habitat. Isolated from the surface of the human body in eczema.

123 a. B. Rheni

Rhine water Bacillus (Burri) : See Frankland, Microörganisms of Water, 1894, 483.

Morphology. Bacilli 0.7 : 2.5–3.5 μ, ends rounded, sometimes slightly bent. In bouillon long filaments and chains. Motility rotatory and progressive.

Gelatin colonies. Deep colonies dirty yellow dots, round, entire, granular. Surface colonies convex, colorless, transparent, becoming yellowish, with an outer zone of clear liquid gelatin. Microscopically, the borders are wavy, and contents rough and irregular.

Gelatin stab. Growth scanty, in depth; liquefaction shallow, funnel-formed; yellow flocculent masses in the liquefied gelatin, and a bluish gray opalescence on the surface of the still solid gelatin.

Glycerin agar slant. Growth thin, glistening, dry, tough, honey-colored.

Potato. Growth moist, glistening, thin, flat: orange- or rust-colored.

Bouillon. Turbid, with a bright, orange-colored pellicle and a yellow deposit.

Milk. Coagulated only partially in the upper part of the tube; the cream layer pale yellow; serum alkaline. No growth at 37°. No gas in glucose bouillon.

Habitat. Rhine River water.

124. B. aurescens Ravenel

l.c., 8.

Morphology. Bacilli short, spindle-shaped, 2–3 times their breadth; occur singly.

Gelatin colonies. Surface colonies minute whitish points, which are granular, brownish, and entire. In 4 days 0.5 mm., yellowish brown, homogeneous, and entire.

Gelatin stab. Slight growth in depth; on the surface a yellow button; in 18–20 days a crateriform dry depression.

Agar slant. Growth thin, yellowish, limited, becoming golden yellow.

Potato. Growth yellow, thick, moist, spreading, becoming orange-yellow.

Bouillon. Slightly turbid, with a dense, flocculent sediment.

Litmus milk. Reduced, no acid, not coagulated. *Indol* negative.

Glucose bouillon. No gas. Grow at 36°.

Habitat. Soil.

125. B. cuticularis Tils

Zeitsch. f. Hygiene, IX, 1890, 282.

Morphology. Bacilli 0.3–0.5 : 2.0–3.0 μ — filaments.

Gelatin colonies. Coli-like. The deep colonies are brown, irregular — entire; the surface colonies have brown centres and colorless borders.

Gelatin stab. Slight growth in depth ; on the surface a yellow membrane.
Potato. Growth scanty, slimy, yellow.
Habitat. Water.

126. B. tremelloides Tils

Zeitsch. f. Hygiene, IX, 1890, 282.

Morphology. Bacilli 0.2 : 1.0 μ.
Gelatin colonies. Raised, which later become spreading.
Potato. Growth coarsely granular, crumpled, yellowish.
Habitat. Water.

127. B. Kornii

B. of liver abscess Korn : Centralblatt f. Bakteriol., XXI, 1897, 438.

Morphology. Bacilli short, thick rods, two times their breadth, or larger forms
 0.4–0.6 : 2–5 μ; often in chains. Not stained by Gram's method.
Gelatin colonies. In 24 hours round, yellowish, granular. Gelatin liquefied
 quickly, with an irregular, ragged nucleus.
Gelatin stab. Liquefaction infundibuliform; an ochre-yellow membrane;
 liquefaction progresses rapidly, with a yellowish fluid and sediment.
Agar slant. At 30° growth, abundant, raised, glistening, yellowish.
Potato. Growth thick, raised, glistening, yellow, becoming brownish.
Milk. Becomes yellowish, syrupy, and alkaline. Pigment insoluble in water,
 slightly so in alcohol.
Pathogenesis. Subcutaneous inoculations of mice with 0.1–0.25 cc. of a bouillon
 culture cause death with tonic cramps ; no change in internal organs, but
 an abscess at point of inoculation.
Habitat. Isolated from a case of liver abscess.

128. B. Schirokikhi

Centralblatt f. Bakteriol., 2te Abt., II, 1896, 205.

Morphology. Bacilli 1.5 times their thickness ; in chains of 2–8. Show polar
 stain. Motility slow.
Gelatin colonies. In 3 days visible ; the gelatin is liquefied to a bright blue
 fluid, with yellowish granules. In 5 days the colonies are 1–2 mm. in
 diameter.
Gelatin stab. Liquefaction rapid, infundibuliform : liquefied gelatin grayish
 to yellowish, with a granular sediment.

Agar colonies. In 2 days like snow crystals, punctiform in centre, surrounded by a bluish toothed zone, becoming round; 2–4 cm.; medium-colored yellowish brown.

Agar slant. Growth white, crumpled, viscid. Nitrates reduced.

Habitat. Isolated from horse manure.

129. B. dianthi Arthur-Bolley

Indiana Ag. Expt. Sta. Bull., 59, 1896.

Morphology. Bacilli oval, occur singly, rarely united, 0.9–1.0 : 1–2 μ. In rich fluid media more united, forming short filaments, afterward forming zoöglœa.

Neutral gelatin colonies. Zoöglœa bodies, with a surrounding irregular area of coalescing forms, which appear as large viscid drops.

Acid gelatin colonies. Zoöglœa make up nearly the entire body of the colony, with a more irregular outline and a lobed, wrinkled appearance of the surface; color, light orange.

Gelatin slant. Growth smooth, limited, creamy, becoming rough, tuberculate, light orange.

Gelatin stab. Slight growth in depth; on the surface, growth arborescent — feathery.

Agar slant. Growth yellowish.

Potato. Growth yellowish.

Habitat. Associated with bacteriosis of carnations.

130. B. Hudsoni Dyar

l.c., 369.

Morphology. Bacilli 0.5–0.6 : 0.7–1.5 μ; occur singly and in pairs.

Potato. Growth thin, translucent, ochreous — orange.

Agar slant. Growth thick, mustard-colored.

Lactose-litmus. Reddened, variable.

Pepton rosolic acid solution. Unchanged. Nitrates not reduced.

Habitat. Air.

131. B. theta Dyar

l.c., 375.

Morphology. Bacilli 0.5–0.7 : 1.0–1.3 μ; in pairs. Motility doubtful.

Agar slant. Growth translucent, ochreous.

Gelatin. Liquefied slowly, after 24 days.

Lactose-litmus. Blue.

Bouillon.   A membrane.
Potato.   Growth thick, spreading, glistening, ochreous — brown.
Pepton rosolic acid solution.   Unchanged.
Habitat.   Air.

132. B. gamma Dyar : l.c.

Morphology.   Bacilli 0.5–0.7 : 1.0–1.5 μ.
Agar slant.   Growth spreading, translucent, ochreous, concentrically marked;
   comes off in pieces under the needle.
Bouillon.   A thick membrane.
Milk.   Slowly peptonized.
Lactose-litmus.   Blue, reduced.
Pepton rosolic acid solution.   Unchanged.
Habitat.   Air.

CLASS VIII.   WITHOUT ENDOSPORES.   AEROBIC AND FACULTATIVE
   ANAEROBIC.   CHROMOGENIC, PIGMENT YELLOWISH.   GELATIN
   NOT LIQUEFIED.

I. Gelatin colonies large, spreading.
  A. Chromogenic function weak, pale yellow.
      133. *B. subflavus* Zimmerman.
  B. Chromogenic function strong, deep or golden yellow.
    1. The gelatin acquires a green fluorescence.
      134. *B. fluorescens* Zimmerman.
    2. Do not cause a fluorescence of the gelatin.
      135. *B. flavus* Adametz.
II. Gelatin colonies aerogenes-like.
      136. *B. aurantiacus* Frankland.
      137. *B. flavescens* Frankland.
      138. *B. Winkleri.*

133. B. subflavus Zimmerman

Bak. Trink u. Nutzwässer, 1890, 62.
B. flavescens Pohl: Centralblatt f. Bakteriol., XI, 1892, 144.

Morphology.   Bacilli 0.8 : 1.5–3.0 μ.
Gelatin colonies. Deep colonies: round, yellowish, white. *Surface colonies:*
   punctiform, becoming spreading; borders irregular.
Gelatin stab.   On the surface, growth thin, yellowish gray.

Agar slant. Growth pale yellow, spreading, becoming darker, pale chrome-yellow to ochre-yellow.

Potato. Growth scanty, clay-yellow.

Habitat. Water.

134. B. fluorescens Zimmerman

B. fluorescens-aureus Zimmerman: l.c., 24.

Morphology. Bacilli 0.7 : 2.0 μ. Optimum temperature 20°.

Gelatin colonies. Deep : round, granular, light yellow. *Surface :* yellowish gray, thickest in the middle.

Gelatin stab. But slight growth in depth ; on the surface, a thin layer, which is yellowish and spreading.

Agar slant. Growth ochreous — golden yellow.

Potato. As in agar slant.

Habitat. Water.

135. B. flavus Adametz

B. aureo-flavus Kruse: Flügge, Die Mikroorganismen, 1896, 310.

Morphology. Bacilli 0.5 : 1.5–4 μ. Grow slowly at room temperatures.

Gelatin colonies. In eight days, small white points, becoming round, yellowish, opaque, 1–4 mm.

Gelatin stab. Slight growth in depth ; on the surface, small round colonies, crowded to make a dark chrome-yellow layer.

Potato. Convex colonies, becoming a chrome-yellow layer, becoming reddish brown in old cultures.

Habitat. Water. Surface of the body in eczema. (Tommasoli) Mon. prak. Dermatol., IX.

136. B. aurantiacus Frankland

Zeitsch. f. Hygiene, VI, 1889, 390.

Morphology. Bacilli short, thick, variable, often in filaments.

Gelatin colonies. Deep colonies : small, round, granular. *Surface colonies :* aerogenes-like, opaque, homogeneous, bright orange.

Gelatin stab. Slight growth in depth ; on the surface, growth glistening, orange.

Agar slant. Growth limited, orange.

Bouillon. Clear, with a thin pellicle, and orange sediment.

Potato. Growth thick, glistening, orange-red. Nitrates slowly reduced.

Habitat. Water.

137. **B. flavescens** Frankland

No name Pohl: Centralblatt f. Bakteriol., XI, 1892, 144.
B. flavescens Frankland: Microörganisms of Water, 1892, 448.

Morphology. Bacilli 0.8 : 2.0 μ. Slightly motile.
Gelatin colonies. Punctiform, yellow, granular. Grow very slowly.
Gelatin stab. Grows in depth, and is spread over the entire surface.
Agar stab. In depth, growth beaded, yellow; on the surface, it spreads
slowly.
Potato. Growth slimy, yellow, spreading.
Habitat. Isolated from marsh water.

138. **B. Winkleri**

B. mesentericus-aureus Winkler: Centralblatt f. Bakteriol., 2te Abt., V, 1899, 577.

Morphology. Bacilli 0.7 : 2–3 μ. Actively motile.
Wort gelatin colonies. In 24 hours, small, snow-white colonies, roundish,
sharp, lobate. Microscopically, cloudy, becoming vitreous drops, becoming golden yellow, becoming slowly liquefied, with a concentric flocculent
zone; outside of this a zone of radiate outgrowths, and outside of this
again, and in the solid medium, are numerous secondary colonies.
Gelatin colonies. Small, coarsely granular, with a central brownish nucleus
and an ameboid border. In 36 hours, colonies 0.25 mm. in diameter.
They are round, entire, with granular borders; centres cloudy, and contain
long, spiral bacterial aggregates, or irregular aggregates variously placed.
Potato. Growth thin, glistening, smooth, yellow.

CLASS IX. WITHOUT ENDOSPORES. AEROBIC AND FACULTATIVE
ANAEROBIC. CHROMOGENIC, PIGMENT REDDISH. GELATIN
LIQUEFIED.

I. Pigment bright carmine or fuchsin-red.
   A. On liquefied gelatin in stab cultures, the production of a membrane.
      1. Pigment carmine-red on agar and potato.
         139. *B. piscatorus* (Lehmann-Neumann).
      2. Pigment brick-red on agar and potato.
         140. *B. indicus* (Koch) Trevisan.
   B. No membrane on the liquefied gelatin in stab cultures.
      1. Pigment granules in the rods.
         141. *B. rubus* Kruse.

2. No pigment granules in the rods.

    142. *B. prodigiosus* (Ehrenberg) Flügge.

    143. *B. fuchsinus* Bökhont-de Vries.

II. Pigment flesh-colored.

    144. *B. carneus* Kruse.

139. **B. piscatorus** (Lehmann-Neumann)

B. ruber-sardinæ Du Bois Saint Sevrin : Ann. Pasteur Inst., 1894, 31.
Bact. piscatorum Lehmann-Neumann : Bak. Diag., 1896, 263.

Morphology. Bacilli short rods, 0.5–0.6 μ, mostly in twos. Gelatin liquefied with a strong membrane of a carmine-red color. Pigment production weaker in bouillon or on agar when grown at 37°.

Potato. Growth of a beautiful carmine-red color. Culture gives a strong odor of trimethylamine.

Habitat. Isolated from sardine oil.

140. **B. indicus** (Koch) Trev.

M. indicus Koch : ref. indt.
B. indicus-ruber Flügge : Die Mikroorganismen, 1886.
Bact. indicum Crookshank : Manual, 1887, 240.
B. indicus Trevisan : Genera, 1889, 17.

Morphology. Bacilli small, very short.

Gelatin colonies. Deep colonies golden yellow, erose. Surface colonies have incised, torn edges.

Gelatin stab. Liquefaction saccate ; on the surface a fragile, red membrane ; sediment white.

Agar slant and *potato*. Growth brick-red.

Pathogenesis. Intravenous inoculations of large doses into rabbits cause death with toxæmia.

Habitat. Isolated from the stomach of an ape.

141. **B. rubus** Kruse

Der rother Bacillus Lustig : Diag. der Bak. des Wassers, 1893, 72.
B. ruber-aquatilis Kruse : Flügge, Die Mikroorganismen, 1896, 303.

Morphology. Bacilli small, 2–3 times their breadth, red pigment granules within the rods.

Gelatin colonies. Erose, reddish, becoming liquefied.

s

Gelatin stab. Liquefaction infundibuliform. No pigment in depth or at 37°.
Agar slant and *potato.* Growth raspberry red.
Habitat. Water.

142. B. prodigiosus (Ehrenberg) Flügge

Serratia marcrescens Bizio: Polent porporp. in Bilb. Ital., XXX, 1823, 288.
Zoogalactina immetropa Sette: Mem. Venezia, 1824, 51.
Monas prodigiosa Ehrenberg: Monatsber. K. Akad., Berlin, 1848.
M. prodigiosus Cohn: Beiträge Biol., I, 1872, 153.
B. prodigiosus Flügge: Die Mikroorganismen, 1886.

Morphology. Bacilli 0.5 : 0.5–1.0; coccoid forms; filaments in acid bouillon.
   Optimum temperature 22°–25°. No pigment at 38°–39°. Flagella peri-
   trichic, 6–8.
Gelatin colonies. Deep colonies round — oval, entire, reddish-brownish; trans-
   lucent borders. Deep colonies irregular, with rough contours; granular,
   gray-brown, liquefied, when the red color of the colony appears.
Agar slant. Growth whitish, becoming reddish.
Gelatin stab. Liquefaction saccate; sediment reddish.
Milk. Coagulated, acid, peptonized, yellowish-reddish.
Potato. Growth rose-red, moist, becoming dark red to purple-red; odor of
   trimethylamine.
Bouillon. Turbid, reddish, with a slight pellicle. Pigment soluble in alcohol
   and ether. The solution becomes orange-yellow on the addition of
   alkalies and violet-red with acids.
Glucose bouillon. Gas production variable; no acid. H$_2$S negative. *Indol*
   slight.
Pathogenesis. Scarcely pathogenic, toxic in very large doses.
Habitat. Commonly present on articles of food, particularly starchy materials;
   also meat, water, etc.

143. B. fuchsinus Bökhout-de Vries

Centralblatt f. Bakteriol., 2te Abt., IV, 1898, 497.

Morphology. Bacilli 0.5–0.7 : 1–1.5 μ. Facultative anaerobic, but oxygen
   necessary for the formation of pigment. Optimum temperature 22°–25°.
   No pigment at 36°.
Agar colonies. White, becoming reddish, thin, metallic, sharp.
Gelatin stab. In depth, growth, but no pigment; on the surface, growth
   bright red.
Agar slant. Growth carmine-red, with a metallic shimmer. Grow best on
   slightly alkaline or neutral media.

Potato. Growth reddish, later shows a bronze shimmer.

Milk. Coagulated, peptonized.

Glucose bouillon. No gas, acid. Pigment soluble in alcohol, chloroform, and carbon bisulphide, less so in ether, and with difficulty in water. The addition of an alkali renders the red-violet solution of the pigment yellow. The most suitable medium is a sodium tartrate pepton agar.

144. **B. carneus** Kruse

Fleischfarbiger Bacillus Tils: Zeitsch. f. Hygiene, IX, 1890, 316.
B. carneus Kruse: Flügge, Die Mikroorganismen, 1896.
B. carnicolor Frankland: Microörganisms of Water.

Morphology. Bacilli 0.5 : 2.0 μ.

Gelatin colonies. In 2 days round, sharp; liquefaction crateriform; centre darker and finely granular; outer zone colorless or concentric.

Gelatin stab. Liquefaction infundibuliform, with a rose-colored sediment.

Agar slant and *potato.* Growth flesh red.

Habitat. Water.

CLASS X. WITHOUT ENDOSPORES. AEROBIC AND FACULTATIVE ANAEROBIC. CHROMOGENIC ; PIGMENT REDDISH. GELATIN NOT LIQUEFIED.

I. Small, slender bacilli.
  146. *B. rubefaciens* Zimmerman.
II. Large, stout bacilli.
  147. *B. rubescens* Jordan.

146. **B. rubefaciens** Zimmerman

Bak. Trink u. Nutzwässer, 1890.

Morphology. Bacilli 0.3 : 0.7–1.6 μ.

Gelatin colonies. *Deep:* round, small, yellowish brown. *Surface:* flat, white-reddish.

Gelatin stab. In depth, a uniform growth ; on the surface, growth grayish white to yellowish, becoming wine-colored.

Agar slant. Growth thick, grayish blue.

Potato. Growth yellow-brown, with flesh-colored edge.

Habitat. Water.

147. B. rubescens Jordan

Mass. State Board of Health, 1890, 368.

Morphology. Bacilli 0.9–4.0 μ; pairs — filaments; rods sometimes slightly curved. Motility slight. Grow at 37°.

Gelatin colonies. Punctiform, porcelain-white, round, entire; later of a brownish cast.

Gelatin stab. Slight growth in depth; on the surface, a porcelain-white bead.

Agar slant. Growth smooth, white, glistening, becoming wrinkled after some time, with a slight pinkish tinge.

Potato. Growth light brown, becoming pink.

Milk. Not coagulated, alkaline; in old cultures a pinkish tinge on the surface.

Bouillon. Turbid; a surface membrane, viscous; becoming clear. Nitrates not reduced.

Habitat. Isolated from sewage.

CLASS XI. WITHOUT ENDOSPORES. AEROBIC AND FACULTATIVE ANAEROBIC. CHROMOGENIC; PIGMENT BROWNISH BLACK OR GRAY ON GELATIN.

I. Gelatin liquefied.

    148. *B. cyanofuscus* (Beijerinck).

    149. *B. aeris.*

    150. *B. ferrugineus* (Rullman).

II. Gelatin not liquefied.

    151. *B. fuscus.*

148. B. cyanofuscus Beijerinck

Bot. Zeit., XLIX, 1891, Nos. 43-47.

Morphology. Bacilli 0.2–0.6 μ, and one-half as thick. Cultures in solutions containing 0.5 per cent of pepton become greenish blue — brownish black. Gelatin colonies surrounded by a black zone in which crystals are formed.

Habitat. Isolated from glue, cheese.

149. B. aeris

B. violaceus-sacchari Ager-Dyar: N. Y. Med. Jour., 1894.

Morphology. Bacilli short, 0.5 : 0.7-1.0 μ; occur singly and in short chains.

Gelatin. Liquefied quickly.

Milk. Coagulated.

Agar slant. Growth thin, translucent; a greenish fluorescence of the medium.

Pepton rosolic acid solution. Decolorized.

Lactose-litmus. Red. Nitrates not reduced. A green fluorescence and a blackish color in the presence of glucose, lactose, and glycerin. A violaceous pigment especially noted in old milk cultures.

Habitat. Air.

150. B. ferrugineus Rullman

Centralblatt f. Bakteriol., XXIV, 1898, 465.

Morphology. Bacilli, in bouillon, 0.8 : 2.0 μ; on potato, 0.5 : 1.4 μ; with or without polar stain.

Gelatin colonies. Deep: slightly brownish. *Surface:* on the first day deep brown. On gelatin media an intense brown pigment penetrates the latter.

Beer-wort agar. Large round colonies with dark brown nuclei and bright brown borders.

Agar slant and *potato.* Growth rusty brown.

Milk. The upper fat layer becomes dark yellow; the milk below slightly colored.

Löffler's blood serum. A liquefaction after some days.

Glucose bouillon. No gas. Reaction of cultures strongly alkaline. Non-pathogenic to mice.

Habitat. Isolated from canal water.

151. B. fuscus

B. fuscus-limbatus Scheibenzuber: Allgemeine Wiener med. Zeitung, 1889, 171.

Morphology. Bacilli short.

Gelatin colonies. As brownish clumps.

Gelatin stab. In depth, growth uniform, with short outgrowths; surface growth, slightly spreading. Gelatin near stab slightly brownish.

Agar slant. A brown color to the medium.

Potato. Growth brownish.

Habitat. Isolated from decayed eggs.

CLASS XII. WITHOUT ENDOSPORES. AEROBIC AND FACULTATIVE ANAEROBIC. CHROMOGENIC; PIGMENT BLUE-VIOLET ON GELATIN OR AGAR.

I. Gelatin liquefied.
   A. Pigment violet on potato.
      1. Pigment violet on agar.
         152. *B. violaceus* Jordan.
      2. Pigment blue-black on agar.
         153. *B. lividus* Flügge-Proskauer.

147. **B.**

Mass. State I

Morphology. Bacilli 0.9-4.0 μ;
curved. Motility slight. C
Gelatin colonies. Punctiform,
brownish cast.
Gelatin stab. Slight growth in
Agar slant. Growth smooth, w
time, with a slight pinkish t
Potato. Growth light brown, b
Milk. Not coagulated, alkaline
Bouillon. Turbid; a surface n
not reduced.
Habitat. Isolated from sewage

CLASS XI. WITHOUT END
ANAEROBIC. CHROMO(
GRAY ON GELATIN.

I. Gelatin liquefied.
148. *B. cyanofuscus* (Beij
149. *B. aeris.*
150. *B. ferrugineus* (Rul'
II. Gelatin not liquefied.
151. *B. fuscus.*

148. **B.**

Bot. Z

Morphology. Bacilli 0.2-0.6 ¡
containing 0.5 per cent o
Gelatin colonies surroun
Habitat. Isolated from glue.

B. violaceus-sacch

Morphology. Bacilli short, ‹
Gelatin. Liquefied quickly.
Milk. Coagulated.
Agar slant. Growth thin, t
Pepton rosolic acid solution.

       b. Much larger rods to coccoid forms, variable.
          157. *B. urinæ.*
          158. *B. Ellingtonii.*
II. Gelatin liquefied quickly.
    A. Gelatin stab cultures arborescent; potato cultures floccose — curled.
        159. *B. lucæmiæ* Lucet.
    B. Cultures on gelatin and potato not characterized as above.
      1. Cultures have a decided aromatic odor.
        160. *B. helvolus.*

155. B. Lesagei Trev.

B. de la diarrhée verte des enfants Lesage: Bull. Acad. Med., Paris, October, 1887.
B. Lesagei Trevisan: Genera, 1889, 14.
B. viridis Kruse: Flügge, Die Mikroorganismen, 1896.

Morphology. Bacilli 0.7 : 1.0–2.4 μ; filaments. Optimum temperature 35°.
Gelatin colonies. Thin, spreading, erose, with green fluorescence.
Gelatin stab. Grow only on the surface.
Potato. Growth dark green, rarely reddish, with odor of old urine.
Pathogenesis. Intravenous inoculations and feeding of rabbits produces "green diarrhœa."
Habitat. Associated with "green diarrhœa" of children.

156. B. smaragdinus Reimann

B. smaragdino-fœtidus Reimann: Phil. Diss., Würzburg, 1887.

Morphology. Bacilli as characterized, somewhat bent. Optimum temperature 37°; scanty or slow growth at 20°.
Gelatin stab. Liquefaction occurs along line of stab; on the surface, a greenish fluorescence.
Agar colonies. Irregular, fluorescent.
Agar slant. Growth dirty yellow.
Potato. Growth chocolate-brown.
Habitat. Isolated from nasal secretions in ozæna.

157. B. urinæ

B. proteus-fluorescens Jäger: Zeitsch. f. Hygiene, XII, 1892, 525.

Morphology. Bacilli variable, cocci — filaments, variable also in thickness.
colonies. Proteus — coli-like, with a green fluorescence. Cultures
as in proteus.
produced.

Agar slant. Warty drops, becoming thick, yellowish white, with a green fluorescence.

Potato. Growth slimy, light yellow, becoming dark brown.

Pathogenesis. Subcutaneous and intraperitoneal inoculations of mice cause death in 3 days to 2 weeks; fatty degeneration of liver and kidneys, spleen enlarged, intestines hemorrhagic, with bacilli in the organs.

Habitat. Isolated from persons in icterus with nephritis; from urinary sediment, etc.

158. B. Ellingtonii

B. No. 21 Conn: l.c., 1893. 52.

Morphology. Bacilli 0.8 : 2.0 μ — chains. Grow at 35°.

Gelatin colonies. Small, opaque, liquefied quickly, becoming large, greenish, granular, with a central granular nucleus.

Gelatin stab. Liquefaction crateriform, becoming a broad funnel; a pellicle on the surface; gelatin clear, green.

Agar slant. Growth thin, white, transparent, slightly spreading; medium becomes greenish.

Potato. Growth thin, moist, brownish.

Milk. Coagulated, alkaline, slowly peptonized.

Bouillon. Turbid, a slight pellicle, medium slightly greenish, later a tenacious pellicle on the surface.

Habitat. Milk, Ellington creamery.

159. B. leucæmiæ Lucet

B. leucæmia-canis Lucet: Baumgarten's Jahresberichte, 1891, 319.

Morphology. Bacilli slender, 3.0 μ long.

Gelatin stab. From the line of stab, lateral outgrowths; a liquefaction and fluorescence of the surface growth.

Potato. Growth abundant, floccose — curled.

Pathogenesis. Inoculations of rabbits cause death in 10 days, with nodular formation on the inner organs, containing bacilli. Guinea pigs immune.

Habitat. Isolated from a dog with leukocythemia.

160. B. helvolus

B. chromo-aromaticus Sternberg: Manual, 1892.

Morphology. Bacilli medium-sized.

Gelatin stab. Growth green, with a yellowish white membrane.

Agar slant. Growth thin, white.

Potato. Growth brown.

Pathogenesis. Intravenous inoculations of rabbits cause death in 2–3 weeks; pneumonia, pleuritis, pericarditis.

Habitat. Isolated from a hog with bronchopneumonia and enteritis.

CLASS XIV. WITHOUT ENDOSPORES. OBLIGATE ANAEROBIC. GELATIN NOT LIQUEFIED.

I. Grow at room temperatures and in nutrient gelatin.

161. B. tumidus

B. I Sanfelice : Zeitsch. f. Hygiene, XIV, 1893.

Morphology. Bacilli of variable length, with bladdery swellings. Motility slight.

Gelatin colonies. Round, glistening, finely granular; edges floccose, or with B. *Zopfi*-like outgrowths.

Agar slant. Growth thick, floccose — tomentous.

Gelatin stab. In death, growth beaded, with gas production.

Habitat. Isolated from putrefying flesh.

II. No growth at room temperatures or in nutrient gelatin.

162. B. thermophilus

Ein neuer anaërober Bacillus des malignen Oedems Novy: Zeitsch. f. Hygiene, XVII. 1894, 209.

B. œdematis-thermophilus Kruse: Flügge, Die Mikroorganismen, 1896, 242.

Morphology. Bacilli 0.8–0.9 : 2.5–5.0. Numerous lateral flagella. Stained by Gram's method.

Gelatin colonies. Floccose.

Glucose bouillon. Gas, without odor, or that of butyric acid. Reduces litmus. No growth below 24°.

Pathogenesis. Very virulent for mice, rats, guinea pigs, rabbits, cats, and pigeons by subcutaneous inoculation of 0.1–0.25 cc. There is an œdema, with much gas; abdominal muscles red with hemorrhagic flecks; pleural and abdominal cavities contain a colorless exudate.

Habitat. Isolated from a guinea pig which was inoculated with a contaminated nuclein culture.

CLASS XV. WITH ENDOSPORES. AEROBIC AND FACULTATIVE ANAEROBIC. RODS NOT SWOLLEN AT SPORULATION. GELATIN LIQUEFIED.

I. Gelatin stab cultures arborescent.
> 164. *B. Prausnitzii* Trevisan.
> 165. *B. aureus* Pansini.

II. Gelatin stab cultures not distinctly arborescent.
> *A.* Liquefaction of the gelatin begins only after 7 days.
>> URO-BACILLUS GROUP.
>> 1. Grow in ordinary culture media only upon the addition of urea or by rendering media distinctly alkaline with NH₄HO.
>>> 166. *B. Pasteuri* (Miquel).
>>> 167. *B. Freudenreichii* (Miquel).
>>> 168. *B. Madoxi* (Miquel).
>> 2. Grow on ordinary culture media.
>>> *a.* Pepton rosolic acid solution decolorized.
>>>> 169. *B. fissuratus* Ravenel.
>>> *b.* Pepton rosolic acid solution not decolorized.
>>>> * No growth on potato.
>>>>> 170. *B. alpha* Dyar.
>>>> ** Growth on potato abundant, but nearly invisible.
>>>>> 171. *B. beta* Dyar.

B. Gelatin liquefied quickly, *i.e.* in less than 7 days.
> 1. Potato cultures becoming crumpled. POTATO BACILLUS GROUP.
>> *a.* Color of potato cultures whitish-yellowish.
>>> * Gelatin liquefied very rapidly.
>>>> † Agar smear cultures white-gray, not yellowish, wrinkled.
>>>>> § Gas generated in glucose bouillon.
>>>>>> 172. *B. Pammelii.*
>>>>> §§ No gas generated in glucose bouillon.
>>>>>> ‡ Milk not coagulated, or at most only slimy; peptonized.
>>>>>>> ♀ Milk shows a slimy fermentation.
>>>>>>> 173. *B. vulgatus* Trev.
>>>>>>> 174. *B. lactis No. 2* Flügge.
>>>>>>> 175. *B. peptonans* Sterling.
>>>>>>> Consistency of milk unchanged.
>>>>>>> 176. *B. megatherium var. Ravenelii.*
>>>>>>> 177. *B. Scheurleni* Sternberg.

    ‡‡ Milk coagulated.

          178. *B. liodermos* Flügge.

  †† Agar smear cultures yellowish-brownish, wrinkled.

          179. *B. mesentericus* Trevisan.

          180. *B. lactis No. 4* Flügge.

  ††† Agar smear cultures smooth or scarcely wrinkled.

          181. *B. tenuis* Trevisan.

  †††† Agar smear cultures not described; broad bacilli up to
    1.7 microns.

          182. *B. tumescens* Zopf.

  ** Gelatin liquefied slowly.

    † Gas produced in saccharose bouillon.

          183. *B. gummosus* Happ.

  †† No gas produced in saccharose bouillon.

          184. *B. stellatus.*

          185. *B. cremoris.*

b. Color of potato cultures becoming eventually brownish.

  * Gelatin colonies round, without radiations.

    † Reduce nitrates to nitrites.

          186. *B. denitrificans* Schirokikh.

    †† Do not reduce nitrates to nitrites.

          187. *B. detrudens* Wright.

          188. *B. maidis* Paltauf-Heider.

  ** Gelatin colonies proteus-like or with outgrowths.

          189. *B. plicatus.*

          190. *B. gangrænæ* Arkövy.

  *** Gelatin colonies filamentous — floccose.

          191. *B. aromaticus.*

  **** Gelatin colonies not described.

          192. *B. magnus.*

          193. *B. Hueppei* Trev.

2. Potato cultures not becoming crumpled. B. SUBTILIS GROUP.

  a. Potato cultures becoming dry, mealy.

    * Potato cultures whitish-gray.

          194. *B. subtilis* (Ehrenberg) Cohn.

          195. *B. lactis-albus* Sternberg.

          196. *B. leptosporus* Klein.

  ** Potato cultures yellowish.

    † Very thick bacilli, approaching 2.5 microns.
>197. *B. megatherium* De Bary.
>198. *B. Ellenbachensis* (alpha) Stutzer-Hartleb.

    †† Bacilli not exceeding 1.0 micron in width.
>199. *B. subtilis-similis* Sternberg.
>200. *B. cereus.*

b. Potato cultures rough, granular.
    * Gelatin liquefied quickly.
>201. *B. capillaceus* Wright.

    ** Gelatin liquefied slowly.
>202. *B. rudis.*

c. Growth on potato thin, scanty, or none, and the color of the medium, *i.e.* white-gray.
    * Gelatin colonies round, entire, without radiations.
>203. *B. circulans* Jordan.

    ** Gelatin colonies with radiating projections or proteus-like.
>204. *B. limosus* Russell.
>205. *B. sputi.*

    *** Gelatin colonies not described.
>206. *B. vacuolosis* Sternberg.

d. Growth on potato thick, white.
    * Colonies of the anthrax type, floccose.
>207. *B. pseudo-anthracis* Kruse.

e. Growth on potato thick, yellowish-brownish.
>208. *B. vaculatus* Ravenel.

f. Growth on potato white-gray, but not otherwise described.
    * Colonies with radiating outgrowths or ciliate margins.
>209. *B. Flüggei.*
>210. *B. crinatus.*

g. Sterilized potato inoculated undergoes a wet rot or putrefactive fermentation.
>211. *B. Krameri.*

164. B. Prausnitzii Trev.

B. ramosus-liquefaciens Flügge: Die Mikroorganismen, 1886, 290.
B. Prausnitzii Trevisan: Genera, 1889, 20.

Morphology. Bacilli large.
Gelatin colonies. Beset with bristles; liquefaction slow.
Gelatin stab. Radiations from all sides.
Habitat. A contamination in cultures.

165. B. aureus Pansini

Virchow's Archiv, CXXII.

Morphology. Bacilli slender threads.

Gelatin colonies. With a Strahlenkranz.

Gelatin stab. With lateral outgrowths, a membrane, and a yellowish sediment.

Agar slant. Growth rugose.

Potato. Growth sulphur — golden yellow. Bouillon turbid, with a membrane.

Habitat. Sputum.

166. B. Pasteuri Miquel

Urobacillus Pasteuri Miquel: Ann. Micrographie, 1889-92.

Morphology. Bacilli large, dimensions variable.

Gelatin colonies. With urea, in 24 hours, minute, surrounded by dumb-bell-shaped crystals; give off odor of NH_3. In urine, an alkaline fermentation and an abundant deposit of ammonia-magnesic-phosphate and alkaline urates. This deposit acquires a blackish color.

Habitat. Urine.

167. B. Freudenreichii Miquel

Urobacillus Freudenreichii Miquel: l.c.

Not clearly differentiated from the preceding.

168. B. Madoxi Miquel

Urobacillus Madoxi Miquel: l.c.

Morphology. Bacilli 1.0 : 3.6 μ.

Gelatin colonies. With urea, small, round, opaque, surrounded by crystals.

Gelatin stab. Fail either to grow, or a scanty development of colonies, with the formation of crystals.

Bouillon. Rendered alkaline, with formation of NH_3; a dense turbidity of the medium.

Habitat. Water and sewage.

169. B. fissuratus Ravenel

l.c., 38.

Morphology. Bacilli small straight rods, with rounded ends; occur singly. Motility slight.

Gelatin colonies. Light yellow, irregular, darker in centre; look like flakes of mineral matter or bits of shell. In 2 days, the *surface colonies* are round, irregular; centre brown; broken and fissured in every direction, 2-3 mm.; gelatin slowly softened.

Agar slant. Growth thin, greenish white, translucent.

Gelatin stab. Slight growth in depth, with later spherical outgrowths; surface growth capitate, sunken in the gelatin; evaporation exceeds liquefaction.

Potato. In 3 days, a bluish discoloration along the line of inoculation, becoming yellowish, moist, glistening.

Bouillon. A slight turbidity, becoming clear.

Pepton rosolic acid solution. Decolorized.

Litmus milk. Alkaline, not coagulated. In two weeks becomes translucent and violet.

Glucose bouillon. No gas. *Indol* negative. Optimum temperature 36°.

Habitat. Soil.

170. B. alpha Dyar

l.c , 366.

Morphology. Bacilli large, 0.8 : 1-2 μ; occur singly and in chains.

Gelatin. Not liquefied for 11 days or more.

Agar slant. Growth translucent, glistening, white.

Litmus milk. Not coagulated; not reddened. Reduction of nitrates negative or slight. *Indol* production slight.

Habitat. Air.

171. B. beta Dyar

l.c., 366.

Morphology. Bacilli short, rounded, 0.6 : 1.5-2.0 μ.

Gelatin. Liquefied slowly. *Surface:* growth thin, feathery.

Agar slant. Growth translucent, glistening, white.

Litmus milk. Not coagulated; not reddened. Reduction of nitrates negative or slight. *Indol* production slight.

Habitat. Air.

172. B. Pammelii

B. gasoformans Pammel : Centralblatt f. Bakteriol., 2te Abt., II, 1896, 642.

Morphology. Bacilli 1.2-1.4 : 3.7-4 μ, variable; in chains of 2-5. Motility slight.

Gelatin colonies. Lenticular.

Gelatin slant. Growth white, limited.

Gelatin stab. Liquefaction infundibuliform.

Agar slant. Growth white, wrinkled.

Bouillon. Good growth, with a pellicle.

Potato. Slightly granular, large wrinkles, spreading.

Milk. Coagulated, peptonized, alkaline.

Glucose and *saccharose bouillon.* Gas. No gas in lactose bouillon.

Habitat. Cheese.

173. B. vulgatus Trevisan.

B. mesentericus-vulgatus Flügge: Die Mikroorganismen, 1886.

B. vulgatus Trevisan: Genera, 1889, 19.

Potato bacillus of various authors.

Morphology. Bacilli 0.8 : 1.6–5.0 μ; somewhat shorter and smaller than
B. subtilis. Stain by Gram's method.

Gelatin colonies. Without radiations like *B. subtilis. Surface colonies:* round,
becoming crateriform, with a grayish white, delicate, crumpled membrane.
Microscopically, in early stage, border of colony segmented.

Gelatin stab. In depth, a narrow tube ; on the surface, a crateriform to stratiform liquefaction, with a surface membrane.

Agar slant. Growth gray white, glistening, becoming crumpled.

Bouillon. Weakly turbid ; a membrane which is firm and not easily broken
by shaking.

Milk. Coagulated slowly, alkaline, slimy, peptonized.

Potato. Growth thin, white, crumpled, becoming light yellowish. *Indol*
negative. H_2S production feeble.

Habitat. Widely distributed.

174. B. lactis No. 2 Flügge

Zeitsch. f. Hygiene, XVII, 1894, 272.

From descriptions, not differentiated from the preceding.

Habitat. Milk.

175. B. peptonans Sterling

B. lactis-peptonans Sterling: Centralblatt f. Bakteriol., 2te Abt., I, 1895, 473.

From descriptions, not differentiated from. 173.

176. B. megatherium var. Ravenelii

B. megatherium Ravenel : l.c., 11.

Morphology. Bacilli thick, rounded, 3–5 times their breadth ; occur in chains,
in which rods are bent on each other ; lengths unequal. Motility slight,
ameboid.

Gelatin colonies. Deep : brown, entire, homogeneous. *Surface :* light brown ; margins irregular ; centres dense, becoming, in 3 days, liquefied, crateriform, 4 mm., with white irregular centres and greenish borders, in which are tangled filaments.

Agar slant. Growth white, glistening, elevated, sometimes yellowish ; odor of sour milk.

Gelatin stab. Liquefaction crateriform, becoming stratiform or infundibuliform.

Potato. Growth elevated, white, moist, glistening, spreading, rugose, becoming smoother ; odor of stale milk.

Bouillon. Turbid, with a flocculent sediment, becoming clear.

Litmus milk. Decolorized slowly, and peptonized ; amphoteric — alkaline.
Indol negative. Grow at 36°.

Habitat. Soil.

177. B. Scheurleni Sternberg

Manual, 1892, 680; Dyar, l.c., 367.

Morphology. Bacilli 0.7-1.0 : 1.0-2.5 μ; occur singly and in short chains.
Agar slant. Growth thin, transparent. Grow slowly at room temperature.
Habitat. Isolated from the surface of the skin, mammæ of woman, air, etc.

178. B. liodermos Flügge

Die Mikroorganismen, 1886.

Morphology. Bacilli thick, rounded, 1.2 : 3.5 μ; in pairs and chains.
Gelatin stab. Liquefaction infundibuliform ; a membrane on the surface.
Agar slant. Growth a white, rosette-like layer.
Potato. A gummy, translucent layer ; gum soluble in water, precipitated by alcohol.
Milk. Coagulated, peptonized ; butyric acid. Related to, perhaps a variety of, *B. vulgatus.*
Habitat. Isolated from potato and from milk.

179. B. mesentericus Trev.

B. mesentericus-fuscus Flügge : Die Mikroorganismen, 1886.
B. mesentericus Trevisan : Genera, 1889, 19.

Morphology. Bacilli 0.8 : 2-4 μ; ends rounded. Stain by Gram's method.
Gelatin colonies. Deep colonies gray-yellow, irregular, with outgrowths. Surface colonies, small, round, gray-white, becoming sunken in the gelatin like *B. subtilis.* Microscopically, centres gray-brown, opaque, with ciliate borders.

Gelatin stab. Liquefaction infundibuliform — saccate, with a white-gray pellicle.
Agar slant. Growth yellow-brown, moist, glistening, crumpled.
Bouillon. Turbid, with a pellicle.
Potato. Growth gray-yellow, moist, glistening, raised, becoming crumpled.
H₂S produced. *Indol* production feeble.
Milk. Coagulated, slightly alkaline.
Habitat. Widely distributed.

VARIETIES. *B. mesentericus-fuscus-granulatus* Dyar.
Agar cultures coarsely granulated ; reduce nitrates vigorously.
B. mesentericus-fuscus-consistens Dyar.
Gelatin colonies approaching in character to those of *B. vulgaris.*

180. B. lactis No. 4 Flügge

Zeitsch. f. Hygiene, XVII, 1894, 294.

From descriptions, not differentiated from the preceding.
Habitat. Milk.

181. B. tenuis Trev.

Tyrothrix tenuis Duclaux: Le Lait, Paris, 1887.
B. tenuis Trevisan: Genera, 1889.

Morphology. Bacilli 0.6 : 3 0 μ — filaments.
Gelatin colonies. Growth on agar, in milk, and in bouillon like *B. subtilis.*
Potato. Growth like *B. vulgatus.*
Habitat. Milk and cheese.

182. B. tumescens Zopf

Spaltpilze, 1885, 82. A. Koch: Bot. Zeitung, 1888, 313.

Morphology. Bacilli short, about 1.7 μ broad ; or filaments, which are bent
and twisted.
Potato. Growth white, viscid, somewhat crumpled.
Habitat. Found in beets.

183. B. gummosus Happ

Philos. Diss., Basel, 1893.

Morphology. Bacilli large, feebly motile.
Potato and *agar slant.* Growth white, crumpled. Ferments cane sugar;
medium viscous, with production of lactic and butyric acids and CO₂.
Habitat. Associated with a slimy fermentation of digitalis infusions, etc.

T

184. B. stellatus

B. lactis No. 9 Flügge: Zeitsch. f. Hygiene, XVII, 1894, 294.

Morphology. Bacilli long.
Gelatin colonies. With radiating outgrowths.
Agar slant. Growth waxy, crumpled.
Potato. Growth whitish yellowish, crumpled.
Bouillon. Turbid, with a crumpled pellicle.
Milk. Peptonized.
Habitat. Milk.

185. B. cremoris

B. lactis No. 10 Flügge: l.c.

Morphology. Bacilli small — filaments. Like the preceding in cultural characters, but with a thick creamy layer on bouillon.
Habitat. Milk.

186. B. denitrificans Schirokikh

Centralblatt f. Bakteriol., 2te Abt., II, 1896, 204.

Morphology. Bacilli 1.5-2 μ times their breadth, ends rounded, in chains of 2-8. Obligate aerobic. Optimum temperature, 37°.
Gelatin colonies. In 2 days, 1-2 mm., round — irregular; liquefied gelatin bluish; containing granules.
Gelatin stab. Liquefaction infundibuliform, with a granular sediment.
Agar colonies. In 24 hours at 37° starred, or like a snow crystal; centres papillate, yellowish, elevated.
Potato. Growth brownish, crumpled, stringy.
Bouillon. Turbid, with a white, crumpled pellicle.
Milk. Peptonized.
Habitat. Isolated from horse manure.

187. B. detrudens Wright

l.c., 452.

Morphology. Bacilli medium-sized, blunt, 2-3 times their breadth. Flagella peritrichic.
Gelatin colonies. *Deep colonies:* round — oval, brownish, entire, or beset with plaques or buds. *Surface colonies:* round, whitish, translucent, entire disks. Microscopically, brownish, granular, entire, with a wide zone of clouded, liquefied gelatin.
Agar slant. Growth creamy white, opaque, and glistening.

Bouillon. Turbid, with a slight pellicle.

Potato. Growth spreading, light brown, reticulately wrinkled.

Litmus milk. Decolorized, amphoteric, coagulated.

Glucose bouillon. No gas. *Indol* production slight or doubtful. Grow at 36°.

Habitat. Water.

188. B. maidis Paltauf-Heider

Wiener Med. Jahrb., 1888, 383.

Not differentiated from *B. vulgatus*, except potato growth is wrinkled, and yellow-brown.

Habitat. Isolated from spoiled corn.

189. B. plicatus

B. lactis No. 7 Flügge: Zeitsch. f. Hygiene, XVII, 1894, 294.

Morphology. Bacilli long.

Agar slant and *potato.* Growth brown, crumpled.

Bouillon. Turbid, slight pellicle.

Milk. Peptonized, toxic.

Habitat. Milk.

190. B. gangrænæ Arkövy

B. gangræna-pulpæ Arkövy: Centralblatt f. Bakteriol., XXIII., 1897, 921.

Morphology. Bacilli 4.0 μ, square-ended; occur singly and in chains. Occasionally as filaments or coccoid forms, with other degeneration forms in old agar cultures. Stain by Gram's method. Spores oval, in the centre of the rods.

Gelatin colonies. In 24 hours, white, mealy. Microscopically, finely granular, golden yellow, round, with outgrowths; in about 30 hours the gelatin is liquefied; on the fluid gelatin a white, crumpled membrane; a stinking, cheesy odor.

Gelatin stab. In 24 hours a small funnel with a white membrane; liquefaction extending to walls of the tube. The membrane becomes dirty brown, the gelatin reddish brown and alkaline.

Agar colonies. In 24-30 hours white, mealy, like gelatin colonies.

Agar slant. Growth limited, crumpled, becoming in 5-6 days ash-gray; the medium a beautiful brownish gray.

Blood serum. A brownish stripe, liquefied.

Bouillon. A membrane; medium-colored, like the liquefied gelatin.

Potato. Growth moist, brown, crumpled.

Milk. Coagulated.

Pathogenesis. By subcutaneous inoculations mice die in 4–12 days with symptoms of diarrhœa; bacilli in the blood. With rabbits there is elevated temperature, and a portion die. Guinea pigs die in a proportion of the inoculations. Inoculation of human teeth results in a gangrenous alteration of the pulp.

Habitat. Associated with gangrene of tooth pulp and caries of teeth, chronic alveolar abscess, etc.

191. B. aromaticus

B. lactis No. 11 Flügge: l.c.

Morphology. Bacilli slender.

Gelatin colonies. An irregular tangle of threads.

Gelatin stab. Show lateral outgrowths, slowly liquefied.

Agar slant and *potato.* Growth thick, leathery, channelled, becoming bright brown.

Milk. Coagulated, slowly peptonized, with an aromatic odor.

Habitat. Milk.

192. B. magnus

B. lactis No. 8 Flügge: l.c.

Morphology. Bacilli large, thick.

Agar slant. Growth white, glistening.

Potato. Growth white — yellow — brown, crumpled.

Bouillon. Clear, with a pellicle.

Milk. Peptonized.

Habitat. Milk.

193. B. Hueppei Trev.

B. butyricus Hueppe: Miteillungen Kaiserlich. Gesundheitsamte, II, 1884, 309.

Clostridium Hueppei Trevisan: Genera, 1889, 22.

B. pseudo-butyricus Kruse: Flügge, Die Mikroorganismen, 1896, 207.

Agar slant. Growth whitish-bluish, smooth.

Potato. Growth brownish, becoming crumpled.

Lactose bouillon. No gas; butyric acid produced.

Milk. Coagulated, not acid, peptonized, with production of leucin, tyrosin, and ammonia.

Habitat. Milk.

194. B. subtilis (Ehrenberg) Cohn

Vibrio subtilis Ehrenberg: Infusionsthierchen als volkommene Organismen, Leipzig, 1838.

B. subtilis Cohn: Beiträge Biol., I, Heft 2, 1875.

Morphology. Bacilli 1.2 : 3–4 μ — filaments — chains; ends rounded. Spore germination equatorial. Flagella peritrichic. Stain by Gram's method.

Gelatin colonies. Crateriform, turbid; microscopically, round, entire, becoming ciliate; dense flocculi in the liquefied gelatin — felted, floccose, densest in the centre.

Gelatin stab. Liquefaction crateriform — saccate — stratiform, with a pellicle on the surface.

Agar colonies. Anthrax-like.

Bouillon. Turbid; a membrane on the surface adhering to the walls of the tube.

Potato. Growth white, thick, mealy.

Agar slant. Growth thick, crumpled.

Milk. Coagulated, peptonized, slightly alkaline. *Indol* negative.

Habitat. Widely distributed.

195. B. lactis-albus Sternberg

Described but not named Löffler: Berliner klin. Wochenschrift, 1887, Nos. 33-34.
B. lactis-albus Sternberg: Manual, 1892, 680.

As far as described like the preceding. In milk, the production of butyric acid, leucin, and tyrosin.

Habitat. Milk.

196. B. leptosporus Klein

Centralblatt f. Bakteriol., VI, 1889, 6.

In fluid media like *B. subtilis.* Characters on solid media not described. Spore germination polar. (See Fig. 6.)

Habitat. A culture contamination.

197. B. megatherium De Bary

Vergl. Morph. Phys. u. Biol. der Pilze, Strassburg, 1884.

Morphology. Bacilli very large, approaching 2-5 microns in thickness; in cultures, after successive generations much smaller, 0.6-0.8 μ, somewhat bent and curved. Numerous peritrichic flagella. Aerobic. Stain by Gram's method.

Gelatin colonies. Deep colonies gray white, translucent, denser in the centre, granular, ciliate. Surface colonies raised, centres dense, borders ciliate.

Gelatin stab. Liquefaction saccate; no membrane.

Agar colonies. Elevated, white, grayish, moist, glistening; microscopically, segmented.

Agar slant and *potato.* Growth like *B. subtilis,* slightly yellowish. *Indol* negative. H_2S positive.

Habitat. Isolated from cabbage infusion.

198. B. Ellenbachensis alpha Stutzer-Hartleb

Centralblatt f. Bakteriol., 2te Abt., IV, 1898, 31.

Morphologically like *B. mycoides* and *B. megatherium*. In long filaments. Spore formation and germination like *B. megatherium*. Spores oval.

Glucose gelatin colonies. Round, with a thick central nucleus and a turbid border; becoming sunken in the liquefied gelatin; border filamentous; odor cheesy.

Agar colonies. In 24 hours at 20°, round, glistening, grayish white. Microscopically, dark gray and granular in the interior; border filamentous — floccose.

Potato. Growth yellowish gray, soft, flat; border indented.

Milk. Weakly acid.

Asparagin agar. Growth soft, with mycelial-like threads from the edge.

Habitat. Isolated from alinite.

199. B. subtilis-similis Sternberg

Manual, 1892, 679.

Morphology. Bacilli 1.0 : 2-4 μ — filaments. Facultative anaerobic. Gelatin liquefied more slowly than *B. subtilis*; a membrane on the surface of the liquefied gelatin.

Gelatin colonies. Like *B. subtilis*.

Agar slant. Growth creamy white.

Potato. Growth dry, yellowish.

Agar stab. Growth arborescent.

Habitat. Isolated from liver of yellow fever cadaver.

200. B. cereus

B. lactis No. 5 Flügge: Zeitsch. f. Hygiene, 1894, 294.

Morphology. Bacilli long, slender.

Gelatin colonies. With outgrowths.

Agar slant. Growth translucent, smooth.

Potato. Growth bright yellow, dry, becoming faintly crumpled.

Bouillon. Clear, with a membrane.

Milk. Peptonized.

Habitat. Milk.

201. **B. capillaceus** Wright

l.c., 456.

Morphology. Bacilli large, ends blunt, in chains.

Gelatin colonies. *Deep colonies:* small, round, hazy; microscopically, dark, granular, irregular — stellate — branched, composed of smaller daughter colonies and filamentous outgrowths. *Surface colonies:* in 2-3 days crateriform, 2 mm, denser and grayish in the centre; microscopically, dense central clumps and ciliate borders.

Gelatin stab. Slight growth in depth; on the surface, a crateriform-stratiform liquefaction and a wrinkled pellicle. Acid gelatin. Slight growth, becoming alkaline.

Agar slant. Growth grayish, frosted, becoming moist, glistening; the agar becomes brownish.

Bouillon. Clear, alkaline.

Potato. Growth thick, rough, slightly granular, purple-pink.

Litmus milk. Coagulated, decolorized, amphoteric.

Glucose bouillon. No gas. *Indol* slight or doubtful. Grow at 36°.

Habitat. Water.

202. **B. rudis**

B. lactis No. 6 Flügge: Zeitsch. f. Hygiene, XVII, 1894.

Morphology. Bacilli slender.

Gelatin colonies. With fine outgrowths.

Agar slant. Growth white, crumpled.

Potato. Growth twisted; surface rough.

Bouillon. Turbid, slight pellicle.

Milk. Coagulated, peptonized.

Habitat. Milk.

203. **B. circulans** Jordan

Report Mass. Board of Health, 1890, 831.

Morphology. Bacilli 1.0 : 2-5 μ, in chains.

Gelatin colonies. In 2-4 days, round, brownish; later, depressions, due to liquefaction.

Gelatin stab. Slight growth in depth, a conical cavity; evaporation exceeds rate of liquefaction.

Agar slant. Growth thin, translucent.

Potato. Growth somewhat scanty, the color of the medium.

Bouillon. In 3-4 days, turbid; no pellicle.

Milk. Slightly acid, coagulated slowly or not at all. Nitrates reduced to nitrites.

Habitat. Water.

204. B. limosus Russell

Zeitsch. f. Hygiene, XI, 1891, 196.

Morphology. Bacilli large, rather slender; plasma granular, 1.2 : 3-4 μ.

Gelatin colonies. Slightly transparent, with slender filaments extending into the gelatin; later, larger liquefied depressions, with thorn-like projections.

Gelatin stab (gelatin made with sea water). Liquefaction infundibuliform — saccate.

Agar slant. Growth moist, glistening, white.

Bouillon. Turbid, with a pellicle.

Potato. Growth thin, grayish white.

Habitat. Sea water.

205. B. sputi

B. No. VI Pansini : Virchow's Archiv, CXII, 1890.

Morphology. Bacilli slender.

Colonies. Proteus-like.

Agar slant. Growth porcelain-white.

Potato. An invisible growth.

Bouillon. A membrane; no odor.

Habitat. Sputum.

206. B. vacuolosis Sternberg

Manual, 1892, 717.

Morphology. Bacilli 1.0 : 1.5-5 — filaments; ends rounded.

Gelatin stab. Liquefaction crateriform; liquefied gelatin viscid; a pellicle on the surface.

Agar slant. Growth creamy white.

Potato. Growth thin, creamy white.

Pathogenesis. Non-pathogenic to rabbits.

Habitat. Isolated from the stomach and intestines of yellow fever cadavers.

207. B. pseudo-anthracis Kruse

Milsbrandähnlicher Bacillus aus Südamerikanischem Fleischfuttermehl Burri : Hyg. Rundschau, 1894, No. 8, ref. Centralblatt f. Bakteriol., XVI, 374.

B. pseudo-anthracis Kruse: Flügge, Die Mikroorganismen, 1896, 233.

Morphology. Bacilli like anthrax, 1 : 3-6 — filaments; slowly motile.

Gelatin colonies. Like anthrax.

Gelatin stab. Liquefaction more rapid than in anthrax, without outgrowths.

Bouillon. Turbid, with a membrane, becoming clear.
Agar slant. Growth gray, smooth, not floccose or crumpled.
Potato. Growth gray-white, soft, becoming moist, glistening.
Milk. Coagulated, amphoteric.
Pathogenesis. Non-pathogenic to mice.
Habitat. Isolated from South American bran.

208. B. vaculatus Ravenel

l.c., 31.

Morphology. Bacilli straight, with rounded ends; occur singly; stain like Klebs-Löffler bacillus.
Gelatin colonies. Deep colonies: yellowish, granular, entire. *Surface colonies:* in 2 days, 0.5 mm., yellowish, entire, granular, mottled; liquefaction crateriform.
Gelatin stab. Surface growth raised; in 3 days, a crateriform liquefaction, becoming stratiform, with a membrane.
Agar slant. Growth thin, translucent, moist, glistening; agar stained a faint greenish.
Potato. Growth light brown, slimy, thick, moist, becoming yellowish brown.
Bouillon. Turbid, becoming clear.
Pepton rosolic acid solution. In 12–14 days, a cherry-red.
Litmus milk. Rendered alkaline, becoming watery, translucent, decolorized, peptonized, alkaline; not coagulated.
Glucose bouillon. No gas. *Indol* positive. Grow at 36°.
Habitat. Soil.

209. B. Flüggei

B. lactis No. 1 Flügge: Zeitsch. f. Hygiene, XVII, 1894.

Morphology. Bacilli thick, short.
Gelatin colonies. With outgrowths.
Agar slant. Growth grayish white.
Potato. Growth gray-white.
Bouillon. Turbid, with a flocculent sediment.
Milk. Peptonized, toxic.
Habitat. Milk.

210. B. crinatus

B. No. V Pansini: Virchow's Archiv, CXXII,

Morphology. Bacilli rather more slender than *B. subtilis.*
Gelatin colonies. Coli-like, with fine, undulating radiations.
Gelatin stab. Liquefaction saccate.

Bouillon. Turbid, with a pellicle.

Agar slant. Growth flat, white.

Potato. Growth flat, white, with odor of rotten cheese.

Habitat. Sputum.

211. B. Krameri

B. der Nassfäule der Kartoffeln Kramer: Österr. landwirtsch. Centralblatt, 1891, 11.

B. of potato rot Sternberg: Manual, 1892, 716.

Morphology. Bacilli 0.7–0.8 : 2.5 μ — chains, filaments.

Gelatin stab. Liquefaction infundibuliform.

Gelatin slant. In 24 hours, a dirty white line of growth, with scalloped margins.

Agar slant. Growth of dirty white, slimy drops.

Milk. Coagulated, but no putrefactive change; dextrin decomposed to butyric acid and CO_2; dissolves starch without decomposing it.

Habitat. Associated with wet rot of potato.

CLASS XVI. WITH ENDOSPORES. AEROBIC AND FACULTATIVE AN-
AEROBIC. RODS NOT SWOLLEN AT SPORULATION. GELATIN
NOT LIQUEFIED.

 I. Pathogenic, animal habitats.

   212. *B. Koubasoffii.*

   213. *B. Afanassieffi* Trevisan.

   214. *B. Cladoi* Trevisan.

 II. Non-pathogenic.

  A. Milk rendered acid.

   215. *B. punctiformis.*

   216. *B. Weigmanni.*

  B. Milk rendered alkaline, or unchanged in reaction.

   217. *B. siccus.*

   218. *B. ginglymus* Ravenel.

  C. Milk cultures not described.

   219. *B. cuticularis* Tataroff.

   220. *B. Foutini.*

212. B. Koubassoffii

B. der krebsartigen Neubildungen Koubassoff: Vortrag. Moskauer Militärärztlichen Verein, No. 22, 1888. Ref. Centralblatt f. Bakteriol., VII, 1890, 317.

B. of Koubassoff Sternberg: Manual, 1892, 405.

Morphology. Bacilli 2–3 times the length of tubercle bacilli, and 3–4 times as thick; ends rounded, or one end pointed. Grow at 36°. Facultative anaerobic.

Glycerin gelatin stab. Growth in depth, slender, jagged; on the surface, a bluish membrane in the form of a funnel.

Agar slant. At 36° growth bluish white.

Potato. Growth at first typhoid-like, becoming a granular membrane.

Pathogenesis. Subcutaneous inoculations of guinea pigs cause death in 1–2 weeks. There is emaciation, paralysis of the sphincter muscles, nodular elevations on the mucous wall of stomach, etc.

Habitat. Isolated from growths on the stomach of a person who died of cancer of the stomach.

213. B. Afanassieffi Trevisan

B. des Keuchhustens Afanassieff: St. Petersburg med. Wochensch., 1887, No. 39–42.
B. Afanassieffi Trevisan: Genera, 1889, 13.

Morphology. Bacilli 0.6–2.2 μ; solitary in pairs and in short chains.

Gelatin colonies. Round — oval, light brown; microscopically, finely granular and dark brown.

Gelatin stab. Slight growth in depth; on the surface the growth is grayish white.

Agar slant. Growth thick, gray, limited.

Potato. Growth yellowish glistening dew-like drops, becoming thicker, brownish, and spreading.

Pathogenesis. Inoculations into the air passages and pulmonary parenchyma of young dogs and rabbits cause bronchial catarrh, bronchopneumonia, and sporadic coughing; bacilli in the bronchial and nasal secretions.

214. B. Cladoi Trevisan

Bacille pedunculé Clado: Bull. Soc. Anat., Paris, 1887, 339.
B. Cladoi Trevisan: Genera, 1889, 14.
B. septicus-vesicæ Sternberg: Manual, 1892, 475.

Morphology. Bacilli 0.5 : 1.6–2.0 μ; never in pairs — chains. Stain by Gram's method. Facultative anaerobic. Grow at 36°.

Gelatin colonies. Do not exceed 1.2 mm., round, oval, transparent, yellowish white — dark gray, with a zone of yellow.

Gelatin stab. Surface growth thin, jagged.

Agar slant. Growth scanty, a grayish yellow stripe.

Potato. Growth flat, dry, brown.

Pathogenesis. Inoculations of mice, guinea pigs, and rabbits cause toxæmia and septicæmia.

Habitat. Isolated from urine of a person with cystitis.

215. B. punctiformis

B. No. 23 Conn : l.c., 1893, 53.

Morphology. Bacilli plump rods, 0.8 : 1.5 μ, in twos, no chains. Facultative anaerobic. Grow at 35°.

Gelatin colonies. Small round transparent beads, 0.5 mm.; in 5 days 1 mm., small, white, glistening.

Gelatin stab. Surface growth thin, rough, somewhat transparent.

Agar slant. Growth white, somewhat elevated, spreading.

Potato. Growth moist, white, glistening, elevated, becoming brownish.

Milk. Not coagulated, thickens to a pasty mass, becoming brownish, acid.

Bouillon. Turbid, with a sediment; no pellicle.

Habitat. Milk.

217. B. siccus

B. No. 25 Conn : l.c., 53.

Morphology. Bacilli 0.7 : 2.0 μ; no chains. Facultative anaerobic. Grow at 36°.

Gelatin colonies. Small, round, with concentric wrinkles, becoming 1 mm., with a central nucleus and a darker rim, separated by a partly clear space; the edge may be rough or lobed.

Gelatin stab. Surface growth thin, spreading, transparent, dry, white.

Agar slant. Growth white, somewhat elevated.

Potato. Growth grayish, dry, elevated, yellowish brown.

Milk. Slightly alkaline.

Bouillon. Turbid, with a pellicle.

Habitat. Milk.

218. B. ginglymus Ravenel

l.c., 37.

Morphology. Bacilli straight, 3-7 times their breadth; occur singly and in chains of 2-3 elements; ends rounded.

Gelatin colonies. Deep colonies: yellowish, granular, 0.5 mm. *Surface colonies:* in 24 hours, minute, white, punctiform, gray, granular; edges irregular; in 36 hours 0.25 mm., white; centre orange-brown, marmorated with an outer gray zone; do not exceed 0.5 mm. In 7 days a colorless border, finely veined, with a brown centre.

Agar slant. A grayish white line, moist, glistening, 1 mm. wide.

Gelatin stab. In depth, 10 days, indistinct globular outgrowths; on the surface a grayish button, 2-3 mm.

Potato. Growth thin, spreading, yellowish, moist, glistening, becoming brownish.

Bouillon. Turbid.

Pepton rosolic acid solution. Becomes lighter; decolorized in 2 weeks.

Litmus milk. Alkaline, not coagulated, becoming translucent, and in 2 weeks violet.

Glucose bouillon. No gas. *Indol* negative. Grow at 36°.

Habitat. Soil.

219. B. cuticularis

B. cuticularis-albus Tataroff: Die Dorpater Wasserbakterien, Dorpat, 1891, 24.

Morphology. Bacilli 3.2 μ long, variously bent filaments, in pairs.

Gelatin colonies. *Deep colonies:* round — oval, entire. *Surface colonies:* irregular, bluish white, opalescent; microscopically, brownish, edge irregular, granular.

Gelatin stab. In depth, beaded; later with ligulate outgrowths; surface growth, irregular — rosette-shaped, white, glistening, spreading.

Agar slant. Growth white, glistening, spreading.

Bouillon. Turbid, white sediment, flocculent particles in the fluid, and a whitish pellicle.

Potato. Growth thick, moist, glistening, brownish; surface irregular, becoming reddish brown — yellowish brown.

Habitat. Water.

220. B. Foutini

Bacillus D Foutin: Centralblatt f. Bakteriol., VII, 1890, 373.

Morphology. Bacilli 1.0 : 5-20 μ, thinner at the poles, which are slightly rounded; one to four spores in a single rod; slightly motile.

Gelatin stab. In depth, growth beaded; on the surface, a nailhead growth.

Agar slant. Growth limited, rather thick, with mother-of-pearl iridescence.

Potato. Growth somewhat raised, yellow, limited.

Pathogenesis. Non-pathogenic.

Habitat. Isolated from hail.

CLASS XVII. WITH ENDOSPORES. AEROBIC AND FACULTATIVE ANAEROBIC. RODS NOT SWOLLEN AT SPORULATION. POTATO CULTURES DEVELOPING A RED PIGMENT, CHROMOGENIC IN PART.

      I. Agar smear cultures yellowish to reddish.

          221. *B. coccineus* Pansini.

      II. Agar smear cultures whitish or grayish.

        1. Milk coagulated.

          223. *B. viscosus.*

        2. Milk not coagulated.

          224. *B. vitalis.*

221. B. coccineus Pansini

Virchow's Archiv, CXXII, 1890.

Morphology. Bacilli large, slightly motile.
Gelatin colonies. Ciliate.
Gelatin stab. Liquefaction infundibuliform; a thin, yellowish membrane, and a white sediment.
Agar slant. Growth yellowish — reddish.
Potato. Red points coalescing, with grayish folds between.
Bouillon. Slightly turbid, with a delicate membrane.
Habitat. Sputum.

223. B. viscosus

Roter Kartoffelbacillus Vogel: Zeitsch. f. Hygiene, XXVI, 1897, 404.

Morphology. Bacilli slender; ends rounded; no chains. Small, glistening spores. Stain by Gram's method.
Gelatin colonies. Flat, crateriform; the dark nucleus shows spiny outgrowths.
Agar colonies. Gray brown; a thick nucleus with delicate outgrowths.
Agar slant. Growth dry, gray white.
Potato. Growth reddish yellow, becoming rose-red, rugose.
Milk. Coagulated.
Bouillon. Clear, with a thick membrane. Optimum temperature 37°.
Habitat. Isolated from stringy bread.

224. B. vitalis

B. mesentericus-ruber Globig: Zeitsch. f. Hygiene, III, 1888, 322.

Morphology. Bacilli 0.4 : 1-4 μ — filaments. Stain by Gram's method. According to Globig, cultures withstand boiling for 5-6 hours. Spores?
Gelatin colonies. Vary in character from those of *B. typhosus* to those of *B. subtilis.*
Gelatin stab. Typhoid-like, later a shallow funnel of liquefied gelatin, with a membrane on the surface.
Bouillon. Clear, with a thick membrane.
Potato. Growth becomes red to reddish brown.
Milk. Not coagulated, slightly alkaline.
Glucose bouillon. No gas. H₂S negative.
Pathogenesis. Non-pathogenic to mice and guinea pigs.
Habitat. Isolated from potato.

CLASS XVIII. WITH ENDOSPORES. AEROBIC AND FACULTATIVE ANAEROBIC. RODS BECOMING SPINDLE-SHAPED AT SPORULATION, OF THE CLOSTRIDIUM TYPE.

I. Gelatin liquefied.
    A. Grow in ordinary nutrient gelatin.
        1. Growth along needle track in gelatin shows filamentous radiations.
            a. Gelatin colonies becoming ameboid or proteus-like.
                225. *B. alvei* Chesire-Cheyne.
                226. *B. licheniformis* Weigmann.
            b. Gelatin colonies seldom with outgrowths.
                227. *B. inflatus* A. Koch.
        2. Growth along needle track in gelatin stab eroded — funnelled.
            228. *B. erodens* Ravenel.
        3. No growth in the depth of the gelatin ; obligate aerobic.
            229. *B. saprogenes* Kramer.
            230. *B. Baccarinii* Macchiati.
    B. Grow in nutrient gelatin only upon the addition of NH_4HO or urea.
        231. *B. Duclauxi* (Miquel).
II. Gelatin not liquefied.
        232. *B. cinctus* Ravenel.
III. Action on gelatin not stated.
        233. *B. catenula* (Duclaux).
        234. *B. urocephalus* (Duclaux).
        235. *B. filiformis* (Duclaux).

225. B. alvei Chesire-Cheyne
Jour. Roy. Mic. Soc., 1885, 582.

Morphology. Bacilli 0.8 : 2.5–5 μ.
Gelatin colonies. Round, entire, becoming ameboid.
Gelatin stab. Growth arborescent, becoming liquefied.
Agar slant. Growth thin, white.
Potato. Growth yellowish.
Milk. Coagulated, slightly acid, peptonized. Pathogenic for bees.
Habitat. Associated with *fowl brood* of bees.

226. B. licheniformis Weigmann
Centralblatt f. Bakteriol., IV, 1898, 822.

Morphology. Bacilli 0.6–0.8 : 1.8–2.6 μ — long filaments. Clostridium — clavate forms at sporulation. Spores 0.45 : 1.3 μ. Spore germination polar. Not stained by Gram's method. Flagella numerous peritrichic.

Gelatin colonies. Cochleate, with filamentous outgrowths like *B. Zopfi.*

Agar colonies. Deep colonies : densely floccose in the centre, loosely filamentous on the border. *Surface colonies :* round, watery, becoming spreading, becoming dry and membranaceous, yellowish gray ; border lobate.

Potato. Growth yellow, flat, spreading, slimy.

Agar slant. A dry, spreading, membranous growth.

Glucose gelatin stab. In 14 days a funnel of liquefaction, with filamentous outgrowths in depth ; on the surface, a thick rugose membrane.

Milk. In 3–4 days at 37° slimy — thick, amphoteric — slightly alkaline ; in 5 days, coagulated, slightly alkaline, becoming slowly peptonized ; a cheesy odor.

Bouillon. A surface growth.

Habitat. Cheese.

227. **B. inflatus** A. Koch

Bot. Zeitung, 1888, 328.

Morphology. Bacilli 0.6–0.8 : 4–5 μ — filaments.

Gelatin colonies. Round, seldom with outgrowths.

Gelatin stab. Short, delicate radiations ; liquefied slowly.

Potato. Growth thin, slimy, brown.

Bouillon. A smooth membrane on the surface.

Habitat. A contamination.

228. **B. erodens** Ravenel

l.c., 35.

Morphology. Bacilli straight, thick rods, 3–7 times their breadth ; occur singly ; ends rounded.

Gelatin colonies. In 24 hours, minute, white, reticulate — moruloid ; in 36 hours, 1.0 mm., round, entire, dark gray ; show swarming movements. In 3 days, 2 mm. ; tunnels run out from edge of colonies, with often curled ends.

Agar slant. Growth thin, spreading, translucent, greenish, with white raised points.

Gelatin stab. In 3 days, a small funnel, curled at bottom, walls of main funnel eroded with minor curled tunnels from the former.

Potato. In 3 days, growth thin, moist, glistening, honey-colored.

Bouillon. Turbid, becoming clear ; no pellicle.

Pepton rosolic acid solution. Unchanged ; in 5–6 weeks slightly deeper in color.

Litmus milk. Color discharged in 12 days, alkaline, not coagulated.
Glucose bouillon. No gas. *Indol* negative. Grow at 36°.
Habitat. Soil.

229. B. saprogenes Kramer

B. saprogenes-vini No. VI Kramer: Bakteriologie Landwirtsch., 1890, 139.

Morphology. Bacilli 1.0 : 2.0 μ. Obligate aerobic.
Gelatin stab. Surface growth dirty white; liquefied rapidly. Ammonia developed in old cultures.
Habitat. Isolated from diseased wine.

230. B. Baccarinii Macchiati

Centralblatt f. Bakteriol., IV, 1898, 332.

Morphology. In gelatin cultures rods 0.7–0.8 : 2.0–3.5 μ, straight — slightly curved; occur singly, in chains, and as zoöglœa. In bouillon, long filaments like Leptothrix. Flagella peritrichic. Rods swollen in the middle at sporulation, oval. Spore germination polar.
Gelatin stab. In 4–5 days, an empty shallow funnel, beginning with an air bubble above, lined with a white growth, with a membrane on the liquefied gelatin. The gelatin of old cultures is colored black, and is fluorescent.
Agar slant. Raised, yellowish colonies, becoming gray, spreading.
Milk. In 3–4 days, coagulated; in 14 days, peptonized, acid.
Potato. Growth light yellow, becoming straw yellow; potato liquefied or softened. Optimum temperature 23–25°, maximum 40°.
Habitat. Associated with *mal-nero* of the vine.

231. B. Duclauxi Miquel

Urobacillus Duclauxi Miquel: Ann. Micrographie, II, 1889, 58.

Morphology. Bacilli 0.6–0.8 : 2–10 μ; chains. Gelatin slowly softened.
Bouillon (made alkaline with NH₄HO). Turbid, becoming viscous; bad odor.
Habitat. Isolated from canal and river water.

232. B. cinctus Ravenel: l.c.

Morphology. Bacilli straight, variable, involution forms. Rods show deeply stained spots.
Gelatin colonies. In 24 hours, minute, yellowish, granular, entire; in 72 hours, 0.5 mm., nucleus surrounded by a yellowish zone, then a gray-veined zone, with irregular margins. In 8 days, 1 mm., grayish white, round, elevated, entire, finely veined — mottled, often with ferny outgrowths.

U

Gelatin stab. Good growth in depth; on the surface, an irregular growth, 2 mm.

Agar slant. Growth thin, glistening, becomes a faint yellow.

Potato. In 3 days, a thin yellow moist glistening growth.

Bouillon. Turbid, becoming clear, the medium a faint greenish.

Pepton rosolic acid solution. In 3–4 days, slightly darker.

Litmus milk. Darker, becoming decolorized, not coagulated, alkaline.

Glucose bouillon. No gas. *Indol* negative. Grow at 36°.

Habitat. Soil.

233. B. catenula (Duclaux)

Tyrothrix catenula Duclaux: Le Lait, Paris, 1887, 249.

Morphology. Bacilli 0.6–1.0 μ thick, in filaments.

Milk. Coagulated, gas produced, peptonized; production of leucin, tyrosin, butyric acid, and ammonia.

Habitat. Cheese.

234. B. urocephalus (Duclaux)

Tyrothrix urocephalus Duclaux: l.c.

Morphology. Bacilli thick, filaments.

Milk. Coagulated at body temperature, at ordinary temperatures scarcely altered; on the surface, a gelatinous mass; production of acid, leucin, and tyrosin.

Habitat. Cheese.

235. B. filiformis (Duclaux)

Tyrothrix filiformis Duclaux: l.c.

Morphology. Bacilli in milk 0.8 μ thick, filaments.

Milk. In 2–3 days, unaltered, becoming peptonized to a turbid fluid; on the surface, a crumpled membrane; production of leucin and tyrosin.

Habitat. Cheese.

CLASS XIX. WITH ENDOSPORES. AEROBIC AND FACULTATIVE AN-
AEROBIC. RODS SWOLLEN AT ONE END AT SPORULATION, OF
THE TETANUS TYPE

I. Gelatin liquefied, at least in gelatin stab cultures.
    A. Grow rapidly in nutrient gelatin.
        236. *B. sublanatus* Wright.
    B. Grow very poorly in nutrient gelatin.
        237. *B. lacteus.*
II. Gelatin not liquefied.
        238. *B. putrificus* Flügge.

236. B. sublanatus Wright : l.c.

Morphology. Bacilli of medium size ; ends rounded ; occur in pairs and in long forms.

Gelatin colonies. Deep colonies : round, brownish, granular, entire. *Surface colonies :* in two days, 1–2 mm., round grayish disks. Microscopically, granular, dense toward centre, margins more translucent and sharp. In 3 days, the colonies are crateriform ; centres yellowish white, borders ciliate.

Gelatin stab. Slight growth in depth ; on the surface, a napiform to stratiform liquefaction, with a white pellicle on the surface.

Agar slant. Growth a translucent thin grayish narrow stripe.

Bouillon. Turbid ; the medium becomes greenish in tint.

Potato. Growth brownish, thin, granular, moist, spreading.

Litmus milk. Not coagulated, decolorized, alkaline, slowly peptonized.

Glucose bouillon. No gas. *Indol* slight or doubtful. Grow at 36°.

Habitat. Water.

237. B. lacteus

B. lactis No. 12 Flügge: Zeitsch. f. Hygiene, XVII, 1894.

Morphology. Bacilli thin, slender.

Gelatin colonies. None visible after 2 days.

Gelatin stab. In 2–3 days, a faint development, and the beginning of liquefaction.

Agar slant. Growth white-gray, slimy.

Bouillon. A thin membrane on the surface, with flocculi in the medium.

Potato. Growth thin, limited, moist, becoming thick, yellowish.

Milk. At 37°, slowly peptonized, becoming bitter and toxic.

Habitat. Milk.

238. B. putrificus Flügge

B. aus Fæces IV Bienstock: Zeitsch. f. klin. Med., VIII.
B. putrificus-coli Flügge: Die Mikroorganismen, 1886.

Morphology. Bacilli slender, delicate, also as filaments. On gelatin, a mother-of-pearl growth, becoming yellowish. Decomposes fibrin.

Habitat. Fæces.

CLASS XX. WITH ENDOSPORES. OBLIGATE ANAEROBIC. RODS NOT
SWOLLEN AT SPORULATION.

MALIGNANT OEDEMA GROUP.

I. Gelatin liquefied.
  A. Gas developed in gelatin or agar media containing sugar.
    1. Colonies on gelatin compact, dense.
      a. Strongly pathogenic.
          239. *B. œdematis* Zopf.
      b. Negatively, or but slightly pathogenic.
          240. *B. pseudœdematis* Kruse.
    2. Colonies on gelatin radiately filamentous or mycelioid.
          241. *B. radiatus* Lüderitz.
          242. *B. thalasophilus* Russell.
  B. No gas development in gelatin or agar media containing sugar.
    1. Gelatin liquefied.
          243. *B. caris.*
    2. Gelatin not liquefied.
      a. Gelatin colonies roundish, dense, entire, not arborescent or
          mossy.
        * Gas produced in nutrient gelatin without sugar (glucose).
          244. *B. amylozyma* Perdrix.
        ** No gas produced in nutrient gelatin without glucose.
          245. *B. solidus* Lüderitz.
          246. *B. tardus.*
      b. Gelatin colonies mossy, with moss-like offshoots.
          247. *B. muscoides* Liborius.

239. **B.** œdematis Zopf

Vibrio septique Pasteur: Compt. rend., LXXXV, 1877.
Œdema Bacillus Koch: Mitteilungen Kaiserlich. Gesundheitsamte, 1881.
B. œdematis-maligni Zopf: Spaltpilze, 1885, 88.

Morphology. Bacilli 0.8–1.0 : 2–10 μ; ends rounded, also forms approaching
    anthrax; occur singly, in chains and in filaments. Not stained by
    Gram's method.
Gelatin colonies. Like *B. subtilis.*
Agar colonies. Composed of a dense network of threads.
Gelatin stab. Below the surface a white line, with short outgrowths, and gas.
Bouillon. Turbid, gas.
Litmus milk. Coagulated slowly or not at all, amphoteric, decolorized in
    depth. *Indol* slight..

Pathogenesis. Somewhat variable; pathogenic to mice, guinea pigs, and rabbits. Subcutaneous inoculations cause a bloody œdema with gas; bacilli present. Sometimes after death the bacilli invade the blood and organs.

Habitat. Associated with malignant œdema; found in earth, dirty water, dust, etc.

240. B. pseudœdematis Kruse

Pseudo-œdema Bacillus Liborius: Zeitsch. f. Hygiene, I, 1886, 115.
Anaerobic No. VII Sanfelice: Zeitsch. f. Hygiene, XIV, 1893, 339.
B. pseudo-œdematis Kruse: Flügge, Die Mikroorganismen, 1896.

Morphology. Bacilli somewhat thicker than the preceding; often many spores in a filament. Cultures like the preceding. Doubtfully or negatively pathogenic.

Habitat. Associated with œdema from earth infection, probably a non-virulent variety of the preceding.

241. B. radiatus Lüderitz

Zeitsch. f. Hygiene, V, 1889, 149.

Morphology. Bacilli 0.8 : 4–7 μ — filaments.
Gelatin colonies. Radiate — mycelioid.
Gelatin stab. Show filamentous outgrowths.
Agar stab. Shows delicate branching, and gas.
Pathogenesis. Non-pathogenic to mice.
Habitat. Soil.

242. B. thalasophilus Russell

Zeitsch. f. Hygiene, XI, 1891, 190.

Morphology. Bacilli slender, of variable length — filaments.
Gelatin stab. Liquefaction saccate; bad-smelling gas.
Gelatin colonies. A thin network of filaments which penetrate the gelatin in all directions.
Agar stab. Scanty growth.
Habitat. Isolated from sea water.

243. B. caris

Anaerobic No. VI Sanfelice: Zeitsch. f. Hygiene, XIV, 1893, 339.

Morphology. Bacilli variable in length.
Gelatin colonies. Branched.
Gelatin stab. A slight turbidity spreading downward; no gas, but a bad odor.

Milk. Coagulated, with a separation of serum.
Litmus milk. Decolorized.
Habitat. Isolated from putrefying flesh.

244. B. amylozyma Perdrix

Ann. Pasteur Inst., 1891, 287.

Morphology. Bacilli 0.5 : 2–3 μ; in twos or short chains.
Gelatin colonies. Small, gas-forming.
Potato. At 37°, white colonies, which soften the medium. In saccharine
  media acetic and butyric acids, with much gas. Starch converted into
  sugar, into ethyl and amyl alcohol, and into butyric acid. Cellulose
  attacked.
Habitat. Water.

245. B. solidus Lüderitz

Zeitsch. f. Hygiene, V, 1889, 149.

Morphology. Bacilli 0.5 : 1–5 μ; not in filaments.
Gelatin stab. Slight growth in depth; no gas.
Glucose gelatin stab. In depth, round colonies, with gas; odor of butyric acid.
Agar colonies. Look like little flocculi of cotton wool.
Bouillon. At 37° turbid; bad gases.
Habitat. Earth.

246. B. tardus

Anaerobic No. III Sanfelice: Zeitsch. f. Hygiene, XIV, 1893.

Morphology. Bacilli short rods, similar to the preceding. Grow very slowly.
Gelatin colonies. Golden yellow, granular, sharp.
Gelatin stab. In depth, isolated colonies.
Habitat. Earth, putrefying fluids, etc.

247. B. muscoides Liborius

Zeitsch. f. Hygiene, I, 1886, 115.

Morphology. Bacilli thick, with slight tendency to form filaments.
Gelatin and *agar colonies.* With delicate, branched, mossy offshoots.
Gelatin stab. Growth arborescent.
Habitat. Earth.

CLASS XXI. WITH ENDOSPORES. OBLIGATE ANAEROBIC. RODS BECOMING LATERALLY SWOLLEN OR SPINDLE-SHAPED AT SPORULATION. A FREQUENT VARIATION IN THIS REGARD IS OFTEN NOTED, IN WHICH THE RODS ARE SWOLLEN NEAR ONE END, APPROACHING THE TETANUS TYPE.

I. Gelatin liquefied. RAUSCHBRAND OR CLOSTRIDIUM GROUP.

   A. Gelatin liquefied slowly, or merely softened by the growth in gelatin stab cultures.

   248. *B. Feseri* (Trevisan) Kitt.
   249. *B. anaerobic V* and *VIII* of Sanfelice.
   250. *B. botulinus* v. Ermengem.

   B. Gelatin liquefied rapidly.
   1. Spores entirely or prevailingly in the centres of the rods, and at sporulation swollen in the middle — typical clostridium types.
      a. Cultures without a bad, putrid odor.
         * Milk coagulated.
         251. *B. butyricus* Botkin.
         ** Milk not coagulated.
         252. *B. amylobacter* v. Tieghem.
      b. Cultures have a bad, putrid odor.
         253. *B. fœtidus* (Liborius).
   2. Spores prevailingly at the ends of the rods, and at sporulation swollen near one end, approaching the tetanus type; often also swollen near the middle, approaching the clostridium type.
      a. Grow in ordinary nutrient gelatin.
         * Colonies in gelatin or glucose gelatin never radiating — filamentous.
            † Milk coagulated.
            254. *B. Kedrowski.*
            †† Milk not coagulated, unchanged in 8 days.
            255. *B. cuneatus.*
            ††† Milk becomes rapidly translucent — transparent, with much gas.
            256. *B. sporogenes* Klein.
         ** Colonies on gelatin radiately filamentous.
            257. *B. spinosus* Lüderitz.
            258. *B. cadaveris.*
      b. Do not grow in ordinary nutrient gelatin.
         259. *B. Weigmanni.*

II. Gelatin not liquefied; rods at sporulation between clostridium and tetanus
types, variable.

    A. Gas produced in media containing milk-sugar.

        260. *B. saccharobutyricus* v. Klecki.

    B. No gas produced in media containing milk-sugar.

      1. Agar colonies compound, moruloid.

        261. *B. polypiformis* Liborius.

      2. Agar colonies simple.

        262. *B. Sanfelicei.*

248. **B. Feseri** (Trevisan) Kitt

Rauschbrand des Rindes Bollinger-Feser: Wochschr. f. Thierheilk., 1878.

B. der Charbon Symptomatique Arloing-Cornevin-Thomas: Compt. rend., XC,
    1880, 1302-5.

Clostridium Feseri Trevisan: Atti Acc. Fis. Med. Stat. di Milano, III, 1885, 116.

B. sarcemphysematis Kitt.

B. carbonis Migula: Die Natürlichen Pflanzenfamilien, 1895.

B. anthracis-symptomatici Kruse: Flügge, Die Mikroorganismen, 1896.

Morphology. Bacilli 3-5 μ in length; in thickness between anthrax and
malignant œdema bacilli. Stained by Gram's method. Flagella
peritrichic.

Gelatin colonies. Round, irregular; surface warty; radiating filaments grow
out into the gelatin.

Gelatin stab. Medium liquefied slowly, with gas production; a turbidity along
the line of inoculation, with outgrowths.

Litmus milk. Decolorized in depth, reddened on the surface.

Milk. Coagulated, slightly acid. *Indol* slight. H_2S positive.

Pathogenesis. Guinea pigs show a bloody gaseous œdema by subcutaneous
inoculation of large doses. Rabbits and mice immune.

Habitat. Associated with Rauschbrand, symptomatic anthrax, quarter-evil, or
black-leg, of sheep, cattle, and goats.

249. **B. anaerobic No. VIII** Sanfelice and **B. anaerobic No. V** Sanfelice

l.c.

Isolated from earth and putrefying flesh, indistinguishable from the preceding,
except that both are non-virulent.

250. **B. botulinus** v. Ermengem

Zeitsch. f. Hygiene, XXVI, 1898, 1. Schneidemühl: Centralblatt f. Bakteriol., XXIV, 582.

Morphology. Bacilli 0.9–1.2 : 4–9 μ; like anthrax and malignant œdema. Clostridium forms, often in chains of two and more. Polar oval spores. Flagella 4–8. Stain by Gram's method. Optimum temperature 20°–30°; growth ceases at 38.5°.

Gelatin colonies. Round, translucent, coarsely granular; gelatin slowly liquefied.

Glucose gelatin colonies. Round, translucent, bright yellowish brown; coarsely granular, with motion of the granules; on the periphery a slight liquefaction, becoming incised and lobed.

Glucose bouillon. Turbid, with a butyric acid odor.

Glucose gelatin stab. In depth, with radiating outgrowths, liquefied slowly, gas, yellowish white sediment.

Potato. No growth even in anaerobic conditions.

Glucose agar slant. As before.

Milk. Not coagulated, and only slight growth. No putrefactive odor in cultures, but a sour smell like rancid butter. No gas in lactose and saccharose bouillon. In ordinary agar and gelatin, without glucose, no gas.

Pathogenesis. Pure cultures given to cats cause the same symptoms as poisonous meat; hyperæmia and small hemorrhages of the liver, kidneys, and central nervous system. In guinea pigs, death in 2 days; in apes, in 30 hours.

Habitat. Isolated from ham which had caused meat poisoning, *botulism.*

251. **B. butyricus** Botkin

Zeitsch. f. Hygiene, XI, 1892, 421.

Morphology. Bacilli 0.5 μ thick, of variable lengths — filaments. Slightly motile. Rods contain granules, which stain with iodine.

Gelatin colonies. Round, with outgrowths.

Agar colonies. Felted — floccose, with filamentous, radiating borders.

Agar stab. Much gas.

Milk. In depth, a clear serum; much gas; coagulated. Produces butyric, propionic, acetic, formic, and lactic acid. Does not decompose cellulose or salts of lactic acid. Gas produced without a bad odor.

Habitat. Isolated from milk, water, earth.

252. B. amylobacter v. Tieghem

Vibrio butyrique Pasteur: Compt. rend., LII, 1861.
B. amylobacter van Tieghem: Compt. rend., LXXXVIII, 1878, LXXXIX, 1879.
B. navicula Reinke-Berthold: Zersetzung Kartoffel durch Pilze, Berlin, 1879.
Clostridium butyricum Prazmowski: Untersuch. über die Entwick. u. Fermentwirk. einiger
    Bact., Leipzig, 1880.

Morphology. Bacilli 1.0 : 3-10 μ — filaments ; contain granulations, which
    stain with iodine. Spores 1 : 1-2.5 μ. Spore germination polar.
Milk. Not coagulated or doubtful ; slowly peptonized. Ferments cellulose.
Habitat. Widely distributed.

253. B. fœtidus (Laborius)

Clostridium fœtidum Liborius: Zeitsch. f. Hygiene, I, 1886, 160.

Morphology. Bacilli 1.0 μ thick, length variable — filaments, like the preced-
    ing. At sporulation, swollen mostly in the middle, now and then at one
    end of the rod.
Agar colonies. Small, yellowish white, with short outgrowths ; irregular
    clumps of variable size ; old colonies show branched outgrowths from all
    sides.
Gelatin colonies. Round, irregular ; liquefaction rapid ; gas, with a bad
    putrid odor.
Habitat. Earth.

254. B. Kedrowskii Migula

Buttersäure Bacillus Kedrowski. Zeitsch. f. Hygiene, XVI, 1894, 445.
B. acidi-butyrici Kruse: Flügge, Die Mikroorganismen, 1896, 256.
B. Kedrowskii Migula: System der Bakterien, 1900.

Morphology. Bacilli quite large.
Glucose gelatin colonies. In 3-4 days, small, delicate, round, sharp ; micro-
    scopically, light yellow, contour irregular, a central nucleus and a fluid
    periphery.
Gelatin stab. In depth, growth beaded.
Agar colonies. Round, elliptical, irregular, grayish white ; border entire —
    irregular ; microscopically, with a dense centre and a reticulately fila-
    mentous border. Gas, with a bad odor.
Milk. Coagulated, becoming acid ; a separation of serum.
Habitat. Isolated from cheese and rancid butter.

255. B. cuneatus

B. Anaerobic No. III Flügge : Zeitsch. f. Hygiene, XVII, 1894, 272.

Morphology. Bacilli long; cuneate at sporulation.
Gelatin colonies. Yellow brown; contour sharp; border irregular.
Agar colonies. Dark brown, irregular, lacerate.
Glucose bouillon. Turbid; gas; rancid odor.
Milk. Unaltered in 8 days.
Habitat. Milk.

256. B. sporogenes Klein

B. enteritidis-sporogenes Klein: Centralblatt f. Bakteriol., XVIII, 1895, 737, XXII, 578, XXIII, 542,913. Report Loc. Gov. Board, Supplement, 1897–1898, 210.

Morphology. Bacilli 0.8 : 1.6–4.8 μ — filaments — chains. Stained by Gram's method. Form oval spores, mostly polar, with clavate enlargement of the rod. Flagella mostly at one side of an end, in bundles.
Glucose gelatin colonies. Uniform, spherical, finely granular, translucent masses of liquefied gelatin.
Agar colonies. Round, gray disks, quite opaque in centre and granular at border.
Blood serum. Growth thin, gray; medium liquefied; fluid turbid, alkaline, and stinking.
Glucose bouillon. Turbid; often no growth.
Potato. Anaerobically, at first no growth; in 8–14 days, a number of small round yellowish colonies.
Formate of soda agar colonies. Small, flat, gray, with dark granular centres and clear sharp borders. Grow best in *milk*, becomes translucent, peptonized, with much gas.
Pathogenesis. Virulent cultures become greatly attenuated in 3–4 generations. Subcutaneous inoculations of 1 cc. of a milk culture into guinea pigs cause death in 18–24 hours. Strong œdema, muscular tissue strongly infiltrated, and a bloody stinking œdematous fluid. Bacilli sparingly present in heart, blood, and spleen.
Habitat. Widely distributed — sewage, water, horse and cow manure, street dust.

257. B. spinosus Lüderitz

Zeitsch. f. Hygiene, V, 1888, 152.

Morphology. Bacilli 0.6 : 4–8 μ. Spores at end of rod where the latter is swollen. No starch reaction with iodine.

Gelatin colonies. Radiately filamentous, like a caterpillar; develops much bad-smelling gas.

Agar colonies. Opaque clumps, reaching 4 mm., felted in centre; filamentous to reticulate on border. Develops much bad-smelling gas. Non-pathogenic to mice and guinea pigs.

Habitat. Earth.

258. B. cadaveris E. Klein

Centralblatt f. Bakteriol., XXV, 1899, 278.

Morphology. Bacilli 2-4 μ long, about as thick as malignant œdema bacillus; ends rounded; filaments and chains. Spores at end, oval exceeding the diameter of the rod. Stain by Gram's method. Flagella peritrichic.

Gelatin colonies (with glucose). Granular, radiately branched colonies; liquefaction begins in 1-2 days; centre dark, granular, from which granular outgrowths proceed. Colonies like the preceding. Klein thinks this organism may be identical with 257.

Agar slant. Anaerobically, at 37°, colonies irregular, ragged, finely granular, with darker centres; later the central colony gives off dark, branched, filamentous, anastomosing outgrowths. Much gas in glucose agar.

Milk. Coagulated, peptonized, amphoteric — alkaline; bad odor.

Blood serum. Liquefied; bad odor.

Bouillon. A slight turbidity.

Pathogenesis. Subcutaneous and intraperitoneal inoculations of guinea pigs negative.

Habitat. Isolated from cadavers.

259. B. Weigmanni

Paraplectrum fœtidum Weigmann: Centralblatt f. Bakteriol., 2te Abt., IV, 1898, 827.

Morphology. Bacilli on agar 0.8-1.3 : 2.2-12.5 μ. Not stained by Gram's method. Motile, but flagella could not be demonstrated. At sporulation clostridium and tetanus forms.

Gelatin colonies. No growth.

Agar (with soda, casein, milk sugar). Growth like *B. licheniformis.* No growth on plain agar.

Glucose agar colonies. Like *B. licheniformis.*

Glucose bouillon. Turbid; a membrane on the surface, becoming clear, with a heavy sediment.

Potato. Growth thin, soft, glistening.

Glucose gelatin stab. A fir tree growth, and a broad liquefied funnel above, or a cylindrical liquefaction along line of stab, with filamentous outgrowths and gas.

Milk. Thick, slimy, cheesy odor, becoming peptonized.

260. B. saccharobutyricus v. Klecki

Centralblatt f. Bakteriol., II, 1896, 169.

Morphology. Bacilli 0.7 : 5–7 μ, straight — slightly bent, often filaments 15 microns long; chains of not more than 2–4 elements. Spores placed at the ends of the rods. Not stained by Gram's method. Rods contain granules which stain violet with iodine.

Lactose gelatin colonies. Oval, sharp, granular in the interior. Ferments milk-sugar, with much gas.

Milk. Production of formic, acetic, and butyric acids. No indol or phenol in milk.

Habitat. Milk.

261. B. polypiformis Liborius.

Zeitsch. f. Hygiene, I, 1886, 162.

B. Anaerobic No. II Sanfelice: Zeitsch. f. Hygiene, XIV, 1893, 369.

Morphology. Bacilli over 1.0 μ thick, of variable length; slightly motile.

Gelatin colonies. Irregular — ameboid, cochleate — multilobular.

Agar colonies. Small, white, irregular, contoured; microscopically, brown, moruloid.

Gelatin stab. Growth arborescent.

Habitat. Isolated from putrefying flesh and earth.

262. B. Sanfelicei

B. solidus Sanfelice: Zeitsch. f. Hygiene, XIV, 1893, 372.

Morphology. Bacilli large.

Gelatin colonies. Small, white points; microscopically, like *Proteus mirabilis* — a more or less rounded colony composed of smaller colonies.

Agar colonies. Round, granular, entire, with a central nucleus and a bright border.

Gelatin stab. In depth, growth beaded.

Milk. Coagulated.

Habitat. Isolated from earth, fæces.

CLASS XXII. WITH ENDOSPORES. OBLIGATE ANAEROBIC. RODS SWOLLEN AT ONE END AT SPORULATION.

TETANUS GROUP.

I. Gelatin liquefied.
  A. Gelatin stab cultures arborescent or with radiating outgrowths.
      263. *B. tetani* Flügge.
      264. *B. pseudotetanicus* Sanfelice.
  B. Gelatin stab cultures not at all arborescent.
      265. *B. cuneatus.*
II. Gelatin not liquefied.
  A. Anaerobic at room temperatures.
      266. *B. Lubinskii* Kruse.
      267. *B. longus.*
  B. Anaerobic only at body temperatures. Will grow with access of air at room temperatures. Non-pathogenic.
      268. *B. pseudotetanicus var. aerobius* Kruse.
III. Do not grow in gelatin; at least cultures therein unsuccessful.
      269. *B. Taveli.*

263. **B. tetani** Flügge

Tetanus Bacillus Nicolaier: Deutsche med. Wochensch., 1884, No. 52.
B. tetani Flügge: Die Mikroorganismen, 1886.
Pacinia Nicolaieri Trevisan: Genera, 1889.

Morphology. Bacilli 0.3–0.5 : 2–4 μ — filaments. Slightly motile. Flagella peritrichic. Stain by Gram's method.

Gelatin colonies. Small, white, punctiform, becoming sunken and surrounded by a zone of liquefied gelatin; microscopically, the centres are yellow brown; borders floccose — fragmented.

Gelatin stab. Growth arborescent, slowly liquefied, with some gas.

Agar stab. A fir tree growth.

Bouillon. A uniform turbidity.

Milk. Not coagulated, amphoteric. H_2S positive. *Indol* slight.

Glucose bouillon. Gas.

Pathogenic. To mice, guinea pigs, rabbits, horses, etc.

Habitat. Associated with tetanus.

264. B. pseudotetanicus Sanfelice

Zeitsch. f. Hygiene, XIV, 1893, 372.

Morphological and cultural characters like the preceding, only differs in its less toxic properties.

Habitat. Isolated from meat infusion and earth.

265. B. cuneatus

B. Anaerobic No. III Flügge: Zeitsch. f. Hygiene, XVII, 1894, 272.

See No. 255.

266. B. Lubinskii Kruse

Ein tetanusähnlicher obligat-anaërober Bacillus Lubinski: Centralblatt f. Bakteriol., XVI, 1894, 771.

B. Lubinskii Kruse: Flügge, Die Mikroorganismen, 1896, 267.

Morphology. Bacilli like *B. tetani.* Stain by Gram's method.

Gelatin colonies. Flat, grayish, radiately crumpled on edges.

Gelatin stab. Growth arborescent.

Agar stab. Much gas.

Pathogenesis. Subcutaneous and intraperitoneal injections of rabbits cause death in 24 hours. There is a necrosis of tissue, a serous exudate, and much gas at the seat of inoculation.

Habitat. Isolated from an abscess.

267. B. longus

B. muscoides-colorabilis Liborius: Ucke, Centralblatt f. Bakteriol., XXIII, 1898, 1001.

Morphology. Bacilli 1.8–2.0 : 4–12 μ — filaments. Polar oval spores, of the tetanus type. Slightly motile. Not stained by Gram's method.

Gelatin colonies. Small, white, punctiform ; grow slowly.

Bouillon. Turbid, with a heavy, white sediment.

Agar slant. Growth bluish white, scarcely visible, with finely erose edges.

Agar stab. In depth, beaded, white colonies, gas.

Glucose agar stab. Abundant gas.

Milk. Not coagulated.

Potato. Growth scarcely visible or very scanty ; no odor.

Litmus media. A reduction, with sugar, acid.

Pathogenesis. Non-pathogenic to guinea pigs.

Habitat. Isolated from garden earth.

268. B. pseudotetanicus var. aerobius Kruse

Flügge, Die Mikroorganismen, 1896, 267.

Morphological and cultural characters like *B. tetani.* Will grow at ordinary temperatures with access of air, at higher temperatures only with the exclusion of air.

Habitat. Isolated from a case of tetanus.

269. B. Taveli

Pseudotetanusbacillus Tavel: Centralblatt f. Bakteriol., XXIII, 1898, 538.

Morphology. Bacilli slender, 0.5 : 5–7 μ, rather more slender than *B. tetani.* Spores of the above oval — those of *B. tetani* round. Flagella peritrichic, ordinarily 4–8 μ. Stained slightly by Gram's method. Gelatin cultures not successful.

Agar stab. Much gas.

Bouillon. Turbid, a white — light gray sediment, becoming clear.

Agar slant. Round discrete colonies, with thin borders, not always entire, but often jagged.

Fluid blood serum. Cultures develop only in a vacuum (with the least trace of oxygen no growth). A strong turbidity, gas, and a bad odor.

Pathogenesis. Non-pathogenic to mice, guinea pigs, and rabbits.

Habitat. Isolated from cases of abscess of the intestines.

CLASS XXIII.   WITHOUT ENDOSPORES.   CHROMOGENIC, PRODUCE PIGMENT ON GELATIN OR AGAR.

I. Aerobic and facultative anaerobic.
   A. Pigment reddish-pink on gelatin; gelatin liquefied.
      1. Rods not swollen at sporulation. *B. subtilis* type.
         270. *B. Lustigi.*
         271. *B. apicum* Kruse.
      2. Rods at sporulation of the tetanus type.
         272. *B. Danteci* Kruse.
   B. Pigment blue-violet.
         273. *B. Berolinensis* Kruse.
         274. *B. Lutetiensis* Kruse.
   C. Pigment brown black; gelatin liquefied.
         275. *B. niger* Biel.

270. B. Lustigi

Der rother Bacillus Lustig: Diag. Bak. des Wassers, 1893, 72.

Morphology. Bacilli small, with rounded ends, generally 2–3 times their breadth, variable, very motile, filaments also motile.

Gelatin colonies. Gray dots with red centres; microscopically; round, granular, edges serrate, centres raspberry-red, becoming liquefied.

Gelatin stab. In depth, growth thin, filiform, liquefied; on the surface, a small, funnel-shaped depression, with pigment, becoming generally liquefied.

Agar slant. Growth at 20°, moist, glistening, spreading, of a crimson lake color; at 37°, growth milky white.

Potato. Growth viscid — slimy, red, spreading.

Bouillon. Turbid; a red pigment at room temperatures.

Habitat. Water.

271. B. apicum Kruse

Described by Canestrini: Atti Soc. Ven. Trent. Sci. Nat., XII, 134.
B. apicum Kruse: Flügge, Die Mikroorganismen, 1896, 233.

Morphology. Bacilli 2 : 4–6 μ, ends rounded; occur singly, in pairs and chains. Grow at 37°. Stained by Gram's method.

Gelatin stab. Liquefied gelatin pink above; sediment white.

Agar slant. Growth whitish.

Blood serum. Liquefied. Bacilli show a capsule, often surrounding a chain of individuals.

Potato. Growth wine-colored.

Pathogenic. To bees, not so to mice and guinea pigs.

Habitat. Isolated from sick bees and their larvæ.

272. B. Danteci Kruse

Bacille du rouge de la morue Dantec: Ann. Pasteur Inst., 1891, 659.
B. Danteci Kruse: Flügge, Die Mikroorganismen, 1896, 270.

Morphology. Bacilli 4–12 μ long, rather thicker than *B. tetani.*

Gelatin colonies. Pale red disks, deeper in color at the periphery; gelatin liquefied slowly.

Potato. A scanty growth.

Bouillon. Turbid; no pigment. Produces red pigment on fish.

Habitat. Isolated from fish.

273. B. Berolinensis Kruse

Described but not named by Plagge-Proskauer: Zeitsch. f. Hygiene, II, 1887, 463.
B. violaceus-Berolinensis Kruse: Flügge, Die Mikroorganismen, 1896, 311.

Morphology. Bacilli 0.8 : 1.7 μ, in twos. Do not grow at 37°.

Gelatin colonies. Irregular, granular, becoming liquefied; centres dark, borders twisted, filamentous.

Gelatin stab. Liquefaction infundibuliform; a violet sediment.

x

Agar slant. Growth smooth, glistening, spreading, deep violet.
Bouillon. Slightly turbid; a violet sediment.
Potato. Growth limited, dark violet. Reduces nitrates.
Milk. Blue.
Habitat. Water.

274. B. Lutetiensis

B. violaceus-Lutetiensis Kruse: Flügge, Die Mikroorganismen, 1896, 311.

Perhaps identical with or a variety of the preceding.
Habitat. Water.

275. B. niger Biel

B. mesentericus-niger Biel: Centralblatt f. Bakteriol., 2te Abt., II, 1896, 137.

Morphology. Bacilli 0.8 : 3.6–5.8 μ, straight, rounded ends; occur singly and
　　in twos. Stain by Gram's method. Spores oval, 1.2–1.3 μ. Obligate
　　aerobic.
Gelatin colonies. Irregular, granular, gray, with long, spiral outgrowths, be-
　　coming more rounded, brownish, granular, surrounded by an irregular,
　　light gray, granular border; in 3 days liquefaction begins.
Gelatin stab. Liquefaction crateriform, with a pellicle.
Acid gelatin. A scanty growth.
Milk. Coagulated, amphoteric, peptonized; a dark brown sediment.
Litmus milk. Bluish-brownish, gas.
Potato. Growth spreading, grayish blue—dark brown, rugose, moist, glisten-
　　ing, or black.
Habitat. Isolated from bread.

PSEUDOMONAS Migula

Cells cylindrical, which now and then form short filaments. Actively motile;
　　flagella attached to the poles. The number of the latter varies in the
　　different species from one to ten, but is more generally three to six. En-
　　dospores known in only a few species.
I. Cells colorless, without a red-colored plasma, and without sulphur granules.
　　A. Grow on ordinary organic culture media.
　　　　1. Without endospores — at least their presence not noted; aerobic
　　　　　　and facultative anaerobic.
　　　　　　a. Without pigment on gelatin or agar.
　　　　　　　　* Gelatin not liquefied. CLASS I, p. 307.
　　　　　　　　** Gelatin liquefied. CLASS II, p. 309.

b. Produce pigment on gelatin or agar. CLASS III, p. 314.

c. Colonies colorless, or colored only slightly yellowish-greenish, but with a yellow green or blue green fluorescence. CLASS IV, p. 320.

2. With endospores. CLASS V, p. 326.

B. Do not grow in nutrient gelatin or other organic media.

CLASS VI, NITROMONAS GROUP, p. 329.

II. Cell plasma with a reddish tint, with also sulphur granules. CLASS VII, p. 329.

CLASS I. WITHOUT ENDOSPORES. AEROBIC AND FACULTATIVE ANAEROBIC. NON–CHROMOGENIC. GELATIN NOT LIQUEFIED.

I. Gas generated in glucose bouillon.

A. *Indol* produced.

1. *Ps. sinuosa* (Wright).

B. No *indol* produced.

2. *Ps. monadiformis* (Kruse).

II. No gas generated in glucose bouillon.

A. Potato cultures whitish-grayish.

3. *Ps. ambigua* (Wright).

4. *Ps. catarrhalis*.

B. Potato cultures brownish.

5. *Ps. nexibilis* (Wright).

1. Ps. sinuosa (Wright)

B. sinuosus Wright: l.c., 440.

Morphology. Bacilli medium-sized, ends rounded, pairs —filaments. A polar flagellum ; some bacilli probably have 2–4 flagella.

Gelatin colonies. In 3 days, 3 mm., thin, delicate, translucent, irregular, sinuous; centres brownish, grained; becoming 6 mm., with radial foldings.

Gelatin slant. Growth grayish white, glistening, translucent.

Agar slant. Growth scanty, limited.

Bouillon. Turbid, with a sediment; no pellicle.

Potato. Growth gray-brown, moist, not thick, rather rough, spreading. *Indol* slight. Grow at 36°.

Milk. Not coagulated.

Habitat. Water.

2. Ps. monadiformis (Kruse)

B. coli-mobilis Messea: Riv. d'igiene, Roma, 1890.
B. monadiformis Kruse: Flügge, Die Mikroorganismen, 1896, 374.

Morphology. Bacilli short rods. Cultural characters like *B. coli.*
Milk. Not coagulated, slightly acid. Gas in lactose, but none in saccharose bouillon. Non-pathogenic to mice.
Habitat. Isolated from typhoid stools.

3. Ps. ambigua (Wright)

B. ambiguus Wright: l.c., 439.

Morphology. Bacilli small, ends rounded, occur singly, and in pairs and filaments. A terminal flagellum.
Gelatin colonies. Deep: round, entire, granular, brownish. *Surface:* in 3-4 days gray, translucent, slightly elevated, rather irregular, 2 mm., sharp. Microscopically, granular, yellowish brown in the centre, with thin translucent margins, finely radiate.
Agar slant. Growth gray, limited, sharply defined.
Bouillon. Turbid; sediment; no pellicle.
Potato. Growth thick, viscid, spreading; gray — creamy.
Litmus milk. Acid, coagulated only after one month, and may not be then. Grow at 36°. *Indol* positive.
Pepton rosolic acid solution. Bleached.
Habitat. Water.

4. Ps. catarrhalis

Der Bacillus der Hundestaupe Jess: Centralblatt f. Bakteriol., XXV, 1899, 541.

Morphology. Bacilli 0.6 : 1.8-2.3, occur singly in cultures, in chains in the animal body. Stain by Gram's method. A polar flagellum.
Agar colonies. In 24 hours, at 37°, dark, entire, sharp, granular.
Agar slant. In 24 hours, at 37°, growth abundant, soft, gray; border entire, sharp; water of condensation turbid.
Glycerin agar. A scanty growth of isolated colonies.
Bouillon. Turbid, with a flocculent sediment.
Potato. Growth white, velvety.
Pathogenesis. The inoculation of pure cultures into dogs and cats produces a pathological picture which the author considers identical with Hundestaupe.
Habitat. Associated with Hundestaupe (*Febris catarrhalis epizootica canum*).

5. Ps. nexibilis (Wright)

B. nexibilis Wright: l.c.

Morphology. Bacilli of medium size, ends rounded, occur in pairs, in long forms, chains, and clumps.

Gelatin colonies. *Deep:* round, irregular, centres brownish, margins faintly radiate, becoming grayish brown. *Surface:* in 3 days, 3 mm., thin, grayish, translucent, opalescent, somewhat sinuous; microscopically, brownish, slightly granular; borders translucent, sinuous — dentate.

Agar slant. Growth thin, translucent, spreading, becoming greenish in time.

Bouillon. Turbid, with a faint greenish tint.

Potato. Growth brown, thick, viscid, spreading.

Litmus milk. Becoming pink, acid. *Indol* positive. No growth at 37°.

Habitat. Water.

CLASS II. WITHOUT ENDOSPORES. AEROBIC AND FACULTATIVE ANAEROBIC. NON–CHROMOGENIC. GELATIN LIQUEFIED.

I. Colonies on gelatin at all stages, round, with no radiations from their edges.
  A. Gelatin liquefied rather quickly.
    1. Gas generated in glucose bouillon.
      a. Gelatin stab cultures crateriform, becoming stratiform.
        6. *Ps. coadunata* (Wright).
      b. Gelatin stab cultures becoming saccate.
        7. *Ps. multistriata* (Wright).
    2. No gas generated in glucose bouillon.
      a. Milk coagulated.
        8. *Ps. Fairmontensis* (Wright).
    3. Gas production in glucose bouillon not stated.
        9. *Ps. liquida* (Frankland).
  B. Gelatin liquefied very slowly.
    1. Gas generated in glucose bouillon.
      10. *Ps. nebulosa* (Wright).
    2. No gas generated in glucose bouillon.
      11. *Ps. cohærea* (Wright).
II. Colonies on gelatin, with filamentous borders or radiate.
  A. Gas generated in glucose bouillon.
      12. *Ps. centrifugans* (Wright).
      13. *Ps. punctata* (Zimmerman).

B. No gas generated in glucose bouillon.
  1. Produce *indol ;* no growth at 36°.
      14. *Ps. fimbriata* (Wright).
  2. Do not produce *indol ;* grow at 36°.
      15. *Ps. geniculata* (Wright).
III. Colonies on gelatin erose, lobed, coli-like.
      16. *Ps. delabens* (Wright).

6. **Ps. coadunata** (Wright)

B. coadunatus Wright : l.c., 460.

Morphology. Bacilli of medium size, ends rounded ; occur in pairs, filaments, and chains. A polar flagellum.

Gelatin colonies. In 3–4 days, round, brownish, dense, less than 1 mm. ; sunken in the liquefied gelatin ; microscopically, with brownish — brownish gray centres, with rough, frayed margins and a zone of liquefied gelatin, in which are scattered granulations.

Gelatin stab. Liquefaction crateriform — stratiform.

Agar slant. Growth translucent, grayish, slightly spreading.

Bouillon. Turbid, with a white sediment and a pellicle ; the medium has a slight greenish tint.

Potato. Growth brown, viscid, moist, glistening, spreading.

Milk. Coagulated, acid. No growth at 37°. *Indol* produced. Nitrates not reduced.

Habitat. Water.

7. **Ps. multistriata** (Wright)

B. multistriatus Wright : l.c., 462.

Morphology. Bacilli of medium size, ends rounded, variable, in pairs.

Gelatin colonies. Deep : brownish, dense, granular, round-oval. *Surface :* round, grayish white, translucent, 1–2 mm. ; microscopically, with dark brownish dense centres, and thinner margins, with radiate brownish lines from the central nucleus.

Gelatin stab. In depth, growth beaded ; on the surface, growth irregular, whitish, gradually sinking into the liquefied gelatin, becoming, in 10 days, saccate.

Agar slant. Growth translucent, narrow.

Potato. Growth grayish to creamy, thick, glistening, spreading, viscid.

Milk. Coagulated, amphoteric. *Indol* negative. No growth at 36°. Nitrates not reduced.

Habitat. Water.

8. Ps. Fairmontensis (Wright)

B. Fairmontensis Wright: l.c., 458.

Morphology. Bacilli of medium size, ends rounded, in pairs and filaments.

Gelatin colonies. Deep: round — oval, dense, granular, dark grayish brown ; in 3 days, surrounded by a zone of liquefied gelatin. *Surface:* in 2 days, round, white, translucent disks, 1–2 mm.; microscopically, with dark centres with a greenish shimmer and thinner edges, and faint radial lines.

Gelatin stab. Liquefaction crateriform, extending to the walls in 2–3 days ; little growth along needle track.

Agar slant. Growth grayish white, glistening.

Potato. Growth granular, elevated, spreading, the color of the medium, becoming brownish, viscid.

Litmus milk. Decolorized. No growth at 36°. *Indol* produced.

Habitat. Water.

9. Ps. liquida (Frankland)

B. liquidus Frankland : Zeitsch. f. Hygiene, VI, 1889, 382.

B. liquefaciens-communis Sternberg : Manual, 1892.

B. aquatilis-communis Kruse : Flügge, Die Mikroorganismen, 1896, 315.

Morphology. Bacilli 0.6 : 1.2–5 μ. Flagella polar.

Gelatin colonies. Round, crateriform, turbid ; edge finely granular, not ciliate.

Agar slant. Growth translucent, gray.

Gelatin stab. In 2 days, a large, saccate liquefaction, turbid, becoming clear.

Potato. Growth yellowish brown, pinkish — flesh-colored. Nitrates reduced.

Habitat. Water.

10. Ps. nebulosa (Wright)

B. nebulosus Wright: l.c.

Morphology. Bacilli of medium size ; flagella polar.

Gelatin colonies. Deep: round, dark, granular. *Surface:* round, thin, gray, translucent, hazy ; centre white, surrounded by a whitish ring ; microscopically, centre dark brownish, granular, surrounded by a thin, transparent zone.

Gelatin slant. Growth viscid, whitish ; lines a shallow furrow, with short, lateral outgrowths.

Agar slant. Growth a thin, translucent stripe.

Bouillon. Turbid, with a sediment.

Potato. A scanty growth, if any.

Litmus milk. Decolorized; casein dissolved; alkaline. *Indol* negative. Grow at 36°.

Milk. Not coagulated.

Habitat. Water.

11. Ps. cohærea (Wright)

B. cohæreus Wright: l.c., 464.

Morphology. Bacilli of medium size, short, ends rounded, in pairs and filaments. · Flagella polar.

Gelatin colonies. Deep: round — oval, granular, brownish, sharp; later a dark brownish tint in the adjacent gelatin; colonies sometimes moruloid. *Surface:* round, elevated, grayish, becoming thicker, denser, and papillate in centre; microscopically, granular, yellowish brown in centre, becoming sunken in the gelatin and crimpled.

Gelatin slant. Growth slightly wrinkled, grayish white, lining a furrow in the gelatin.

Agar slant. Growth elevated, grayish white, translucent, glistening.

Bouillon. Turbid, with a wrinkled membrane; becoming clear.

Potato. Growth elevated, granular, the color of the medium.

Litmus milk. Decolorized, viscid; coagulated; alkaline, becoming brownish. *Indol* negative. But slight growth at 36°.

Habitat. Water.

12. Ps. centrifugans (Wright)

B. centrifugans Wright: l.c., 462.

Morphology. Bacilli of medium size, in pairs and filaments. Flagella polar.

Gelatin colonies. Deep: round, dark, granular, with a greenish shimmer, soon surrounded by a zone of liquefied gelatin. *Surface:* in 24–48 hours, round, · crateriform, 3 mm., turbid, flocculi in centre; microscopically granular, a circulating motion; margin fringed with short hairs.

Gelatin stab. Liquefaction saccate, a pellicle, greenish below and alkaline.

Agar slant. Growth translucent, glistening, grayish, thin, becoming brownish to greenish brown.

Bouillon. Turbid; a slight pellicle; later a brown green tint.

Potato. Growth thick, spreading, gray-pinkish; sometimes a rough granular surface.

Litmus milk. Coagulated, decolorized, amphoteric — acid. *Indol* produced. Nitrates not reduced. Grow at 35°-36°.

Habitat. Water.

13. Ps. punctata (Zimmerman)

B. punctatus Zimmerman: Bak. Nutz u. Trinkwässer, Chemnitz, 1890, 38.
Bact. punctatum Lehmann-Neumann: Bak. Diag., 1896, 238.

Morphology. Bacilli 0.5 : 0.8 μ — filaments. A polar flagellum. Not stained by Gram's method. '

Gelatin colonies. Round, punctiform, entire, becoming erose, then filamentous — ciliate, then sunken and liquefied like cholera colonies.

Milk. Coagulated, becoming fluid.

Glucose bouillon. Gas. H_2S produced. *Indol* slight.

Habitat. Water.

B. annulatus Zimmerman : l.c., 1894. Probably identical with the preceding.

14. Ps. fimbriata (Wright)

B. fimbriatus Wright: l.c., 463.

Morphology. Bacilli medium-sized, ends blunt, short — long forms — chains. Flagella several.

Gelatin colonies. Deep : round, granular, sharp, grayish in centre. *Surface :* rounded, yellowish white, sunken, 1-2 mm., sometimes surrounded by a clouded liquid zone ; microscopically, dark brownish, granular in centre ; edge a delicate fringe.

Gelatin stab. Liquefaction napiform, a slight pellicle, iridescent, becoming slightly greenish.

Agar slant. Growth smooth, dark gray, glistening ; agar becomes brownish.

Bouillon. Turbid ; a slight pellicle, becoming a dark greenish tint.

Potato. Growth grayish — light brownish, slightly rough, spreading.

Litmus milk. Coagulated, decolorized, amphoteric. *Indol* produced. Nitrates not reduced. Grow at 36°.

Habitat. Water.

15. Ps. geniculata (Wright)

B. geniculatus Wright: l.c., 459.

Morphology. Bacilli medium-sized, in pairs and filaments.

Gelatin colonies. Deep : yellowish, round, sharp, slightly granular ; studded with small plaques and buds. *Surface :* round, translucent, whitish, somewhat depressed : microscopically, brownish, granular centre, thin margin, entire — undulate with radiating fibrils, becoming liquefied, crateriform ; centres gray white — yellowish, with an outer zone of radiating fibrils.

Gelatin stab. Liquefaction infundibuliform, an air space above; sediment whitish-pinkish.

Agar slant. Growth grayish, glistening, limited, translucent, brownish gray.

Bouillon. Turbid, flocculi in suspension, a slight pellicle, and a slight greenish tint to the medium.

Potato. Growth thin, viscid, moist, glistening, brownish.

Litmus milk. Coagulated, decolorized, alkaline. *Indol* not produced. No growth at 35°-36°.

Habitat. Water.

16. Ps. delabens (Wright)

B. delabens Wright: l.c., 456.

Morphology. Bacilli small, short, and long forms.

Gelatin colonies. Deep: round — irregular, slightly granular, brownish yellow. *Surface:* thin, translucent, glistening, wavy — irregular, centre grayish; microscopically, thin, translucent, marmorated, brownish centre; later the growth sinks in the slowly liquefied gelatin.

Gelatin slant. Growth a gray white stripe which sinks into the gelatin.

Gelatin stab. In depth, growth beaded, brownish gray; on the surface, growth thin, white, irregular, which sinks slowly into the liquefied gelatin.

Agar slant. Growth whitish, translucent, glistening, slightly spreading; the agar becomes greenish.

Bouillon: Turbid; a slight pellicle, and a slight greenish tint.

Potato. Growth brownish, viscid, thick, spreading.

Litmus milk. Decolorized, not coagulated, alkaline; a tough membrane. *Indol* slight. Nitrates not reduced. No growth at 36°.

Habitat. Water.

CLASS III. WITHOUT ENDOSPORES. AEROBIC AND FACULTATIVE ANAEROBIC. CHROMOGENIC, PIGMENT ON GELATIN OR AGAR.

I. Pigment yellowish.

  A. Gelatin liquefied; no gas in glucose bouillon.

    1. Milk coagulated.

        a. Grow well at 35°-36° C.

            17. *Ps. pullulans* (Wright).

        b. Do not grow at 35°-36° C.

            18. *Ps. annulata* (Wright).

    2. Milk not coagulated.

            19. *Ps. ochracea* (Zimmerman).

            20. *Ps. campestris* (Pammel) Smith.

B. Gelatin not liquefied.
> 21. *Ps. turcosa* (Zimmerman).

II. Pigment blue-violet on gelatin or agar.
> *A.* Gelatin liquefied.
>> 1. On potato, pigment violet.
>>> 22. *Ps. janthina* (Zopf).
>> 2. On potato, growth grayish blue — blue green.
>>> 23. *Ps. cærulea* (Voges).
>> 3. On potato, growth dark blue — blue black.
>>> 24. *Ps. Smithii.*
> *B.* Gelatin not liquefied.
>>> 25. *Ps. indigofera* (Voges).
>>> 26. *Ps. Berolinensis* (Classen) Migula.

17. **Ps. pullulans** (Wright)

B. pullulans Wright: l.c., 445.

Morphology. Bacilli small, short, in pairs. Flagella several, polar.

Gelatin colonies. In 2-3 days 2 mm., yellowish gray, slightly elevated, translucent. Microscopically, with yellowish centres and colorless margins, sausage-shaped granules on lower side of colony, surrounded by a zone of liquefied gelatin.

Gelatin stab. Surface growth raised; liquefaction napiform.

Agar slant. Growth yellowish, translucent.

Glucose bouillon. No gas. Nitrates not reduced. Grow at 36°. *Indol* produced.

Habitat. Water.

18. **Ps. annulata** (Wright)

B. annulatus Wright: l.c., 443.

Morphology. Bacilli small; occur singly, in pairs, and as long forms. Flagella several, at one or both ends.

Gelatin colonies. In 3-4 days, 2-3 mm., round; centres yellowish, edges indistinct — somewhat fringed; liquefaction saccate.

Gelatin stab. Liquefaction crateriform, yellowish flocculi, a slight pellicle.

Agar slant. Growth yellowish, translucent, glistening.

Bouillon. Turbid, with a yellow pellicle.

Litmus milk. Coagulated, decolorized. *Indol* slight. No growth at 36°.

Habitat. Water.

19. **Ps. ochracea** (Zimmerman)

B. ochraceus Zimmerman: Bak. Nutz u. Trinkwässer, Chemnitz, 1890.

Morphology. Bacilli 0.5–0.8 : 1.2–3.6 μ. Stain by Gram's method. Flagella polar.

Gelatin colonies. *Deep:* round, small, light yellowish. *Surface:* at first coli-like, becoming slightly fringed on borders, becoming liquefied and depressed ; a pellicle on the surface of a grayish yellow or deep yellow color, often with a reticulate structure. Microscopically, brownish, granular, often warty.

Gelatin stab. On the surface a yellowish — yellowish gray layer, which sinks, becoming a funnel and later a cylindrical liquefaction, with a pale yellow or ochreous sediment.

Agar slant. A thin yellowish gray — ochreous expansion.

Bouillon. Weakly turbid, with a slight pellicle, and much sediment. *Indol* positive. H₂S produced.

Milk. Not coagulated, somewhat slimy.

Glucose bouillon. No gas.

Potato. Growth ochre-yellow.

Habitat. Water.

20. **Ps. campestris** (Pammel) Smith

B. campestris Pammel: Bull. 27, Iowa Ag. Expt. Sta., 1895, 130.
Ps. campestris Smith: Centralblatt f. Bakteriol., 2te Abt., III, 1897.
B. campestris Russell: Wis. Ag. Expt. Sta., Bull. 65, 1898.

Morphology. Bacilli 0.4–0.6 : 1–2 μ ; occur singly and in chains. Motility active. Stain uniformly with Löffler's methylene blue ; stain irregularly with Ziehl's carbol-fuchsin. Decolorized by Gram's method.

Gelatin colonies. *Deep:* round, sharp, darker than surface growths, concentric. *Surface:* in 3–4 days 1 mm., moist, glistening, raised, light yellow. Microscopically, round, entire, centres darker, finely granular, sometimes radiately streaked. In 12–15 days the surface colonies begin to liquefy.

Agar colonies. *Deep:* lenticular — irregular, dark, granular, concentric. *Surface:* 1–4 mm., thin, moist, glistening, translucent, light yellow — olive. Microscopically, thicker and denser in centre, concentric.

Agar slant. At first an elevated streak, becoming spreading, thinner, and somewhat translucent ; rich golden yellow.

Gelatin stab. Slight growth in depth, spreading slightly at the surface. In 7–10 days a slight liquefaction under the latter, and a pit due to evaporation. Liquefaction extends laterally, very slowly, becoming stratiform.

Potato. Growth moist, glistening, pasty, light yellow, becoming cadmium-yellow to golden brown.

Bouillon. A slight turbidity, and a yellow deposit adhering to walls at surface, becoming clear, with a yellow granular precipitate.

Litmus milk. In 10 days pink; no true curd, but a layer of whey on top. The casein is gradually digested. No gas in glucose and lactose bouillon; growth only in the open end of fermentation tubes.

Pigment. Soluble in ethyl and methyl alcohol; color destroyed by mineral acids. Bouillon rendered slightly alkaline. Non-pathogenic to rabbits.

Habitat. Associated with a bacterial rot of cabbages and allied plants.

21. Ps. turcosa (Zimmerman)

B. turcosus Zimmerman: Bak. Nutz u. Trinkwässer, Chemnitz, II Theil, 1894, 32.

Morphology. Bacilli 0.2–0.3 : 0.3–1.5 μ. A single polar flagellum.

Gelatin colonies. Small, punctiform, translucent, intense yellow. Microscopically, amorphous.

Gelatin stab. Surface growth small, slowly growing, round, convex, intense yellow, with a slight greenish tint, becoming gradually sunken, without liquefaction.

Agar slant. A scanty growth, intense yellow.

Potato. A scanty growth, greenish yellow, dry, with a soft, glistening appearance.

Bouillon. Weakly turbid. H₂S negative. *Indol* negative.

Glucose bouillon. No gas.

Milk. Not coagulated.

Habitat. Water.

22. Ps. janthina (Zopf)

B. janthinus Zopf: Spaltpilze, 1885, 68.
B. violaceus Macé: Ann. d'Hyg. publ. et de Méd. leg. XVII, 1887.
B. violaceus Frankland: Zeitsch. f. Hygiene, 1889, 394.
B. violaceus-laurentius Jordan: State Board of Health, Mass., 1890, 838.

Morphology. Bacilli 0.5–0.8 : 1.5–5.0 μ, ends rounded. One or two polar flagella; according to Lehmann-Neumann it may show 3–4 peritrichic flagella. Stain by Gram's method.

Gelatin colonies. Small yellow points, becoming violet; becoming liquefied and depressed, with grayish centres and violet borders. The unliquefied colonies may be yellowish violet, with ragged, erose, lobular borders. Microscopically, fragmental — grumose, dark yellow-brownish.

Gelatin stab. Old cultures show on the surface a white expansion, becoming violet-blue; after a time the growth sinks, and the gelatin becomes slowly liquefied. Freshly isolated cultures may show a funnel-formed-cylindrical liquefaction, whose contents are gray violet.

Agar slant. Growth moist, glistening, yellowish — brownish white, becoming deep violet.

Potato. Growth violet — violet-black, spreading.

Bouillon. Turbid, with a membrane, slightly violet.

Milk. Slowly coagulated, ordinarily not coagulated; colored violet, at least the cream layer.

Glucose bouillon. Slightly acid; no gas. H_2S produced. *Indol* produced.

Habitat. Water.

23. **Ps. cœrulea** (Voges)

B. cœruleus Voges: Centralblatt f. Bakteriol., XIV, 1893, 303.

Morphology. Bacilli 0.8 : 1.0–1.4 μ. One polar flagellum. At 37° good growth, with pigment production.

Gelatin colonies. Typhoid-like, becoming grayish blue; slowly liquefied.

Gelatin stab. Slight growth.

Bouillon. A gray membrane on the surface.

Milk. Coagulated; the cream layer sky-blue.

Potato. Growth grayish blue — blue green, becoming darker, coarsely granular.

Habitat. Water.

24. **Ps. Smithii**

B. cœruleus Smith: Med. News, II, 1887, 758.

Morphology. Bacilli 0.5 : 2.0–2.5; frequently form leptothrix-like threads.

Gelatin colonies. Form cup-like liquid depressions. No color in the depth of the gelatin, but the surface colonies exhibit a faint blue tint.

Gelatin stab. A funnel; in depth, a few colonies.

Agar slant. Growth bluish.

Potato. Growth dark blue, becoming intense blue black. According to Wright (l.c., 451).

Gelatin colonies. *Deep:* irregular — oval, finely granular, yellowish-brownish. *Surface:* thin, translucent, slate-blue; microscopically, finely granular, with irregular outlines. Later the surface colonies are bluish gray masses within depressions of liquefied gelatin which microscopically are dense, brown, opaque, coarsely granular, with ragged margins.

Gelatin stab. Liquefaction napiform, with a bluish gray membrane and a bluish sediment; little growth along the line of stab.

Agar slant. Growth glistening, limited, slate-blue, becoming gray.

Bouillon. Clouded, whitish flocculi, and a few bluish flocculi on the surface, or a bluish ring.

Potato. Growth slate-blue, dense, becoming dirty brown.

Litmus milk. Coagulated, decolorized; serum bluish.

Glucose gelatin stab. A fair growth; no gas. *Indol* produced. Grow at 36°. The species seems subject to considerable variations, and may be identical with, or a variety of, *B. janthinus.*

Habitat. Water.

25. Ps. indigofera (Voges)

B. indigoferus Voges: Centralblatt f. Bakteriol., XIV, 1893, 307.

Morphology. Bacilli 0.6–1.8 μ; occur singly. A polar flagellum. Not stained by Gram's method. Grow at 37°.

Gelatin colonies. Flat, spreading, iridescent, blue.

Gelatin stab. Surface growth flat, glistening.

Bouillon. A delicate membrane of a blue color.

Agar slant. Growth dark blue.

Potato. Growth greenish blue. Non-pathogenic. Pigment soluble in H_2SO_4 (brown); in HNO_3 (yellowish); in HCl (bluish). Addition of NH_4HO has no effect on the pigment.

Habitat. Water.

26. Ps. Berolinensis (Classen) Migula

B. Berolinensis-indicus Classen; Centralblatt f. Bakteriol., VII, 1890, 13.

Morphology. Bacilli slender, with rounded ends, like *B. typhosus*; occur singly, in pairs, in threes, or in packets. The bacillus is surrounded by a delicate capsule.

Gelatin colonies. Grayish white punctiform, becoming indigo-blue; borders irregular, typhoid-like, colorless.

Gelatin stab. On the surface, growth punctiform, deep indigo-blue, slowly spreading; contour irregular. The color does not penetrate into the gelatin.

Agar slant. Growth thick, moist, glistening, deep indigo-blue.

Potato (acid). Growth deep blue; on alkaline potato it is dirty green, glistening.

Bouillon. Turbid, with flocculent particles; no color produced. Do not grow at 37°; grow best at 15°.

Habitat. Water.

CLASS IV. WITHOUT ENDOSPORES. FLUORESCENT BACTERIA.

I. Gelatin liquefied.
 A. Gelatin liquefied slowly.
   27. *Ps. Schuylkilliensis* (Wright).
 B. Gelatin liquefied quickly.
  1. Milk coagulated.
   28. *Ps. pyocyanea.*
   29. *Ps. capsulata* (Pottien).
  2. Milk not coagulated.
   30. *Ps. fluorescens* (Flügge) Migula.
II. Gelatin not liquefied.
 A. Milk coagulated.
   31. *Ps. rugosa* (Wright).
 B. Milk not coagulated.
  1. Milk rendered slightly acid, or litmus milk becomes slightly pinkish.
   32. *Ps. incognita* (Wright).
   33. *Ps. foliacea* (Wright).
  2. Milk rendered alkaline in reaction.
   a. Gelatin surface colonies thin, flat.
    * Do not grow at body temperatures.
     34. *Ps. syncyanea* (Ehrenberg) Migula.
    ** Grow at body temperature.
     † Gelatin surface colonies become filamentous on their borders.
      35. *Ps. striata* (Ravenel).
     †† Gelatin surface colonies sharp, entire.
      36. *Ps. ovalis* (Ravenel).
   b. Gelatin surface colonies convex.
    37. *Ps. convexa* (Wright).
  3. Milk reaction not changed.
   a. Reddish granules in the rods.
    38. *Ps. erythrospora* (Cohn) Migula.
   b. Rods not characterized as before.
    39. *Ps. putrida* (Flügge) Migula.

27. Ps. Schuylkilliensis (Wright)

B. fluorescens-Schuylkilliensis Wright: l.c., 448.

Morphology. Bacilli small, short, ends rounded, in pairs and filaments. A polar flagellum.

Gelatin colonies. In 2 days, 1.5 mm., grayish white, translucent; microscopically, with brownish centres, borders thin, with radiate structure; later show a greenish white — blue green fluorescence.

Gelatin stab. Liquefaction slow, crateriform, with a blue green fluorescence.

Agar slant. Growth grayish, translucent; agar fluorescent.

Bouillon. Turbid, with a slight pellicle and a blue green fluorescence.

Potato. Growth brownish, elevated, spreading.

Litmus milk. Coagulated slowly, decolorized slowly.

Glucose bouillon. No gas. *Indol* doubtful.

Habitat. Water.

28. **Ps. pyocyanea** (Gessard) Migula

B. pyocyaneus Gessard: De la pyocyanine et son microbe, Thèse de Paris, 1882.
Ps. pyocyanea Migula: Die Natürlichen Pflanzenfam., 1896.

Morphology. Bacilli ·0.4: 1.4-6 μ, variable. A polar flagellum. Stain by Gram's method.

Gelatin colonies. Deep: round, oval, yellowish white — greenish yellow, often moruloid. *Surface:* round, entire, yellowish; later, the borders become irregular and coli-like; often beset with hairs, or fringed; becoming sunken, with an irregular, variable structure, gray or greenish gray, irregular, ragged, or coarsely granular.

Gelatin stab. Liquefaction crateriform, becoming saccate, with a greenish yellow-blue green fluorescence; contents granular, flocculent.

Agar colonies. Deep: round, oval, entire, undulate, granular, yellow — greenish yellow. *Surface:* round, entire, glistening. greenish white — yellowish; microscopically, round, entire, granular, moruloid, yellow — greenish yellow.

Agar slant. Growth soft, glistening, spreading. yellowish green — greenish; the agar shows a strong yellow green fluorescence.

Bouillon. Turbid, with a pellicle, and a strong yellow green fluorescence.

Milk. Coagulated, becoming fluid and alkaline, with a yellow green fluorescence.

Potato. Growth yellowish, moist, glistening, slightly raised, becoming brownish yellow — brown. *Indol* negative. H_2S negative.

Glucose bouillon. Acid, no gas. Nitrates reduced to nitrites.

Pathogenesis. Subcutaneous inoculation of 1 cc. of a virulent bouillon culture into guinea pigs and rabbits causes a purulent infiltration and inflammatory œdema at the point of inoculation, and death in some cases in 24 hours. Intraperitoneal inoculations cause peritonitis and death. With small doses only local inflammation, and recovery.

Y

Habitat. In the mouth, intestines, and on the surface of the body; in suppurating wounds; sometimes associated with peritonitis, appendicitis, in
phlegmons, otitis media, bronchopneumonia, etc.

VARIETIES (non-pathogenic). *B. fluorescens-mutabilis* Wright: l.c., 449.
Habitat, water.

B. No. 21 Conn.: l.c., 1893, 52. *Habitat,* milk.

These are probably only non-pathogenic varieties of the preceding, except
that the chromogenic function is weaker. Cultural characters otherwise
within the range of probable normal variations.

29. Ps. capsulata (Pottien)

B. fluorescens-capsulatus Pottien: Zeitsch. f. Hygiene, XXII, 1896, 146.

Morphology. Bacilli small, with rounded ends, scarcely one-half the length of
tubercle bacilli, and somewhat thicker; often in twos, somewhat variable
in size, rarely coccoid. Show a polar stain. Not stained by Gram's
method. A capsule in the body and on media. A long, undulating polar
flagellum. Optimum temperature 37°.

Gelatin colonies. Deep: grayish, greenish, concentric, often with radiate
fibrils. *Surface:* brownish, granular, becoming liquefied in 4 days; a
greenish shimmer, with sunken surface growth; microscopically, concentric, the outer zone has a *B. subtilis* character. An odor in the liquefied
gelatin of musty cheese.

Gelatin stab. Liquefaction saccate — infundibuliform, becoming stratiform;
the liquid gelatin becomes yellowish green — bluish red.

Agar slant. Growth moist, glistening, slimy; a bluish shimmer, with a gray
green — blue green fluorescence.

Bouillon. A thick, slimy membrane, and a green fluorescence.

Glucose bouillon. No gas.

Milk. 37°, a green fluorescence at the surface; at 20°, coagulated in 48
hours.

Potato. Growth grayish green — yellowish green, slimy. Grow in acid gelatin.
Indol positive or scanty. Reduction of nitrates to nitrites negative or
scanty.

Pathogenesis. Subcutaneous inoculation of small doses into mice cause death
in 1–3 days, with cramps, œdema at the point of inoculation, and inflammation of the internal tissues. Bacilli by culture from blood and organs.

Habitat. Isolated from a case of cholera-nostras.

30. Ps. fluorescens (Flügge) Migula

B. fluorescens-liquefaciens Flügge: Die Mikroorganismen, 1886, 289.
B. viscosus Frankland: Zeitsch. f. Hygiene, VI, 1887, 39.
B. fluorescens-nivalis Schmelck: Centralblatt f. Bakteriol., IV, 1888, 544.

Morphology. Bacilli 0.5 : 1.0–1.5 μ; occur chiefly in pairs. A polar bundle of 3–6 flagella. Not stained by Gram's method. Cultural characters like No. 28, but according to Lehmann and Neumann, milk is not coagulated, and cultures show a weak indol reaction.

Ruzicka (Archiv f. Hygiene, XXXIV, 1898, 149) shows that the present species and No. 28 are subject to such variations in cultural characters that no sharp lines between the two can be drawn. The student should consult the above paper by Ruzicka.

Habitat. Widely distributed, water, etc.

31. Ps. rugosa Wright

B. rugosus Wright: l.c., 438.

Morphology. Bacilli of medium size, ends rounded, in pairs, chains, and filaments. Flagella 1–4, polar.

Gelatin colonies. In 3 days, 3–4 mm., translucent, grayish, slightly elevated, irregular, sinuous, sharp, radiately rugose, with a smooth border.

Gelatin slant. Growth grayish green, dense, limited, delicately wrinkled, reticulate; gelatin a faint green. Grow in acid gelatin.

Agar slant. Growth translucent, somewhat limited, grayish — grayish white, with delicate wrinkles. The agar becomes greenish.

Bouillon. Turbid, with a pellicle.

Potato. Growth moist, glistening, brown.

Litmus milk. Coagulated, acid.

Glucose bouillon. No gas. *Indol* slight. Nitrates not reduced to nitrites. No growth at 36°.

Habitat. Water.

32. Ps. incognita (Wright)

B. fluorescens-incognitus Wright: l.c., 436.

Morphology. Bacilli of medium size, short, ends rounded, in pairs, chains, and filaments. A polar flagellum.

Gelatin colonies. Deep: round, oval, yellow brown, granular. *Surface:* in 2–3 days, thin, translucent; edge irregular — wavy, coli-like; microscopically, slightly granular, slightly yellowish brown; centre with a yellow brown nucleus, marmorated; older colonies 6–8 mm., with a greenish tint, and the gelatin acquires a blue green fluorescence.

Gelatin slant. Growth thin, translucent, slightly greenish, limited.

Agar slant. Growth thin, moist, translucent; acquires a greenish color.

Bouillon. Turbid, with a pellicle; becomes greenish.

Potato. Growth moist, glistening, brown, spreading.

Litmus milk. Slightly decolorized (after a month or so), reaction acid.

Habitat. Water.

33. Ps. foliacea (Wright)

B. fluorescens-foliaceus Wright: l.c., 439.

Dr. Wright doubtless has good grounds for making this and the preceding distinct species; a careful study of the descriptions, however, do not reveal upon what specific differentiation can be well based. The features of distinction appear to be in the character of the growth on gelatin slant, which shows a central furrow, with also laterals, giving a leaf-like etching of the medium. The surface gelatin colonies show heavy, brown, radial stripes.

Litmus milk. A deeper blue, alkaline, becoming acid, pink. *Indol* slight.

Habitat. Water.

34. Ps. syncyanea (Ehrenberg) Migula

Vibrio syncyaneus Ehrenberg: Gurlt u. Hertwig's Magaz. f. ges. Thierheilk., VII, 1841.

B. cyanogenes Flügge: Die Mikroorganismen, 1886.

Ps. syncyanea Migula: Die Natürlichen Pflanzenfam., 1892, 29.

Morphology. Bacilli 0.3–0.5 : 1.0–4.0 μ. Flagella polar, in bundles of 1–5, rarely bipolar.

Gelatin colonies. *Deep:* small, round, yellowish, granular. *Surface:* large, thin, spreading, with erose edges like *B. coli.* Colonies rarely of the aerogenes type.

Gelatin stab. Slight growth in depth; on the surface, growth white — bluish gray.

Potato. Growth slimy, bluish gray — brown.

Agar slant. Growth white; agar variously colored.

Bouillon. Turbid, gray green — bluish green, with a pellicle.

Glucose bouillon. No gas, slightly acid. *Indol* a trace. H_2S negative.

Milk. Alkaline, blue.

Habitat. Isolated from blue milk.

35. Ps. striata (Ravenel)

B. striatus-viridis Ravenel: l.c., 22.

Morphology. Bacilli slender, of variable lengths; rods irregular; stain like diphtheria; occur singly and in pairs. Flagella polar.

Gelatin colonies. Deep: round, yellowish, granular, entire, becoming brownish. *Surface:* in 5 days 1 mm., zoned; border filamentous.

Agar slant. Growth thin; agar becomes slightly green — yellowish green.

Gelatin stab. Slight growth in depth; on the surface; growth white, elevated.

Potato. Growth moist, glistening, becoming chocolate-brown.

Bouillon. Turbid; becoming in 7 days slightly greenish.

Pepton rosolic acid. Unchanged.

Litmus milk. Alkaline, becoming decolorized.

Glucose bouillon. No gas. *Indol* negative.

Habitat. Soil.

36. Ps. ovalis (Ravenel)

B. fluorescens-ovalis Ravenel; l.c., 9.

Morphology. Bacilli short, rounded, 2–3 times their breadth; occur singly. Flagella polar.

Gelatin colonies. Deep: pale gray, slightly granular, entire. *Surface:* round, light gray, slightly granular, entire, becoming blue white — yellowish green.

Gelatin stab. In depth, growth filiform; on the surface, a white button, with irregular leafy margins.

Agar slant. Growth thin, greenish white, limited; the agar becomes a faint green.

Potato. Growth thin, moist, honey-yellow, becoming yellowish brown.

Bouillon. Turbid, a flaky pellicle on the surface; the medium has a greenish tint.

Litmus milk. Deep blue, alkaline.

Glucose bouillon. No gas. *Indol* negative.

Habitat. Soil.

37. Ps. convexa (Wright)

B. fluorescens-convexus Wright: l.c., 438.

Morphology. Bacilli of medium size, short, thick, rounded. A polar flagellum.

Gelatin colonies. Deep: round, sharp, slightly granular. *Surface:* round, convex, glistening, light greenish, translucent; the gelatin acquires a blue green fluorescence.

Gelatin stab. Surface growth elevated, glistening, light green; the medium becomes blue green, fluorescent. Grow in acid gelatin; no fluorescence.

Agar slant. Growth translucent, moist, glistening, light greenish; agar becomes greenish.

Bouillon. Turbid, greenish.

Potato. Growth pale brown, spreading.

Litmus milk. Not coagulated, alkaline; color deepened.

Glucose bouillon. No gas. *Indol* doubtful. At 36° little or no growth.

Habitat. Water.

38. Ps. erythrospora (Cohn) Migula

B. erythrosporus Cohn: Beiträge Biol., III, 1879, Heft I, 128.

Ps. erythrospora Migula: Die Natürlichen Pflanzenfam., 1896.

Morphology. Bacilli slender threads. Cultures at 20° show in every rod 2-8 reddish ovoid granules. Flagella polar, in bundles of 3-6.

Gelatin colonies. Deep: lobed, undulately channelled, with a green fluorescence. *Surface:* round — irregular, brownish, with a faint radial striping.

Gelatin stab. In depth, a uniform growth; surface growth flat.

Potato. Growth slightly spreading, reddish — nut-brown.

Habitat. Isolated from flesh infusion, and from water.

39. Ps. putida (Flügge) Migula

B. fluorescens-putidus Flügge.

Morphology. Bacilli, small, short, ends rounded.

Gelatin colonies. Deep: very small, dark. *Surface:* round, strongly refracting, yellowish; borders bright gray, finely granular. Later with eroselobate borders, with a greenish shimmer.

Gelatin stab. In depth, a weak, gray, milky turbidity; surface growth spreading; a greenish coloration of the gelatin; not liquefied.

Potato. Growth thin, grayish-brownish. In putrefying solutions, an odor of trimethylamin.

Habitat. Isolated from putrefying substances, water, etc.

CLASS V. WITH ENDOSPORES.

I. Aerobic and facultative anaerobic.
    A. Non-chromogenic.
        1. **Rods not swollen at sporulation.** *B. subtilis* type.
            40. *Ps. rosea* (Bordas).
        2. Rods swollen at one end at sporulation. Tetanus type.
            41. *Ps. Tromelschlägel* (Ravenel).

B. Cultures show a greenish fluorescence.
   1. Gelatin liquefied.
      42. *Ps. viridescens* (Ravenel).
   2. Gelatin not liquefied.
      43. *Ps. undulata* (Ravenel).

40. **Ps. rosea** (Bordas)

B. roseus-vini Bordas-Joulin-Rackowski: Compt. rend., CXXVI, 1898, 1550.

Morphology. Bacilli 0.8 : 8–12 μ; filaments. A brush of polar flagella. Stain by Gram's method.

Gelatin colonies. Large, soft, white; gelatin not liquefied. Reduce nitrates to nitrites.

Milk. Coagulated. *Indol* negative. Grows in media containing 0.3 per cent tartaric acid. Acts feebly on glycerin and glucose, producing succinic acid. No action on alcohol. Does not ferment saccharose. In solutions containing glucose, a production of acetic and butyric acids. In yeast-water glucose solutions there is a rose coloration.

Habitat. Cultures in wine become cloudy, lose color, and in 20 days there is a sensible diminution of tartar and glucose, and a slight increase of acidity.

41. **Ps. Tromelschlägel** (Ravenel)

B. Tromelschlägel Ravenel: l.c.

Morphology. Bacilli straight, 5–7 times their breadth; occur singly and in chains. Flagella at the poles.

Gelatin colonies. *Deep:* round — irregular, yellowish, granular, entire. *Surface:* colonies show a nucleus, with a yellowish granular and an outer grayish granular, veined, irregular border. In 3 days 1 mm., white, elevated, entire. Microscopically, dense, gray, irregular, with veined margins.

Gelatin stab. Surface growth thin, white, irregular; in 2 weeks liquefaction crateriform to a depth of 2 mm.

Agar slant. Growth thin, translucent, grayish, becoming yellowish-gray, then brownish.

Potato. Growth thin, yellow, becoming brown, dry, with a metallic lustre.

Bouillon. Slightly turbid, becoming clear.

Litmus milk. Becomes darker, then is decolorized, alkaline, peptonized, watery, translucent.

Glucose bouillon. No gas. *Indol* negative. Grow at 36°.

Habitat. Soil.

42. Ps. viridescens (Ravenel)

B. viridescens-liquefaciens Ravenel: l.c., 24.

Morphology. Bacilli small, straight, ends rounded, 3–5 times their breadth, occur singly. Flagella polar.

Gelatin colonies. Deep: gray, granular, irregular margins, becoming lobed and fissured at edges. *Surface:* in 24 hours, gray, granular, with irregular margins; centres of grayish cloudy masses; liquefaction crateriform; liquefaction proceeds rapidly and colonies lose their characters quickly.

Gelatin stab. Good growth in depth; liquefaction infundibuliform, becoming in 3–4 days nearly stratiform to the depth of the stab; no green color to the gelatin.

Agar slant. Growth smooth, elevated, greenish white, becoming spreading, with a green tint; edges thin, yellowish green.

Potato. Growth yellow, moist, glistening, not very thick, becoming brownish.

Bouillon. Turbid, medium greenish.

Pepton rosolic acid solution. A thin film on the surface, unchanged.

Litmus milk. Acid, not coagulated.

Glucose bouillon. No gas. *Indol* negative. Grow at 36°.

Habitat. Soil.

43. Ps. undulata (Ravenel)

B. fluorescens-undulatus Ravenel: l.c., 20.

Morphology. Bacilli slender, straight, ends rounded, 7–10 times their breadth, in chains. Flagella polar.

Gelatin colonies. Deep: dense, gray brown, with margins of fine wavy lines; do not exceed 0.25 mm. *Surface:* like drops of moisture, greenish gray; microscopically, gray, granular, with a nucleus and an outer zone of fine hairs. In 3 days, 1 mm., elevated, dense, iridescent; microscopically, dense, with a finely striate border; the gelatin becomes faint green.

Gelatin stab. Slight growth in depth; on the surface, a button, white yellow; the gelatin tinged green near the surface.

Agar slant. Growth thin, translucent; agar tinged green.

Potato. Growth yellow, moist, becoming dirty yellow brown, thin.

Bouillon. Turbid, becoming clear; the medium becomes a clear green.

Pepton rosolic acid solution. Slight growth, becoming in 3 weeks darker.

Litmus milk. Rendered alkaline, not coagulated, consistency unchanged.

Glucose bouillon. No gas. *Indol* negative. Grow at 36°.

Habitat. Soil.

CLASS VI. NITROMONAS GROUP OF WINOGRADSKY

Do not grow in nutrient gelatin or other organic media.

44. Ps. Europæa (Winogradsky) Migula

Nitromonas Europæa Winogradsky : Archiv Sci. biol. St. Pétersbourg I, 1892, Nos. 1-2.
Ps. Europæa Migula : Die Natürlichen Pflanzenfam., 1892, 29.

Morphology. Bacilli 0.9-1.0 : 1.1-1.8 μ; occur singly, rarely in chains of 3-4 elements, coccoid forms — short rods. A single polar flagellum 2-3 times the length of a rod, or rarely one at either end. Grow readily in a fluid medium composed of tap water, 1 litre ; ammonium sulphate, 1 gramme ; potassium phosphate, 1 gramme ; and basic carbonate of magnesia, 10 grammes. The organisms united in zoöglœa masses adherent to the particles of magnesium carbonate in the bottom of the flask. On Kühne's gelatinous silica medium (Zeitsch. f. Biol., XXVII, 1890, Heft 1 ; Migula System der Bakterien, I, 1897, 263). Compact, sharply defined colonies of a brownish color; after 10-14 days the colonies spread somewhat and are bright uncolored masses with differently formed outgrowths. Ammonium salts converted into nitrites.

Habitat. Isolated from soils from Europe, Africa, and Japan.

45. Ps. Javanensis (Winogradsky) Migula

Nitromonas Javanensis Winogradsky : l.c.
Ps. Javanensis Migula : l.c.

Morphology. Bacilli small, ovoid, 0.2 : 0.5 μ. A polar flagellum 20 times as long as the rod. In fluid inorganic media as in (44) ; minute flocculi or scales adherent to the walls of the flask; no turbidity. On Kühne's silica medium the colonies are round — elliptical. Nitrites are converted into nitrates.

Habitat. Isolated from Quito soil.

CLASS VII. CELLS WITH REDDISH PLASMA AND WITH SULPHUR GRANULES. *CHROMATIUM* PERTY.

46. Ps. Okenii (Ehrenberg) Migula

Monas Okenii Ehrenberg : Die Infusionstierchen, 1838.
Ps. Okenii Migula : Die Natürlichen Pflanzenfam., 1892.

Bacilli short, ovoid forms, 0.5 : 8.0-15.0 μ, with 1-3 polar flagella.

47. Ps. rosea Migula : l.c.

Bacilli long, cylindrical, 2.0 : 8.0-1.2 μ, with 1-3 polar flagella.

It is doubtful whether the two last species should be included in this genus.

SPIRILLACEÆ

Cells more or less spirally curved. Division in one direction of space, at right angles to the longer axis of the cell. Generally without endospores, which are, however, present in a few species. With or without flagella. The flagella, when present, are attached to the poles, usually in bundles.

I. Cells stiff, not flexile.
 A. Without flagella. Spirosoma Migula.
 B. With flagella.
  1. With 1, rarely with 2–3 polar flagella. Microspira Schröter.
  2. With a bundle of polar flagella. Spirillum Ehrenberg.
II. Cells flexile. Spirochæta Ehrenberg.

SPIROSOMA Migula

Cells comma-formed to spiral filaments, not flexile, stiff; non-motile, flagella absent; occur singly or commonly united in zoöglœa; endospores not yet discovered in any of the species.

I. Non-chromogenic, without pigment on gelatin or agar.
 A. Gelatin colonies floccose.
  1. *Spirosoma linguale* (Weibel) Migula.
 B. Gelatin colonies finely granular.
  2. *Spirosoma nasale* (Weibel) Migula.
II. Chromogenic, produce pigment on gelatin or agar.
 A. Chromogenic function weak, pale yellows.
  3. *Spirosoma flavescens* (Weibel).
  4. *Spirosoma flava* (Weibel).
 B. Chromogenic function stronger, golden yellows.
  5. *Spirosoma aurea* (Weibel).

1. Spirosoma linguale (Weibel) Migula

Vibrio lingualis Weibel: Centralblatt f. Bakteriol., IV., 1888, 227.
Mpma. linguale Migula: Die Natürlichen Pflanzenfam., 1892, 3.

Morphology. Bacilli short, curved, cholera-like rods, or slightly undulate filaments; no true spirals.
Gelatin colonies. Floccose, anthrax-like.
Gelatin stab. In depth, growth filiform; on the surface, no growth; gelatin not liquefied.

Bouillon. Slight turbidity, with a flocculent sediment.
Agar slant. Growth dirty white, finely granular. Non-pathogenic.
Habitat. Isolated from the tongue.

2. Spirosoma nasale (Weibel) Migula

Vibrio nasalis Weibel: l.c.
Mpma. nasale Migula: l.c.

Morphology. Bacilli from the nasal mucus as thick vibrios; in bouillon cultures as straight rods; on agar as spirals.
Gelatin colonies. Small, entire, yellowish brown, finely granular; gelatin not liquefied.
Potato. No growth. No odor in cultures. Non-pathogenic.
Habitat. Isolated from nasal mucus.

3. Spirosoma flavescens (Weibel)

Vibrio flavescens Weibel: l.c.

Morphology. Bacilli 1.5 times as thick as cholera bacteria, comma and S forms, longer and shorter forms; tendency to form long spirals not evident.
Gelatin colonies. *Deep:* ovoid, granular. *Surface:* round, entire, dirty grayish yellow.
Gelatin stab. Surface growth flat, with lobular edges, not depressed.
Agar slant. Growth dull yellow, with grayish spots.
Potato. Growth thick, pasty, dull yellow.
Bouillon. Turbid; no pellicle.
Habitat. Sewer mud.

4. Spirosoma flava (Weibel)

Vibrio flavus Weibel: l.c.

Morphology. Bacilli like the preceding.
Gelatin colonies. *Deep:* golden yellow, finely granular. *Surface:* light yellow, mottled, with a white zoned border.
Agar slant. An ochre-yellow layer.
Bouillon. Turbid; no pellicle.
Habitat. Sewer mud.

5. Spirosoma aurea (Weibel): l.c.

Morphology. Bacilli as in No. 3.
Gelatin colonies. *Deep:* ovoid, granular; centre golden yellow, with a lighter border. *Surface:* round, entire, granular, golden yellow, with a lighter border.

Gelatin stab. In depth, growth thick, granular, yellow; on the surface, a yellowish expansion, with a bowl-shaped depression.

Agar slant. Growth dirty white, becoming thicker, golden yellow, pasty.

Potato. Growth thick, pasty, golden yellow.

Bouillon. Turbid; no pellicle.

Habitat. Sewer mud.

MICROSPIRA Migula

Cells mostly small, weakly curved comma forms or short spirals, occasionally longer spiral filaments, which, with iodine staining, may show segmentation into comma elements. Every cell bears, as a rule, one polar flagellum, and, less commonly, 2–3. Immediately before division there may be flagella at both poles. Endospores not known.

I. Cultures show a bluish to a silvery phosphorescence.
MARINE BACTERIA.

    1. *Microspira phosphorescens* (Fischer).

    2. *Microspira Fischeri* (Beijerinck).

    3. *Microspira luminosa* (Beijerinck).

II. Cultures do not show phosphorescent properties.

  A. Gelatin liquefied.

    1. Cultures show the nitro-indol reaction.

      a. Very pathogenic to pigeons.

        4. *Microspira Metschnikovi* (Gamaleïa) Migula.

        5. *Microspira Schuylkilliensis* (Abbott).

      b. Not distinctly pathogenic to pigeons.

        * Milk coagulated.

         † Gelatin liquefied rather slowly.

         6. *Microspira comma* (Koch) Schröter.

        †† Gelatin liquefied rapidly.

         7. *Microspira danubica* (Heider). ,

        ** Milk not coagulated.

         8. *Microspira Berolinensis* (Neisser) Migula..

    2. Nitro-indol reaction negative or very weak, at least after 24 hours.

      a. Grow on potato and in neutral bouillon.

        9. *Microspira protea.*

        10. *Microspira Gindha* (Kruse).

      b. No growth on potato.

        * Liquefaction of the gelatin takes place only at the surface, crateriform.

        11. *Microspira aquatilis* (Günther).

** Liquefaction of the gelatin takes place in depth, cholera-like.

        12. *Microspira tyrogena* (Dencke) Migula.

3. Indol reaction not stated.

        Microspira marina Russell, see No. 20.

B. Gelatin not liquefied, or only slightly so in No. 20.

    1. A slight liquefaction of the gelatin at the surface in stab cultures.

        13. *Microspira choleroides* (Bujwid).

    2. Absolutely no liquefaction of the gelatin.

      a. Growth in gelatin plates slow, colonies minute.

        14. *Microspira Weibeli.*

        15. *Microspira denitrificans* (Sewerin).

      b. Colonies on gelatin plates of average size.

        * Potato cultures becoming brownish.

          16. *Microspira saprophile* (Weibel).

          17. *Microspira cloaca.*

      ** Potato cultures yellowish white.

          18. *Microspira terrigena* (Günther).

1. **Microspira phosphorescens** (Fischer)

B. phosphorescens Fischer: Zeitsch. f. Hygiene, II, 1887.

Photobact. indicum Beijerinck: Akademie van Wetenschappen Afdeeling Natuurkunde 2. Reeks, VII, Amsterdam, 1890.

B. phosphorescens-indicus Kruse: Flügge, Die Mikroorganismen, 1896, 330.

Morphology. Bacilli 0.6–0.8 : 2.0 μ — bent filaments.

Gelatin colonies. Deep : round, entire, bluish — sea green. *Surface :* granular, brownish, borders undulate, sinking slowly in the liquid gelatin.

Gelatin stab. Slight growth in depth ; on the surface, liquefaction napiform, with an air bubble.

Agar slant. Growth grayish white.

Potato. No growth except when cooked in salt water. Phosphorescence bluish, disappearing in old cultures. Grow well in sea water on fish, meat, blood, and egg, with a bluish phosphorescence. Non-pathogenic.

Habitat. Isolated from phosphorescent sea water, West Indies.

2. **Microspira Fischeri** (Beijerinck)

Photobact. Fischeri Beijerinck: l.c.

B. phosphorescens-indigenus Kruse: Flügge, Die Mikroorganismen, 1896, 331.

Cultural characters as before. Differs in liquefying gelatin rather more slowly, phosphorescence less intense, the latter absent when grown on fresh media, as above.

Habitat. Sea water.

3. Microspira luminosa (Beijerinck)

Photobact. luminosum Beijerinck: Archives Néelandaises, XXIII, 1889, 104.

Morphology. Bacilli 0.6 : 2.0 μ, bent — curved rods — threads, variable.

Gelatin stab. Liquefaction rapid ; in later generations it occurs more slowly. Grow best in 3.5 per cent NaCl gelatin. Phosphorescence weak, silvery ; does not appear on meat and potato cultures, evident in sea water, on fish, and in 3.5 per cent NaCl sugar-free gelatin.

The last three species are doubtfully placed in this genus.

4. Microspira Metschnikovi (Gamaleïa) Migula

Vibrio Metschnikovi Gamaleïa : Annales Pasteur Institut, 1888, 482.

Msp. Metschnikovi Migula : Die Natürlichen Pflanzenfam., 1892, 33.

Morphology. Bacilli like *Msp. comma*, but somewhat shorter and thicker. A long spiral polar flagellum.

Gelatin colonies. Quite like *Msp. comma.*

Gelatin stab. Liquefied more rapidly than *Msp. comma*, but otherwise identical.

Bouillon. Strongly turbid ; a strong, white membrane on the surface ; in 24 hours a strong nitro-indol reaction.

Potato. Slight growth at 20° ; at 37° a delicate brownish layer.

Pathogenesis. Pigeons inoculated into the pectoral muscle die in 24 hours. Muscles greatly swollen and yellowish as if cooked, and infiltrated with a serous fluid containing many bacilli ; numerous bacilli in the blood and organs. Intestines pale and filled with a grayish yellow fluid, with only a few bacilli present. Subcutaneous inoculations of guinea pigs cause death in 24 hours. There is great bloody œdema with bacilli in the heart blood.

Habitat. Isolated by Gamaleïa from an epizoötic of fowl, and by Pfuhl (Zeitsch. f. Hygiene, 1894) from water.

5. Microspira Schuylkilliensis (Abbott)

Vibrio Schuylkilliensis Abbott: Jour. Expt. Med., 1896, I, No. 3.

Morphology. Bacilli in fresh cultures, rather plump commas, often with a decided curve, or nearly straight, ends rounded — slightly pointed, long spiral filaments uncommon, involution forms in old cultures. A single polar flagellum. Spores absent. Stain irregularly ; not stained by Gram's method.

Gelatin colonies. Round, sharply defined, slightly granular, with fine irregular lining as if creased, or concentric becoming moruloid, or again distinctly concentric, with a dark central portion; later ragged granular clumps in pits of liquefied gelatin, with granular or ciliate borders.   Plates completely liquefied in 36-40 hours.   Colonies quite like the preceding (4). The liquefied gelatin becomes decidedly alkaline.

Gelatin stab.   Quite like *Msp. comma*, but liquefaction rather more rapid.

Agar slant (neutral to phenolphthalein).   In 24 hours, colonies 1.5-2 mm., smooth, glistening, opaque; in 48 hours, growth dryer, wrinkled.

Löffler's blood serum.   In 24 hours, at 37°, a depressed line, not spreading due to liquefaction, becoming dirty brown.

Bouillon.   Turbid, with a pellicle, alkaline.

Potato.   In 48 hours, at 37°, very slight growth, dirty yellow — brownish, not spreading.

Litmus milk.   Acid, coagulated, sometimes not coagulated.

Pepton solution.   In 24 hours, a strong indol reaction, alkaline to litmus.

Pepton rosolic acid solution.   Color slightly intensified, or orange-red.

Glucose bouillon.   No gas.   Grow more slowly under anaerobic conditions.

Pathogenesis.   Very pathogenic to pigeons and guinea pigs.   Pigeons, 0.2-0.3 cc. of an agar suspension into pectoral muscle causes death in 16-18 hours.   Muscle swollen with œdema of subcutaneous tissue; muscle marked by yellow striations and red necrotic areas; bacilli present in large numbers.   Blood clots in the heart; lungs, kidneys, and spleen quite normal; intestines scarcely affected.   Subcutaneous inoculations of 0.3-0.5 cc. of an agar suspension into guinea pigs causes death in 18 hours.   Tissues at the point of inoculation much injected. The subcutaneous œdema may be widespread or less so.   Lymph glands red and enlarged.   Peritoneum red, or a general peritonitis with perhaps much fibro-purulent exudate.   The liver may be pale and mottled or normal.   Kidneys and adrenal bodies usually acutely congested; lungs and spleen normal.   Bacilli in the fluid at the point of inoculation, and sparingly in the blood and organs.

Habitat.   Isolated from Schuylkill River water.

The preceding may be a variety of *Msp. Metschnikovi*, but Abbott claims a distinct species.

6. Microspira comma (Koch) Schröter

Comma Bacillus Koch: Berliner klin. Wochensch., 1884, 31-32.
Spirillum choleræ-asiaticæ Zopf: Spaltpilze, 1885.
Microspira comma Schröter: Pilze Schles., 1886, 168.
Vibrio choleræ Lehmann-Neumann: Bak. Diag., 1896, 317.

Morphology.   Bacilli 0.4 : 2.0 μ, slightly curved — semicircular forms, also spirals and involution forms.   One or two spiral flagella.   Not stained by Gram's method.

Gelatin colonies. Small, yellowish white, becoming sunken in the liquefied
   gelatin; zone of liquefied gelatin turbid — flocculent, later becoming concentric. Microscopically, in 16–18 hours small, round, yellowish, coarsely
   granular; the central colony becomes irregular — brecciated.

Gelatin stab. A napiform liquefaction, with an air-bubble depression at the
   top, becoming saccate; gelatin turbid — granulated or shows yellowish
   white masses.

Agar colonies. Deep: round — irregular — naviculate, entire, knobbed or
   granular, becoming darker, or a brownish nucleus, with a grayish zone.
   Surface: round, whitish-brownish, moist, glistening, entire, translucent,
   slightly raised. Microscopically round, light yellowish, translucent,
   punctate — granulated.

Agar slant. Growth brownish gray, moist, glistening, or yellowish brown.

Blood serum. Slowly liquefied.

Bouillon. Slightly turbid; a wrinkled fragile membrane; alkaline; becoming
   clear.

Milk. Coagulated in 2 days at 37°, acid.

Potato (acid). No growth; on alkaline potato growth dirty white — yellowish, scarcely raised, moist, glistening, becoming brownish red, spreading.
   H_2S positive. *Indol* positive. No gas in glucose, lactose, or saccharose
   bouillons. In glucose bouillon the production of left-handed lactic acid.

Litmus milk. On the surface, a blue pellicle, below this a red discoloration.

Pathogenesis. Inoculations *per os* and by ingestion gave positive results with
   young cats and rabbits (see text-books). Intraperitoneal inoculations of
   guinea pigs cause distention of the abdomen, subnormal temperatures,
   and general collapse. Bacilli in the peritoneal cavity, or, with large doses,
   in the blood and small intestines. Intravenous injections of rabbits cause
   toxæmia and death in 18 hours; with smaller doses a true cholera picture
   may follow.

Habitat. Isolated from the alvine discharges and intestinal contents of cholera patients, and in water supplies during epidemics.

PROBABLE VARIETIES OF *Msp. comma.*

(*A*) *Pseudo-cholera-spirillum* Rénan : Annales Pasteur Institut, VI, 1892, 621.
Bacilli much larger than *Msp. comma*; commas and S forms.

Gelatin colonies. Small, lenticular, with outgrowths : centres darker and yellowish, becoming in 4 days surrounded by a liquefied zone.

Gelatin stab. Like *Msp. comma*, but grows more rapidly.

Agar slant. Growth thick, creamy white, limited.

Bouillon. Turbid, with a thin pellicle.

Pathogenesis. Non-pathogenic to guinea pigs.

Habitat. Well water.

(*B*) *Vibrio choleroides*, *β*. Bujwid : Centralblatt f. Bakteriol., XIII, 1893, 120.

Morphological and cultural characters like *Msp. comma*, but more anaerobic in habit, and forms a deeper liquefied funnel.

Habitat. Water.

(*C*) *Vibrio Ivanoff* Kruse : Flügge, Die Mikroorganismen, 1896, 592. Noted by Ivanoff: Zeitsch. f. Hygiene, XV.

Bacilli fine and elongated, slightly curved comma forms ; otherwise identical with *Msp. comma*. Ivanoff considers it only a morphological variety.

Habitat. Isolated from the dejecta of typhoid patients.

7. Microspira danubica (Heider)

Vibrio danubicus Heider: Centralblatt f. Bakteriol., XIV, 1893, 341.

Morphology. Bacilli like *Msp. comma*. Gelatin rapidly liquefied, crateriform — funnel-formed.

Gelatin colonies. When the gelatin layer is thick, colonies like *Msp. comma*; when thin, surface colonies flat, spreading, irregular, with undulate — coarsely lobate borders.

Milk. Coagulated.

Potato. At 37° growth brownish, scanty. *Indol* positive. Pathogenic for guinea pigs, slightly so for pigeons. A negative reaction with Pfeiffer's serum.

Habitat. Isolated from canal water.

8. Microspira Berolinensis (Neisser) Migula

Vibrio Berolinensis Neisser: Archiv f. Hygiene, XIX, 1893.

Morphology. Bacilli like *Msp. comma*, but somewhat smaller. A single polar flagellum. Not stained by Gram's method.

Gelatin colonies. Like *Msp. comma*, but not much granulated or fragmented. Liquefaction of the gelatin very slight. *Deep colonies:* small, round, entire, colorless, transparent, slightly granular. *Surface colonies:* small, thin, membranous, with a round central nucleus ; no depression is formed, and the edges remain sharp.

Gelatin stab. Growth like *Msp. comma*, but a slower growth,

Agar slant and *potato.* Growth like *Msp. comma*.

z

Bouillon. Growth more rapid than *Msp. comma*; medium rendered alkaline.

Milk. Not coagulated, not rendered acid.

Habitat. Isolated from filtered river Spree water.

9. **Microspira maaseï** (v. Hoff)

Spirillum maaseï v. Hoff: Centralblatt f. Bakteriol., XXI, 1897, 797.

Morphology. Bacilli short, thick, 1.0–1.2 μ. One or two polar flagella. Spirals of 1–2 turns.

Agar slant. Growth milky white.

Milk. Not coagulated.

Bouillon. A membrane on the surface; no acid production. *Indol* produced.

Habitat. Isolated from Rotterdam water.

10. **Microspira protea** (Buchner)

Vibrio der cholera-nostras Finkler-Prior: Centralblatt f. allg. Ges. Ergänzungshefte, Bd. I, 1884.

Vibrio proteus Buchner: Sitzungsber. d. Ges. f. Morph. u. Physiol., München, I, 21.

Morphology. Bacilli 0.4–0.6 : 2.4 μ, more or less curved, rather longer and thicker than *Msp. comma*, and often pointed at the ends and thicker in the centre. S forms and spirals less common than in the case of *Msp. comma*.

Gelatin colonies. In early stages like cholera, but grow rather more vigorously, and the gelatin liquefied more rapidly. Microscopically, round, entire, yellow, finely granular, often concentrically zoned. Colonies retain their regular form and finer structure better than with cholera.

Gelatin stab. Gelatin liquefied much more rapidly than with cholera; a saccate liquefaction in 24 hours, with turbid contents.

Milk. Coagulated, becoming liquefied (Lehmann-Neumann); unchanged according to other authors; slightly acid.

Potato. Growth at room temperature, within 48 hours, slimy, grayish yellow. (*Msp. comma* does not grow on potato at room temperatures.)

Glucose bouillon. No gas. H_2S production slight. Cultures have a fœtid odor.

Agar slant. Growth dirty yellowish (cholera whitish).

Plover's egg albumen. A bright yellow growth (cholera whitish). See Pearmain-Moor, "Applied Bacteriology," 1892, 205. No indol reaction after 3 days.

Pathogenesis. On experimental animals action like *Msp. comma*, but somewhat milder.

Habitat. Isolated from stools in cholera-nostras, cholera infantum, etc.

VARIETIES. —

(*A*) *Vibrio Lissabonensis* Pestana-Bettencourt : Centralblatt f. Bakteriol., XVI, 1894, 401.

According to Chantemesse identical, or nearly so, with the preceding. From descriptions, indistinguishable from the latter.

Habitat. Isolated from a case of choleraic disease in Lisbon.

(*B*) *Vibrio helicogenes* Fischer: Centralblatt f. Bakteriol., XIV, 1894, 73. From descriptions, indistinguishable from *Msp. protea.*

Habitat. Isolated from the stools of a sick woman.

11. **Microspira Gindha** (Kruse)

Vibrio Gindha Kruse: Flügge, Die Mikroorganismen, 1896, 590.

Morphology. Bacilli rather long curved rods, rather smaller than *Msp. comma.* A single polar flagellum.

Gelatin colonies. Cholera-like. In pepton solution in 24 hours, at 37°, a negative or only a faint indol reaction.

Pathogenesis. Subcutaneous inoculations into pigeons and guinea pigs negative. Living or dead cultures show toxic properties when inoculated intraperitoneally into guinea pigs.

Habitat. Isolated by Pasquale from water.

12. **Microspira aquatilis** (Günther)

Vibrio aquatilis Günther: Deutsche med. Wochenschrift. 1892, 1124.

Morphology. Bacilli like *Msp. comma.* A single polar flagellum.

Gelatin colonies. Round, entire, brownish, finely granular.

Gelatin stab. Growth only on the surface, becoming a crateriform liquefaction.

Agar slant. Growth like cholera.

Potato. No growth either at 20° or 37°.

Bouillon. No growth in alkaline or amphoteric bouillon at 37°; at 20°, only a faint growth after some weeks. *Indol* negative. A strong odor of H$_2$S. Non-pathogenic.

Habitat. Isolated from water of river Spree.

13. **Microspira tyrogena** (Dencke) Migula

Spirillum tyrogenum Dencke: Deutsche med. Wochenschrift, 1885, No. 3.
Msp. tyrogena Migula: l.c.

Morphology. Bacilli like *Msp. comma,* but rather smaller and more slender, often very long spirals with close windings.

Gelatin colonies. In 24 hours, small white points; microscopically, round, entire, dark greenish brown, becoming dark yellow in centre with lighter ·borders, becoming liquefied. Colonies generally retain their sharp contour.

Gelatin stab. Grow like cholera, but the gelatin liquefied more rapidly.

Agar slant. Growth yellowish white.

Potato. No growth either at 20° or 37°. In bouillon or pepton solution, no indol reaction.

Pathogenesis. Very feebly pathogenic (Koch), or not at all so.

Habitat. Isolated from cheese.

14. **Microspira choleroides** (Bujwid)

Vibrio choleroides Bujwid: Centralblatt f. Bakteriol., XIII, 1893, 120.

Morphology. Bacilli like *Msp. comma*, but movements not so rapid.

Gelatin colonies. At low temperatures, cholera-like colonies; at higher temperatures colonies larger, and do not sink deeply into the gelatin; microscopically the contour is more regular than cholera, and only finely granular.

Gelatin stab. Growth only on the surface, with slight liquefaction, or an air bubble may be produced.

Agar slant. A good growth, with an odor of methyl-mercaptan.

Bouillon. Slightly turbid; no pellicle.

Habitat. Water.

15. **Microspira Weibeli**

Vibrio saprophiles β Weibel: Centralblatt f. Bakteriol., II, 1887, 469.

Morphology. Bacilli slender curved rods of about the thickness of tubercle bacilli and about 2.0 μ long; ends blunt, commas — S forms; long filaments do not occur. Grow slowly at room temperatures.

Gelatin colonies. Do not exceed 0.3 mm.; microscopically, round, yellowish brown.

Gelatin stab. A slight growth in depth; on the surface, growth thin, white, spreading.

Agar slant. Growth creamy, yellowish white, viscid.

Potato. Growth thin, glistening, varnish-like, of a dirty brownish gray color.

Habitat. Isolated from putrefying hay infusion.

16. **Microspira denitrificans** (Sewerin)

B. denitrificans Sewerin: Centralblatt f. Bakteriol., 2te Abt., I, 1895, 162, and IV, 451.

Morphology. Bacilli on agar 0.5 : 2–4 μ; on nitrate bouillon, comma forms — longer vibrios and Spirillum forms. Older cultures show branched individuals and other involution forms.

Gelatin colonies. Small, white, entire.
Gelatin stab. In depth, growth pearly.
Agar colonies. In 2 days, at 30°, 1-2 mm., bluish white, slimy.
Agar slant. In 2 days, growth grayish white, limited, becoming spreading.
Nitrate bouillon. Turbid, with a slight membrane; nitrates reduced.
Habitat. Isolated from horse manure.

17. **Microspira saprophile** (Weibel)

Vibrio saprophiles A. Weibel: Centralblatt f. Bakteriol., II, 1887, 469.

Morphology. Bacilli bent rods 3.0 μ long to S forms, rarely filaments.
Gelatin colonies. Deep : round, entire, yellowish brown, concentric, edges
serrate. *Surface :* flat, yellowish white; microscopically, with dark yellowish gray centres, paler at borders, finely granular.
Gelatin stab. In depth, growth thin, veily; on the surface, growth thin,
spreading, white.
Agar slant. Growth dirty white, spreading; agar colored below.
Potato. Growth slimy, pasty, yellowish red, becoming chocolate-brown.
Bouillon. Turbid, with a yellowish granular sediment.
Habitat. Isolated from putrid hay infusion and sewer mud.

18. **Microspira cloaca**

Vibrio saprophiles Y Weibel: l.c.

Bacilli like the preceding, but twice as large, rarely forms long-twisted threads;
a great tendency to produce involution forms.
Gelatin colonies. Deep : ovoid, entire, granular; centres orange-colored, with
light yellow outer zones. *Surface:* flat, dirty white — opalescent, with
prominent white centres; microscopically, like *B. coli;* borders irregular
— lobed, marmorated — spotted.
Gelatin stab. In depth, growth filiform; on the surface, a whitish expansion.
Agar slant. Growth spreading, dirty white.
Potato. Growth yellowish brown — brown, moist, glistening, becoming dry,
tough and dark brown.
Bouillon. Turbid, with a thick firm pellicle.
Habitat. Isolated from sewer mud.

19. **Microspira terrigena** (Günther)

Vibrio terrigenus Günther: Centralblatt f. Bakteriol., XVI, 1894, 746.

Morphology. Bacilli show at each end one or several flagella.
Gelatin colonies. Entire, small, structureless.

Potato. Growth yellowish white.

Glucose bouillon. No gas.

Milk. Not coagulated. Non-pathogenic. Strongly aerobic.

Habitat. Soil.

ADDENDA

20. **Microspira marina** (Russell)

B. marinus Russell: Zeitsch. f. Hygiene, XI, 1891, 198.

Morphology. Bacilli small, straight — curved — spirals. Motion rotatory and progressive.

Gelatin colonies. Small, radially striped, becoming liquefied and rougher, with flocculi in the liquefied gelatin.

Gelatin stab. A rapid liquefaction, the liquid gelatin turbid, and a thin pellicle on the surface.

Agar slant. A moist whitish expansion, like pus.

Potato. Growth thick, waxy, spreading.

Bouillon, with sea water, turbid, with a smooth white pellicle.

Habitat. Isolated from sea water and sea mud.

SPIRILLUM Ehrenberg

Spirally curved or corkscrew forms of variable thickness. Endospores present in a few species. Cells actively motile, and possess at one or both poles bundles containing four or more flagella.

I. Cell contents colorless — EUSPIRILLUM of Migula.

    A. Grow in nutrient gelatin or other culture media.

        1. Aerobic, *i.e.* grow in the presence of atmospheric oxygen.

           a. Non-chromogenic; do not produce pigment on gelatin or agar.

                * Cell plasma contains numerous black granules, very large spirilla, 1.5–2.0 μ broad and 20–30 μ long.

                    1. *Spirillum volutans* Ehrenberg.

              ** Cell plasma does not contain dark granules. Spirilla smaller.

                † Gelatin liquefied, sometimes very slowly.

                    § Pathogenic to pigeons and guinea pigs. Comma-like.

                      2. *Spirillum Massauah* Kruse.

                  §§ Terrestrial forms probably not pathogenic.

                    3. *Spirillum serpens* (Müller) Winter.

                    4. *Spirillum tenerrimum* Lehmann-Neumann.

†† Gelatin not liquefied.

     5. *Spirillum undula* (Müller) Ehrenberg.

     6. *Spirillum concentricum* Kitasato.

b. Produce pigment on gelatin or agar.

   * On gelatin, growth yellowish.

     7. *Spirillum tenue* Ehrenberg.

  ** On gelatin and agar, growth reddish.

     8. *Spirillum rubrum* v. Esmarch.

2. Anaerobic; do not grow in the presence of atmospheric oxygen.

     9. *Spirillum rugula* (Müller) Winter.

B. Do not grow in nutrient gelatin or other culture media.

     10. *Spirillum sputigenum* Kruse.

II. Cell plasma reddish.   THIOSPIRILLUM Migula.

A. Cells 3–3.5 microns thick and 10–40 long.

     11. *Spirillum jenense* (Ehrenberg) Winter.

     12. *Spirillum sanguineum* (Ehrenberg) Cohn.

B. Cells scarcely exceed one micron in thickness.

     13. *Spirillum rufum* Perty.

1. **Spirillum volutans** Ehrenberg

Die Infusionstierchen als vollkommene Organismen, 1838.

Morphology. Bacilli 2–3 : 13–30–50 μ, ends slightly attenuated. Spirals of 2–5 elements, amplitude 10–15 μ. A polar bundle of flagella (Kutscher),. 3–8. Dark granules in the interior of the rods, supposed to be of sulphur.

Gelatin colonies. Coli-like.

Agar colonies. Like diphtheria.

Gelatin stab. Slight growth in depth; on the surface, growth porcelain-white, crumpled.

Potato. Growth dry.

Bouillon. Turbid; no membrane.

Habitat. Isolated from stagnant and marsh water.

2. **Spirillum Massauah** Kruse

Flügge, Die Mikroorganismen, 1896, 589.
Noted by Pasquale: Baumgarten's Jahresberichte, 1891, 336.

Morphology. Bacilli similar to *Msp. comma*, with as many as 4 polar flagella.

Gelatin colonies. Completely round, entire, yellowish; only a trace of lique-faction. *Indol* produced in 24 hours in pepton solution.

Pathogenesis. Subcutaneous and intramuscular inoculations of pigeons and guinea pigs fatal.

Habitat. Isolated from dejecta of cholera patients.

3. **Spirillum serpens** (Müller) Winter

Vibrio serpens Müller: Animalia infusoria, 1786.
Spirillum serpens Winter: Die Pilze, 1884. Zettnow: Centralblatt f. Bakteriol., X, 689.

Morphology. Bacilli rigid filaments, with 2–3 wave-like undulations, 0.8–1.0 : 10–30 μ. Polar flagella in bundles of 14. Spores absent.

Gelatin colonies. Typhi- or coli-like.

Gelatin stab. Growth coli-like; surface growth becomes sunken, with the formation of a bubble.

Agar slant and *potato.* Growth like *B. coli.*

Bouillon. Turbid, with a delicate membrane.

Habitat. Isolated from stagnant water.

4. **Spirillum tenerrimum** Lehmann-Neumann

Spirillum I Kutscher: Zeitsch. f. Hygiene, XX, 1895, 47.
Spirillum tenerrimum Lehmann-Neumann : Bak. Diag., 1896, 346.

Morphology. Bacilli short forms; as a rule, thin. Spirilla of 3–4 elements.

Gelatin colonies. Show a compact centre, a thin, finely granular zone, and an outer filamentous border.

Gelatin stab. Growth like that of the bacillus of mouse septicæmia, liquefaction slow.

Agar slant. Growth of dewdrop colonies.

Bouillon. Slightly turbid — no membrane. The above may be a Microspira.

Habitat. Sewage.

5. **Spirillum undula** (Müller) Ehrenberg

Vibrio undula Müller: Historia Vermium terrestrium et fluviat Hauniæ, 1773.
Spirillum undula Ehrenberg: Abhandl. Ber. Akad., 1830, 38.

Morphology. Bacilli stout threads, 1.2–1.5 : 8–16 μ, one-half to three turns. Wave length, 4–5 μ. Bundles of 3–9 flagella at both poles. Spores present (Migula, l.c.).

Gelatin colonies. Deep: small, entire, granular; no growth on the surface.

Gelatin stab. A development in the upper part of the stab; on the surface, growth thin, white, somewhat rugose.

Potato. No growth.

Bouillon. Turbid; no membrane.

Habitat. Putrid and stagnant water.

VARIETIES. — *Spirillum undula-minus* and *Spirillum undula-majus* of Zett-now-Kutscher: Centralblatt f. Bakteriol., XVIII, 614; XIX, 393. One-third larger than the above. Grow well on gelatin and agar.

6. Spirillum concentricum Kitasato

Centralblatt f. Bakteriol., III, 72, 1888.

Morphology. Bacilli 0.5 : 1.0–8.0 μ; spiral forms, ends pointed. Bundles of polar flagella at one or both poles.

Gelatin colonies. Round, grayish, concentric.

Gelatin stab. Slight growth in depth; on the surface, growth, thin, veily, spreading.

Agar slant. Growth thin, adherent.

Potato. No growth at 20°–37°.

Bouillon. Turbid; becoming clear, with a sediment.

Milk. Not coagulated. H_2S negative. *Indol* negative.

Habitat. Isolated from putrid blood.

7. Spirillum tenue Ehrenberg

Die Infusionsthierchen als vollkommene Organismen, 1838.

Morphology. Bacilli 0.8 μ wide and 4–15 μ long, of 2–5 undulations, wave length and amplitude, 2–3 μ. Bundles of flagella at either pole. Spores negative.

Gelatin colonies. Deep: round, entire, finely granular, yellowish. *Surface:* round, thin, yellowish.

Gelatin stab. In depth, growth filiform; on the surface, a yellowish layer, and a slow liquefaction, with a bubble of gas.

Potato. No growth.

Bouillon. Turbid, with a thick membrane. See Bonhoff, Hyg. Rundschau, VI, 351. Kutscher, Centralblatt f. Bakteriol., 2te Abt., I.

Habitat. Putrefying vegetable infusions, etc.

8. Spirillum rubrum v. Esmarch

Centralblatt f. Bakteriol., I, 1887, 225.

Morphology. Bacilli short spirals 0.6–0.8 : 1.0–3.2 μ, with 1 — 2 — 3 screw twists when grown in gelatin or agar; in bouillon, longer forms which are twice as thick as *Msp. comma*. The shorter spirilla are very motile, the longer ones only slightly so. There are glistening spots in the rods like spores, but which do not stain. Stained by Gram's method. Flagella in bundles at the poles.

Gelatin colonies. A very slow growth, visible in 8 days. *Deep colonies :* gray — pale red, entire, becoming wine-red. The *surface colonies* develop but little pigment.

Gelatin stab. In depth, wine-red colonies ; on the surface no color. *Indol,* a slight trace.

Agar slant and *blood serum.* Growth moist, glistening, grayish white, limited, wine-red where thicker.

Potato. Growth of small, deep-red colonies.

Bouillon. Slightly turbid, with a reddish sediment. H₂S negative. No gas produced.

Habitat. Water.

9. Spirillum rugula (Müller) Winter

Vibrio rugula Müller : Cohn's Beiträge, I, 1875.
Spr. rugula Winter : Die Pilze, 1884.

Morphology. Bacilli 1.5–2.0 : 8–16 μ, rods curved, spiral, or in long chains. A bundle of polar flagella. Spores at ends of swollen rods; according to Bonhoff, not present.

Gelatin colonies. At 20°–22° yellowish white, round, becoming liquefied.

Agar slant. Growth at 37° white, somewhat rugose.

Potato. Growth at 37° white-yellowish, spreading, rugose.

Blood serum. Growth at 37° thin, white ; medium liquefied. A penetrating fæcal odor in cultures.

Habitat. Isolated from the mouth and putrefying fluids.

10. Spirillum sputigenum Kruse

Flügge, Die Mikroorganismen, 1896, 594.
Described but not named by Miller : Deutsche med. Wochenschrift, 1884, Nos. 34 and 48.

Morphology. Bacilli curved rods, commas, and S forms, also spiral filaments. According to Lehmann-Neumann (Bak. Diag., 1896, 344), the flagella are in bundles at one side of the end of a rod. Do not grow in the ordinary culture media.

Habitat. Isolated from tartar on teeth and from saliva.

11. Spirillum jenense (Ehrenberg) Winter

Ophidomonas jenensis Ehrenberg : l.c.
Spr. jenense Winter : Die Pilze, 1884, 65.

A large species of a dirty green red — brownish green color. Sulphur granules in the plasma and at the poles. Flagella bundles of 3–9 elements, very long and stout. The flat spirals are 40 long and 3.5 μ thick.

12. **Spirillum sanguineum** (Ehrenb.) Cohn

Ophidomonas sanguinea Ehrenberg: l.c.
Spr. sanguineum Cohn: Beiträge Biol., I, 1875, Heft 3, 169.

Morphology. Bacilli 3.0 : 10-30 μ. The spirals have an amplitude of 9-12 μ, and a length of 6-10 μ. Color pale red. Sulphur granules in the plasma.

Habitat. Found in brackish water containing putrefying marine algæ.

13. **Spirillum rufum** Perty

Zur Kenntniss Kleinsten Lebensformen, Berne, 1852.

Morphology. Bacilli as filaments 8-16 μ long and 1-1.2 μ thick, contents slightly reddish, 1-4 spiral turns, not broken up into segments. At both poles bundles of 6-18 flagella. Spores absent.

Habitat. From well water, forming on sides of well red mucus-like spots.

SPIROCHÆTA Ehrenberg

Long, slender, closely coiled filaments; cells flexile. Show undulatory or snake-like movements which are not progressive, or a turning upon the longer axis. Flagella not known, endospores apparently absent.

I. PATHOGENIC SPECIES.

1. **Spirochæta Obermeieri** Cohn

Beiträge Biol., I, 1875, Heft 3.

Morphology. Bacilli very slender flexile spiral or wavy filaments with pointed ends, 0.1 : 16-40 μ. Stain easily with analine colors. Not stained by Gram's method. Has never been cultivated in artificial media.

Pathogenesis. Inoculations into apes positive, into mice, rabbits, sheep, and swine negative.

Habitat. Found in the spleen and in the blood in relapsing fever.

2. **Spirochæta febris**

Ueber einen aus dem Körper einer Rekurrenskranken erhaltenen Bacillus
Afanassieff: Centralblatt f. Bakteriol., XXV, 1899, 405.

Morphology. From preparations of the first day of the disease, bacilli small, 0.3 : 1-1.5 μ, with rounded ends; occur singly, in clumps and chains. In preparations on the following day bacilli 5-6 or 10-14 μ long, as commas or S forms. In overstained preparations an uncolored capsule is demonstrated. Cultures made from the blood.

Bouillon. Turbid; bacilli very actively motile, 1–6 : 0.3 μ; a scum on the surface.

Gelatin colonies. Very small, white; do not increase in size; borders erose, granulose; no liquefaction.

Gelatin slant. A thin thread-like growth, 0.5 mm. wide; here the bacilli are undulate with spindle-shaped thickenings.

Agar slant. In 24 hours at 37° a delicate partly transparent whitish growth.

Blood serum. Growth as on gelatin.

Lactose-litmus agar. Color unchanged.

Potato. At first a watery growth which disappears later. Grows in an atmosphere of hydrogen in absence of oxygen.

Milk. Not coagulated. *Indol* produced. Gas in glucose bouillon at 37°.

Pathogenesis. Subcutaneous inoculation of rabbits with 0.2–0.5 cc. of a bouillon culture causes a progressive elevation of temperature, followed after 5–10 days by a fall to normal. Similar inoculations of 0.1–0.2 cc. into man causes a chill and elevation of temperature. Bacilli found in the blood.

3. Spirochæta anserina Sakharoff

Centralblatt f. Bakteriol., XI, 1892, 203.

Morphology. Resembles No. 1. Has never been cultivated in artificial media.

Habitat. From the blood of geese in septicæmia. Inoculations into geese positive.

II. NON-PATHOGENIC SPECIES.

4. Spirochæta plicatilis Ehrenberg

Abhandl. Berlin Akad., 1833, 313.

Morphology. Bacilli very thin flexible filaments, 0.5 : 100–200 μ, ends rounded, undulations close and regular.

Habitat. Stagnant water containing decomposing vegetable matter.

5. Spirochæta dentium Cohn

Beiträge Biol., I, Heft II, 180, Heft III, 197, 1875.

Morphology. Bacilli long flexible spiral filaments of unequal thickness, and irregular spiral windings 8–25 μ long. Do not grow on culture media as far as known.

Habitat. From the mouth of healthy individuals.

MYCOBACTERIACEÆ

Cells either short or long, cylindrical — clavate — cuneate in form, which at times may show true branching, or as long-branched mycelial-like filaments. Filaments not surrounded by a sheath as in *Chlamydobacteriaceæ*. Without endospores, but with the formation of gonidia-like bodies due to a segmentation of the cells. Division at right angles to the axis of a rod or filament.

A. Cells in their ordinary form short cylindrical rods, often bent and irregularly swollen, clavate or cuneate. At times Y-shaped forms or longer filaments with true branchings. May produce short coccoid elements, perhaps gonidia *Mycobacterium* (Lehmann-Neumann), including *Corynebacterium* (Lehmann-Neumann).

B. Cells in their ordinary form as long-branched filaments. Produce gonidia-like bodies. Cultures generally have a mouldy appearance due to the development of aerial hyphæ. *Streptothrix* (Cohn); *Oöspora* (Lehmann-Neumann).

MYCOBACTERIUM Lehmann-Neumann

Characters emend. including *Corynebacterium* Lehmann-Neumann: Bak. Diag., 1896.

Cells in their ordinary forms as short cylindrical rods, which are often bent irregularly, swollen, clavate or cuneate, and which also at times may show Y-shaped forms or longer filaments, with true branchings. Without endospores. Without flagella. May produce short coccoid elements, perhaps gonidia.

I. Stain with aqueous solutions of basic aniline colors, and are easily decolorized by mineral acids when stained with Ziehl's carbol fuchsin.

    A. Cells slender, straight or bent, generally cylindrical, rod-like; rarely show branched forms.

        1. Do not grow on ordinary nutrient gelatin.

            a. Have not been cultivated on any known artificial media.

                1. *Mycobact. lepræ* (Hansen) Lehmann-Neumann.

                2. *Mycobact. syphilidis* (Schröter).

            b. Grow only on special blood media.

                3. *Mycobact. influenzæ* (Pfeiffer). INFLUENZA GROUP.

                4. *Mycobact. Elmassian.*

2. Grow on ordinary nutrient gelatin. SWINE ERYSIPELAS GROUP.
   a. Stain by Gram's method.
        5. *Mycobact. rhusiopathiæ* (Kitt).
        6. *Mycobact. murisepticum* (Flügge).
   b. Not stained by Gram's method.
        7. *Mycobact. malei* (Löffler) Migula.
B. Cells commonly irregularly swollen, or clavate — cuneate ; rarely show
   branched forms. DIPHTHERIA GROUP.
   1. On Löffler's blood serum a decided yellow growth.
        8. *Mycobact. lactis.*
   2. On Löffler's blood serum growth whitish, not pigmented.
   a. Stain by Gram's method.
        9. *Mycobact. diphtheriæ* (Klebs).
       10. *Mycobact. pseudodiphthericum* (Kruse).
       11. *Mycobact. pseudotuberculosis* (Kutscher).
   b. Do not stain by Gram's method.
       12. *Mycobact. hastilis* (Seitz).
II. Not stained with aqueous solutions of basic aniline colors ; not easily de-
    colorized by mineral acids when stained with Ziehl's carbol fuchsin.
                                              TUBERCLE GROUP.
A. Do not grow in nutrient gelatin at room temperatures.
   1. Not decolorized by alcohol when stained with Ziehl's carbol fuchsin.
   a. Growth on glycerin agar slow, dry, rough, warty, or fragmented.
       13. *Mycobact. tuberculosis.*
   b. Growth on glycerin agar visible in about 8 days ; at 37°, flatter
      and more watery.
       14. *Mycobact. avium.*
   2. Decolorized by alcohol when stained with Ziehl's carbol fuchsin.
       15. *Mycobact. smegmatis.*
B. Grow at least feebly in nutrient gelatin at room temperatures.
   1. Growth on agar or glycerin agar becomes a deep yellowish or orange.
       16. *Mycobact. butyri.*
   2. Growth on agar or glycerin agar whitish, or only a pale yellow.
   a. On the surface of bouillon a yellowish membrane.
       17. *Mycobact. Moëlleri.*
   b. On the surface of bouillon a whitish membrane.
      * No odor in bouillon cultures.
       18. *Mycobact. graminis.*
     ** Bouillon cultures have a bad odor.
       19. *Mycobact. friburgensis* (Korn).

1. **Mycobact. lepræ** (Hansen) Lehmann-Neumann

B. lepræ Hansen: Ueber die Aetiol. des Aussatzes, in Norsk. Magaz. for Laegeve-
densk, Christiania, 1874, Heft IX.
Mycobact. lepræ Lehmann-Neumann: Bak: Diag., 1896, 372.

Morphology. Bacilli thin rods of about the same size as tubercle bacilli,
straight — slightly curved, occur singly or in twos, often tapered at one or
both ends. Stain uniformly or irregularly, and by Gram's method.

Habitat. Found in large numbers in leprous lesions, in the round cells of
granulation tissue of tuberculous nodules, in lymphatic spaces, in endo-
therial cells, and in the walls of blood vessels.

2. **Mycobact. syphilidis** (Schröter)

Syphilis bacillus Lustgarten: Med. Jahrb. der K. K. Gesellsch. der Aerzte in
Wien, 1885.
B. syphilidis Schröter: Pilze Schles., 1886.

Morphology. Bacilli similar to tubercle bacilli, 0.2-0.3 : 3-7 μ, often bent, S-
formed, or clavately swollen or irregular, with irregular staining. In the
tissues occur singly or in clumps. Stain by Gram's method; resist decolorization with alcohol. Cultures not known.

Habitat. Found in the lesions of syphilis.

3. **Mycobact. influenzæ** (Pfeiffer)

B. influenzæ R. Pfeiffer: Zeitsch. f. Hygiene, XIII, 1893.

Morphology. Bacilli 0.2 : 3-5 μ, commonly in twos. Stain with Löffler's
alkaline blue and carbol fuchsin. Not stained by Gram's method. On
agar moistened with blood, in 24-48 hours, small glassy drops; older
colonies have yellowish-brownish centres.

Nastiukoff's solution.[1] In 24 hours, at 37°, small white flecks at the bottom
of the tube composed of chains of bacilli.

Nastiukoff's agar.[1] Colonies as small gray points, which microscopically
are round yellow and translucent.

For the differential diagnosis make (1) cover-glass preparations from bronchial secretions, sputum, etc., and (2) smear cultures on agar moistened
with blood, and plate cultures with Nastiukoff's agar.

Habitat. Isolated from nasal and bronchial secretions and urine of man
affected with influenza.

---

[1] For the preparation of Nastiukoff's media see Centralblatt f. Bakteriol., XVII,
492.

4. Mycobact. Elmassiani

Bacille analogue au Bacille de Pfeiffer Elmassian: Annales Pasteur Inst.,
XIII, 1899, 625.

Morphology. Bacilli in size like *Bact. conjunctivitis*, but a little thicker, and
with a slight constriction in the middle; ends tapered or rounded, others
distinctly rod-like, others like *Bact. pneumoniæ*. Not stained by Gram's
method. In a medium containing two parts of gelatin and one of blood
serum in slanted tubes inoculated with washed sputum from a case of
whooping-cough, and incubated at 37°, round transparent punctiform
colonies, 0.25–0.5 mm. in diameter, developed. No growth in ordinary
bouillon or on gelatin. In *serum bouillon* in 48 hours, at 37° C., a uniform turbidity, becoming clear.

Pathogenesis. Intravenous inoculation of large doses into pigeons and guinea
pigs, negative. Intraperitoneal inoculations of 2–4 cc. into guinea pigs
cause death in 24 hours. Abdomen soft, painful; animal immobile; elevation of temperature, peritonitis, and a sero-fibrinous exudate.

Habitat. Isolated from sputum in whooping-cough, tuberculosis, pneumonia,
and la grippe.

5. Mycobact. rhusiopathiæ (Kitt)

Bacillus des Schweinerotlaufs Löffler: Arbeiten Kaiserl. Gesundheitsamte, Bd. I,
1886, 46.
B. rhusiopathiæ-suis Kitt: Bakterienkunde u. path. Mikroskopie, 1893, 284.
Bact. erysipelatus-suis Migula: Die Natürlichen Pflanzenfam., 1895.

Morphology. Bacilli very small, slender, bent or curved, also filaments,
0.2 : 0.6–1.8 μ.

Gelatin colonies. Thin, veily, which under the microscope show a fine
filamentous structure.

Gelatin stab. In depth, gray cloudy radiating outgrowths; after some time
the gelatin is softened.

Agar slant. A delicate layer.

Bouillon. Turbid; later a gray white sediment.

Potato. No growth. *Indol* negative.

Pathogenesis. Inoculations of mice, white rats, and pigeons cause death in
3–4 days with septicæmia; bacilli in the blood and enclosed within
leucocytes. Mice die in a sitting posture with the eyes sealed by a
secretion.

Habitat. Associated with swine erysipelas, Schweinerotlauf.

6. Mycobact. murisepticum (Flügge)

B. murisepticus Flügge: Die Mikroorganismen, 1886.
B. Septikamie bei Mäusen Koch: Aetiol. Wundinfectionsk., 1878.
B. marinus Schröter: Kryptogamenflora Schlesien, III, 1886, 162.

Probably identical with the preceding.

7. Mycobact. malei (Löffler) Migula

B. mallei Löffler: Arbeiten Kaiserlich. Gesundheitsamte, I, 1886, 141.

Morphology. Bacilli small, slender, bent, 0.25–0.4 : 1.5–3 μ; may occur as coccoid elements. According to Marx (Centralblatt f. Bakteriol., XXV, 274), may show diphtheria-like and branched forms. Stain badly. Grow best on glycerin agar; a scanty growth on blood serum.

Agar slant. In 24–48 hours at 37° C. whitish translucent watery colonies.

Gelatin colonies. After weeks the gelatin begins to soften and small funnels are formed. Grows with an acid reaction of the medium.

Potato. Growth yellowish — reddish brown; show pleomorphic, anthrax-like threads which become swollen involution forms. *Indol* doubtful.

Pathogenesis. Rabbits but slightly affected, white and gray mice immune. Intraperitoneal inoculations of 1–2 cc. into male guinea pigs cause death in 12–15 days; testicles swollen and reddened, tubercles on tunica vaginalis, suppurating organ contains the bacilli from which pure cultures can be made.

Habitat. Secretions and ulcers and tubercles in glanders in men, horses, cats; sheep, goats, dogs, and rarely swine; cattle and birds immune.

8. Mycobact. lactis

A Bacillus resembling B. diphtheriæ found in milk and American cheese Park-Beebe-Williams: Sci. Bull. No. 2, Health Dept., New York City, 1895.

Morphology. Bacilli more or less regular in shape and size resembling diphtheria, but slightly thicker and of more variable length. On Löffler's blood serum bacilli like diphtheria, except slightly thicker and a little more irregular.

Gelatin colonies. In 24 hours minute, punctiform; microscopically, round, entire, sharp, yellow, granular. No liquefaction.

Gelatin stab. In depth, growth finely granular, white; on the surface, growth abundant, light yellow, becoming deeper yellow.

Agar colonies. In 24 hours at 20° small, punctiform, cream-colored, which microscopically are coarsely granular, entire, grayish; later the surface colonies are a little larger and have a decided yellow color.

2 A

Agar slant. Growth abundant, moist, creamy, becoming a deep yellow.

Löffler's blood serum. Growth in 24 hours at 37° abundant, moist, light cream-colored, becoming a decided yellow.

Bouillon. Slightly turbid, with a granular sediment, becoming clear and tinged with yellow ; no pellicle.

Glucose bouillon. Rendered decidedly acid.

Milk. Not coagulated ; rendered a deeper cream color.

Potato. Growth a narrow stripe, becoming raised, dry, granular, and deep yellow. Non-pathogenic to guinea pigs.

Habitat. Found commonly in milk and cheese.

See Klein Jour. Path. and Bact., 1894. 441. and Henrici Arbeiten aus dem Bact. Inst. Tech. Hochsch. Karlsruhe, 1894, Heft 1.

9. **Mycobact. diphtheriæ** (Klebs)

B. diphtheriæ Klebs : Verhandl. Congr. für inneren Medicin, 1883. 143.

Bacillus bei Diphtherie des Menschen Löffler : Mitteilungen Kaiserliche Gesundheit-samte, II, 1886, 421.

Corynebacterium diphtheriæ Lehmann-Neumann : Bak. Diag., 1896, 350.

Morphology. Bacilli slender, rather long, straight or somewhat bent, commonly swollen at one or both ends, wedge-shaped (cuneate) or clavate, or various irregular forms, rarely those which show true branching. With alkaline methylene blue a beaded appearance due to irregular staining. Stained by Gram's method. Grow best in the presence of oxygen, and at body temperatures.

Glycerin gelatin colonies. *Deep:* round, light yellow, granular, border entire — rough. *Surface:* delicate, grayish white — light yellowish, translucent, darker and finely granular in the centre, coarsely granular on the border.

Gelatin stab. A scanty development ; no liquefaction.

Glycerin agar colonies. *Deep:* round — oval, dark gray — greenish, entire, amorphous. *Surface:* delicate, grayish white, translucent ; microscopically, round, entire, yellowish, translucent, granular.

Agar slant. But scanty development.

Glycerin agar slant. Growth delicate, white — yellowish white.

Blood serum or *Löffler's blood serum.* Opaque whitish colonies, or a dull whitish, granular streak.

Bouillon. Turbid, due to granular particles, either immersed or often forming a film on the surface. Medium at first acid, then alkaline.

Milk. Not coagulated, amphoteric.

Potato. On acid potato growth scanty, when medium is rendered alkaline a delicate glistening growth which can be raised by the needle and which has the color of the potato.

Glucose bouillon. Acid; no gas. H₂S slight. *Indol* positive.

Pathogenesis. Subcutaneous inoculations of guinea pigs cause death in 36–72 hours. There is œdema, hemorrhage, and a fibro-purulent exudation about the point of inoculation, hemorrhagic enlargement of lymph glands, congestion of the lungs and other organs, hemorrhage of supra-renal capsules. There is generally only a local development of bacteria.

Habitat. Associated with diphtheria and present in the throats of persons who have been exposed to infection.

10. **Mocobact. pseudodiphthericum** (Kruse)

Pseudodiphtheria Bacillus Löffler: Centralblatt f. Bakteriol., II, 1887, 105.
Xerose Bacillus Neisser-Kuschbert: Centralblatt f. Bakteriol., I, 178.
B. pseudodiphthericus Kruse: Flügge, Die Mikroorganismen, 1896.

This may be only a non-virulent variety of the preceding. In the more marked types it may be distinguished from the preceding by being shorter and thicker. On *glycerin agar* and in *bouillon* it grows rather more abundantly. According to Escherich and others, the pseudo-diphtheria bacillus during the first 2–3 days of growth causes an increased alkalinity of the medium (bouillon), while the true diphtheria bacillus causes a diminished alkalinity.

Habitat. Isolated by Hoffmann, etc., from the healthy mouth and throat; by Neisser-Kuschbert, etc., in xerosis and other affections of the conjunctiva.

11. **Mycobact. pseudotuberculosis** (Kutscher)

B. pseudotuberculosis-murium Kutscher: Zeitsch. f. Hygiene, XVIII, 1894.
B. pseudotuberculosis-ovis Preisz: Annales Pasteur Inst. 1895.

Morphology. Bacilli like those of diphtheria. Stain by Gram's method.
Gelatin colonies. Granular, with erose borders.
Potato. No growth.
Milk. Unchanged.
Bouillon. Slight growth.
Pathogenesis. Subcutaneous inoculation of mice result in abscesses at the point of injection, with a general inflection and death in 5–8 days. Intra-peritoneal and intrapulmonary injections cause death, with pseudotuber-cular lesions. Subcutaneous and intraperitoneal inoculations of guinea pigs cause pseudotuberculosis of the abdominal viscera, and death in 2–10–35 days.
Habitat. Found by Kutscher in cheesy nodules in lung and pleura of a mouse; by Preisz and Guinard in pseudotuberculosis of sheep; and by Kitt from a cheesy pneumonia of cattle.

12. Mycobact. hastilis (Seitz)

B. hastilis Seitz: Zeitsch. f. Hygiene, XXX, 1899, Heft 1.

Morphology. Bacilli slender or rather broad rods, pointed at one or both
ends, and here and there slightly thickened in the middle; straight or
somewhat bent; occur singly, in twos or short-long chains. Not stained
by Gram's method. With ordinary analine colors, often a beaded staining. No growth on blood serum, but a growth in the water of condensation. Grow in ordinary bouillon with the generation of gas, and a foul
odor like carious teeth.

Habitat. Isolated from the mouth.

13. Mycobact. tuberculosis (Koch)

B. tuberculosis Koch: Die Aetiologie des Tuberculose, Berliner, klin. Wochensch.,
1882, No. 15.
Mycobact. tuberculosis Lehmann-Neumann: Bak. Diag., 1896, 363.

Morphology. Bacilli mostly slender, straight or curved or bent rods, 0.4 : 1.5–4 μ.
Occasionally longer filamentous forms, with true branching, have been
noted (Coppen Jones, Centralblatt f. Bakteriol., XVII, 1). With carbol
fuchsin, an irregular, beaded staining. May also show deeply stained
bodies, which Coppen Jones thinks homologous with chlamydospores.

Glycerin agar slant. Growth whitish, dry, rough — warty, with commonly a
faint pinkish or flesh color. The cultures have a peculiar yeast-like odor.

Blood serum. Growth white, dry, scaly — granular, which is friable but
coherent.

Potato. An abundant raised growth.

Glycerin bouillon. A whitish grayish, membranous, rugose growth on the
surface, which readily sinks; medium clear. *Indol* negative. H_2S
negative.

Pathogenesis. Subcutaneous inoculation of rabbits and guinea pigs cause a
generalized tuberculosis, with death in 2–3 months.

Habitat. Associated with tuberculosis in man and the lower animals.

14. Mycobact. avium (Kruse) Lehmann-Neumann

Bacillus der Huhner oder Geflügeltuberkulose Maffucci: Zeitsch. f. Hygiene, XI, 1892, 445.
B. tuberculosis-avium Kruse: Flügge, Die Mikroorganismen, 1896, 506.
Mycobact. tuberculosis-avium Lehmann-Neumann: Bak. Diag., 1896, 370.

Morphology. Bacilli like the preceding, but somewhat longer and more
slender, with a greater tendency to form branched and clavate forms.
Staining reactions as in No. 13. Grow at 43° C. *B. tuberculosis* does not
grow above 42°.

On *blood serum* and *glycerin agar* growth softer, flatter, and more watery; also grows more rapidly, *i.e.* a visible growth in about eight days.

Pathogenesis. Highly pathogenic to fowls; not truly pathogenic to guinea pigs, rabbits. and apes.

Habitat. Associated with avian tuberculosis.

15. **Mycobact. smegmatis** (Kruse)

Smegmabacillus Tavel-Alvarez: Bull. de l'Acad. de Médecine, 1885.
B. smegmatis Kruse: Flügge, Die Mikroorganismen, 1896, 517.

Morphology. Bacilli like *Mycobact. syphilidis*, but shows a greater variability in size and form. Stain with difficulty, and not decolorized with potassium permanganate, and also retain the stain after long treatment with mineral acids in contradistinction to syphilis, but are easily decolorized with alcohol. For differential diagnosis, stain with hot carbol fuchsin, and immerse in a saturated alcoholic solution of methylene blue, when the smegma bacilli are stained blue and the tubercle bacilli red.

Habitat. Found on the mucous membranes of the urino-genital tract in man and the lower animals, on the mamma, and in urine.

16. **Mycobact. butyri**

Tuberkelähnlicher Bacillus Rabinowitsch: Zeitsch. f. Hygiene, XXVI, 1897, 101.
Tuberclebacillen in Butter u. Milch Petri: Arb. a. d. Kais. Gesundheitsamte, XIV, 1.

Morphology. Bacilli identical with tubercle. Stain with carbol fuchsin, and resist decolorization after 4 minutes' immersion in 6 per cent H_2SO_4.

Agar colonies. Deep, round — oval, gray, granular. Surface colonies have gray granular centres, and clear crumpled borders, and are often dry and cupped.

Gelatin stab. Slow growth; in depth, small disjointed colonies along the line of stab; medium not liquefied.

Agar slant. Freshly isolated from the body, the growth is thick, moist, and creamy; in old cultures, a crumpled membrane, of an orange or copper color. By repeated passage through animals, the cultures on agar or glycerin agar are dry and fragmented or crumpled, closely simulating true tubercle bacilli.

Potato. A moist gray layer.

Bouillon and *glycerin bouillon.* A crumpled membrane on the surface; media clear. The cultures have an ammoniacal odor, and lack the characteristic yeasty odor of true tubercle cultures. The medium is rendered alkaline. Grow in acid bouillon. *Indol* is produced.

Pathogenesis. Intraperitoneal inoculations into guinea pigs, of butter containing the preceding bacilli, result, in 3-4 weeks, in peritonitis, with abundant
tubercles on the peritoneum, which have cheesy, purulent centres. Similar
tubercles may be found on the pleura and in the spleen. These tubercles
lack the microscopic characters of true tubercles in the absence of giant
cells, but more closely simulate glanders nodules. They contain the
bacilli in large numbers, from which pure cultures can be readily obtained.

Habitat. Found by Rabinowitsch, Petri, and others in butter and milk.

17. Mycobact. Moëlleri

Timothee Bacillus or *Grass Bacillus I* Moëller: Wiener med. Wochenschr., 1898, p. 2358.

Morphology. Bacilli 0.2-0.5; 1.4 μ; often bent; often in chains of 2-3, or in
clumps; also as filaments with clubbed ends, or branched. Show irregular staining like tubercle bacilli, and retain the red color of carbol
fuchsin when treated with acid and acid alcohol.

Glycerin agar. At 37°, after several days, grayish white, dry, scaly colonies.

Bouillon. Small granular masses on the walls and bottom of the tube; on
the surface, a yellowish membrane.

Milk. On the surface, yellowish spots or colonies, or a yellowish ring adherent
to the walls.

Potato. A warty layer like tubercle bacilli.

Pathogenesis. Pathogenic to rabbits and guinea pigs, producing a pseudotuberculosis as in *Mycobact. butyri.*

Habitat. Isolated by Moëller from infusions of timothy grass.

The MIST BACILLUS of Moëller, isolated from cow manure, is a closely related
organism (Berliner thierärztl. Wochenschr., 1898, 100). In morphology it
simulates the tubercle bacillus; also occurs as long filaments; also clubbed
at one or both ends, and without branching. Does not grow in milk.
Grows well on glycerin agar.

18. Mycobact. graminis

Grass Bacillus II Moëller: Centralblatt f. Bakteriol., XXV, 1899, 369.

Morphology. Bacilli in fluid media mostly as rods whose morphology and
staining properties simulate tubercle bacilli. In old cultures, often filaments and branched forms. On solid media, at first only rods and coccoid
forms; later, filaments showing true branching. Rods 0.2-0.4: 1-5 μ.
Stain by Gram's method

Glycerin agar slant. In 2 days, at 37°, small, delicate, watery colonies, becoming confluent. Growth rather raised, often with a yellowish tinge. In the water of condensation, which is clear, small particles which sink.

Potato. At 37° a thick grayish white growth along the line of inoculation.

Milk. Becomes acid in 2–3 days.

Bouillon. Medium clear; a grayish white membrane and a stringy sediment; no odor.

Gelatin slant. In 4–5 days, at 20°, a grayish white thick growth along the line of inoculation; no liquefaction.

Gelatin stab. A good growth along the line of stab.

Pathogenesis. Intraperitoneal inoculations of guinea pigs cause death in 4–6 weeks, with the same macro-pathological picture as in tuberculosis.

Habitat. Found in hay dust in lofts.

19. Mycobact. friburgensis (Korn)

B. friburgensis Korn : Centralblatt f. Bakteriol., XXV, 1899, 532.

Morphology. Bacilli in bouillon vary from those simulating *B. coli* to longer and slightly bent rods. On agar the rods are rather thinner; on old agar and serum cultures, coccothrix forms. On potato, cocci, diplococci, and short rods or partly bent bacilli of variable thickness, also clavate forms. In the animal body, tubercle and coli forms; also branched individuals. Stain poorly with ordinary aniline colors, well with aniline and carbol fuchsin, and withstand decolorization for one minute in 10 per cent H_2SO_4. Stain by Gram's method.

Agar colonies. Deep: round — elongated, granular. *Surface:* round, grayish white, becoming sunken in the centre.

Gelatin stab. In depth, growth slight, uniform; on the surface, growth white, and rather flat and lobate.

Glycerin agar slant. In 2–3 days, at 37°, growth thick, glistening, rough, crumpled, with a membrane on the water of condensation.

Bouillon. On the surface, a thick membrane. Cultures have a bad odor. No growth in acid bouillon.

Glucose bouillon. Grow only in the open end; no gas. *Indol* slight, and produced only in glycerin bouillon.

Milk. In 6 days unaltered; after a longer time the milk becomes rather gray, with a slight sediment; not coagulated or peptonized. Whole milk or cream assumes a copper color.

Glycerin blood serum (horse). In 6–8 days an orange-colored growth.

Potato. A soft flat whitish growth, becoming brownish. In an atmosphere of hydrogen a slight growth at 37°, none at 20°.

Pathogenesis. Subcutaneous inoculation of guinea pigs causes abscesses at the point of injection containing the bacilli, but without fatal issue. Intraperitoneal inoculations of mice cause death after some days. There is a serous fluid in the peritoneal cavity; the peritoneum is studded with submilliary nodules. Spleen swollen, and studded with knots. Knots in liver and kidneys. Bacilli in the lesions.

STREPTOTHRIX

Cells in their ordinary form as long branched filaments. Cultures on solid media raised. Growth coherent, dry, rough, or crumpled, often with a mouldy appearance, due to the formation of aerial hyphæ. Without endospores, but by a multiple segmentation of a filament, the production of short, gonidia-like bodies.

SYNOPSIS OF THE GENUS

I. In the animal body the radially arranged filaments show a clavate enlargement of their ends.
 A. Cultures show abundant branched filaments.
  1. Grow at room temperatures, 20°–22°, and on potato.
   1. *Streptothrix bovis* (Harz).
  2. No growth below 22° C.
   a. Grow on potato.
    2. *Streptothrix flava.*
   b. Do not grow on potato.
    3. *Streptothrix Hofmanni* (Gruber) Kruse.
 B. On ordinary media, viz. agar, no branched filaments, but diphtheria-like forms.
   4. *Streptothrix Israeli* Kruse.
   5. *Streptothrix Krausei.*
II. Filaments do not show a clavate enlargement of their ends.
 A. Do not grow in nutrient gelatin; grow on blood serum or blood-serum agar.
   6. *Streptothrix necrophorus* (Löffler) Schmorl.
 B. Grow in nutrient gelatin.
  1. Gelatin liquefied.
   a. Non-chromogenic, colonies whitish; no pigment on gelatin or agar.

* Gelatin liquefied very slowly or imperfectly.
> 7. *Streptothrix Rosenbachii* Kruse.
** Gelatin liquefied rather rapidly.
> 8. *Streptothrix invulnerabilis* (Acosta-Grande-Rossi) Kruse.
> 9. *Streptothrix Foersteri* Cohn.
b. Chromogenic; produce pigment on gelatin.
* Gelatin colonies yellowish.
> 10. *Streptothrix albido* Rossi-Doria.
** Gelatin colonies brownish; gelatin stained a deep brown.
> 11. *Streptothrix chromogena* Gasperini.
*** Gelatin colonies violet.
> 12. *Streptothrix violacea* Rossi-Doria.
**** Gelatin colonies, grown anaerobically, brick-red.
> 13. *Streptothrix rubra* Kruse.
2. Gelatin not liquefied.
> *a.* No distinct pigment on gelatin or agar; growth white — gray, or at most, yellowish white.
> > 14. *Streptothrix farcinica* (Trevisan) Rossi-Doria.
> *b.* Chromogenic; a pigment on gelatin or agar.
> * On gelatin, colonies or growth yellowish — orange.
> > 15. *Streptothrix aurantiaca* Rossi-Doria.
> > 16. *Streptothrix asteroides* (Eppinger).
> ** On gelatin or agar, colonies or growths become reddish.
> > 17. *Streptothrix carnea* Rossi-Doria.
> > 18. *Streptothrix maduræ* Vincent.

1. Streptothrix bovis (Harz)

Actinomyces bovis Harz: Jahresb. Münch. Central Thierarzneischule, 1877–78.
Discomyces bovis Rivolta: Sul. cosi detto mal del rospo della Trutta e sul' Actinomyces bovis di Harz, 1878.

Morphology. In the body of man and animals, as foci of granulation tissue, containing the fungus, which is composed of interlacing filaments, and which on the periphery show a radial arrangement or clavate enlargement of their ends. Filaments 0.4–0.6 μ in diameter, either long and unsegmented, or breaking up into longer or shorter rods. The filaments often show a segmentation of their plasma and the formation of gonidial elements, which later are set free and are presumably gonidia. In cultures, filaments showing true branching are numerous. The clavate enlargements of the ends of the filaments, due to a capsular thickening, are found in the deeper portions of the culture. Filaments, but not the capsular enlargements, stained by Gram's method.

Gelatin colonies. Irregular, yellowish gray, glistening; microscopically, dark yellowish gray, homogeneous or slightly concentric; border dark, with a fine filamentous structure.

Gelatin stab. In depth, small yellowish white spheres, which become bristly; on the surface, growth yellowish white, flat, raised, soft, glistening, rather tough, becoming sunken, due to a slight liquefaction; liquefied gelatin syrupy, brownish.

Agar slant. Growth of delicate colonies, becoming whitish to whitish yellow, soft, glistening, raised, warty, becoming sunken and brownish in color. The growth penetrates deep into the medium.

Blood serum. Isolated colonies becoming spreading and thicker, rather dry; the lower surface of the growth in contact with the medium, orange-yellow — brick-red.

Bouillon. Remains clear; at the bottom globular masses.

Milk. Unchanged in 8 days.

Potato. Growth warty, yellowish white, very adherent, limited. Odor of cultures weak, not mouldy.

Glucose bouillon. No gas, acid. H_2S negative.

Pathogenesis. Intraperitoneal inoculations of guinea pigs and rabbits result, in 30 days, in the formation of nodules on the peritoneum containing the fungus.

Habitat. Associated with antinomycosis in man and cattle.

2. Streptothrix flava

Described by Bruns: Centralblatt f. Bakteriol., XXVI, 1899, 11.

Morphology. Filaments 1-2 μ broad and reaching 100 μ in length, often with clavate swellings. Filaments show branching, also shorter forms like *Mycobact. diphtheriæ.* Stain by Gram's method.

On *agar*, after 3-4 weeks, a colony 0.75-1.0 cm., yellowish. Surface irregular, adherent, not easily fragile.

In *bouillon*, a whitish yellow, fragmentary growth on the bottom; no growth on the surface; fluid clear. Grows on blood serum and potato, but not so well on agar. On gelatin, at 24°-26°, a slight growth after 4 weeks; no liquefaction. Optimum temperature, 35°-38°; minimum, 25.5° C.

Pathogenesis. Non-pathogenic to mice, guinea pigs, and rabbits.

3. **Streptothrix Hofmanni** (Gruber) Kruse

Mikromyces Hofmanni Gruber: Archiv f. Hygiene, XVI, 1893, 34.

Morphology. Branched filaments 1.0 μ thick. Do not produce aerial hyphæ. The contents of older filaments become segmented into coccoid gonidia. In animal body, clavate enlargements; also in old, 3 months, bouillon cultures. Aerobic. No growth below 22° C. Optimum temperature, 37° C.

No growth on *gelatin* and *potato*; on *agar* and *blood serum* a scanty development.

Glycerin agar slant. A raised, rugose, dull, grayish white to brownish growth.

Bouillon. Clear, with a granular sediment, and often with a surface membrane. Grows well in fluid media containing 0.5–3 per cent of sugar with the production of acetic acid and alcohol.

Pathogenesis. Subcutaneous inoculations of rather large doses into rabbits cause a fibro-purulent inflammation at the point of inoculation, with abscess formation, which remains localized. There appears to be no development of the fungus.

Habitat. Isolated from the air.

4. **Streptothrix Israeli** Kruse

Flügge, Die Mikroorganismen, 1896, 56.
Noted by Wolff-Israel: Virchow's Archiv, CXXVI.

Anaerobic, grow poorly in the presence of air. An absence of branched filaments in cultures. In anaerobic (Buchner's method) cultures on agar at 37° C., fine, dew-like drops or convex colonies, which generally remain discrete.

Bouillon. Growth composed of small scaly particles.

No growth on *gelatin*.

Agar slant. Cultures show only rods greatly similar to diphtheria, with but little tendency to form filaments. Egg cultures show typical filaments.

Pathogenesis. Intraperitoneal inoculations of rabbits and guinea pigs result in (4–7 weeks) the formation on the peritoneal viscera of nodules, varying in size from that of a millet seed to that of a plum, in the smaller of which typical actinomyces kernels are found. These contain branched filaments with clavate ends.

Habitat. Isolated by Wolff and Israel from two cases of human antinomycosis.

5. Streptothrix Krausei

Streptothrix Krause: Centralblatt f. Bakteriol., XXVI, 1899, 209.

Morphology. Short and long rods and clavate forms like diphtheria. Stain by Gram's method. Grow best at 37°; no growth at 22° C.

Glycerin agar colonies. In 4 days, small, in 8 days, slightly yellowish, 2–3 mm.; borders erose or rosette-like; adherent to the medium.

Bouillon. Clear, but a sediment of colony clumps.

No growth on *gelatin* or *potato.* No gas produced. No *indol*, and no H_2S produced. Grows better aerobicly than anaerobicly. Non-pathogenic to mice, guinea pigs, and rabbits.

Habitat. Isolated from a case of actinomycosis in man.

6. Streptothrix necrophorus (Löffler) Schmorl

B. necrophorus Löffler: Mitteilungen a. d. Kaiserlich. Gesundheitsamte, II, 1884, 493.
B. diphtheriæ-vituorum Löffler: l.c.
Streptothrix cuniculi Schmorl: Zeitsch. f. Tiermed., XVII, 1891.

Anaerobic. Grows best on blood serum and on blood-serum agar at the body temperature. Noted by Schmorl in an infectious disease of rabbits, characterized by a progressive necrosis of the subcutaneous tissue, also by a fibrinous inflammation of the serous membranes, etc. By Bang and Löffler, in diphtheria of calves, etc. The organisms are found at the periphery of the necrotic areas, where it forms thick tufts and filaments, in which true branching is not certainly demonstrated. The filaments now and then break up into rods.

Pathogenesis. Subcutaneous inoculation of mice and rabbits causes a local necrosis, with multiple necrotic foci in the inner organs.

7. Streptothrix Rosenbachii Kruse

Flügge, Die Mikroorganismen, 1896, 61.
Discovered by Rosenbach: Archiv f. Chirurgie (Langenbeck), 1887.

Morphology. Very fine branched filaments, breaking up into short rods or coccoid forms. The filaments often end in a thick point. In cultural characters like the bacillus of mouse septicæmia. Old cultures become brownish. Grow best at 20° C., and badly at 37° C.

Pathogenesis. By inoculations into man, Rosenbach produced erysipeloid lesions.

Habitat. Associated with erythema exudativum multiforme.

8. Streptothrix invulnerabilis (Acosta-Grande-Rossi) Kruse

Cladothrix invulnerabilis Acosta-Grande-Rossi: Centralblatt f. Bakteriol., XV, 1893, 1.
Streptothrix invulnerabilis Kruse: Flügge, Die Mikroorganismen, 1896, 64.

Morphology. Mycelium branched with aerial hyphæ.
Gelatin colonies. Tough, with a whitish bloom, crumpled; gelatin not
 liquefied. Grows in the absence of air.
Potato. Stained black around the growth. Causes a cloudiness in water.
Habitat. Water, etc.

9. Streptothrix Fœrsteri Cohn

Streptothrix Fœrsteri Cohn: Beiträge Biol., Bd. I, Heft. 3.
Streptothrix alba Rossi-Doria: Annal. dell' Ist. d' Ig. di Roma, 1891.
Streptothrix I–II Almquist: Zeitsch. f. Hygiene, VIII, 1890.
Actinomyces albus Gasperini: Ann. Micrographie, IX, 1890, 449.
Cladothrix liquefaciens Hesse-Garten: Zeitsch. f. Chirurgie, 34 and 41.

Morphology. A branched mycelium with abundant aerial hyphæ, giving
 cultures a white appearance. Aerobic.
On *gelatin*, a broadly folded membrane with a white bloom and oil drops.
Milk. Peptonized. Grows well on cooked vegetables.
Pathogenesis. Non-pathogenic (Rossi-Doria). According to Gasperini, may
 produce actinomyces in cattle. Almquist found this species in a case of
 meningitis.
Cladothrix liquefaciens of Hesse-Gartner was isolated from a lesion in man,
 simulating actinomycosis. The various forms classed under this head
 are not sufficiently described to closely differentiate them.
Habitat. Air and water; also assuming pathogenic rôles as already intimated.

10. Streptothrix albido Rossi-Doria

Annal. dell' Ist. d' Ig. di Roma, 1891.

Morphology. Mycelium branched, with aerial hyphæ sparingly. Do not
 grow with the exclusion of air. Color of colonies yellowish. Surface
 growth channelled or fluted. Gelatin liquefied slowly.
Milk. Growth as islands on the surface, peptonized. Non-pathogenic.
Habitat. Air.

11. Streptothrix chromogena Gasperini

Streptothrix nigra Rossi-Doria: l.c., 1891.

Morphology. Branched filaments often evidently septate, composed of long
 and short elements. In the aerial hyphæ by segmentation short coccoid

gonidia. Stain by Gram's method. Grow at 20°, but best at 37° C. Aerobic.

Gelatin colonies. Round, slightly raised, brownish, but with a whitish dry chalky appearance in the centre, becoming concentric. Gelatin around the colony dark brown and slowly liquefied, leaving a chalky crust on the surface of the liquefied gelatin. Microscopically, filamentous, tangled, becoming opaque in centre with a filamentous border.

Gelatin stab. In depth, short radiate bundle-like outgrowths after some time; on the surface like gelatin colonies, gelatin slowly liquefied beneath.

Agar stab. In depth, bristly outgrowths; on the surface, growth moist, yellowish, glistening, raised, becoming dry, warty; agar stained a deep brown.

Agar slant. Growth brownish, slightly spreading, becoming whitish, chalky.

Bouillon. On the surface, a delicate and later a tough membrane.

Glucose bouillon. Radiate masses at the bottom; medium brownish.

Milk. A tough yellowish brown growth on the surface; medium rendered alkaline and peptonized.

Potato. Growth yellowish — yellowish brown, becoming chalky. The medium is stained a deep brown or black. The culture has an intense mouldy odor.

Habitat. Air, water, and stomach contents.

12. **Streptothrix violacea** Rossi-Doria

l.c., 1891.

Morphology. Typical Streptothrix, with branched mycelium and spore formation and aerial hyphæ. Stain by Gram's method. No anaerobic growth.

Gelatin colonies. Violet, isolated, becoming united and forming a rugose membrane; also the medium stained violet.

Bouillon. Scanty growth; compact nodules at the bottom and isolated colonies on the surface; medium colored a faint wine-red.

Potato. Red violet colonies, with a whitish bloom and a brownish discoloration of the medium.

Milk. Violet points of growth; medium slowly peptonized.

Pathogenesis. Negative, or injection of 2 cc. intraperitoneally into test animals may cause a pseudotuberculosis of the mesentery, spleen, liver, and lungs.

Habitat. Air and water.

13. **Streptothrix rubra** Kruse

Flügge, Die Mikroorganismen, 1896, 63.
Noted by Casabó: Centralblatt f. Bakteriol., XVII, Nos. 13–14, 1895.

Morphology. Thick-branched filaments with spores. A good growth under anaerobic conditions with the formation of brick-red colonies. Non-pathogenic.

Habitat. Isolated from sputum.

14. **Streptothrix farcinica**

Bacillus du Farcin Nocard: Ann. Pasteur Inst., II, 1888, 293.
Nocardia farcinica Trevisan: Genera, 1889, 9.
Streptothrix farcinica Rossi-Doria: l.c., 1891.

Morphology. Filaments 0.25 μ thick, branched, or short jointed. Stain by Gram's method. Slight growth at 20°, good growth at 37°.

Gelatin colonies. Grow slowly; in 10 days small round transparent glistening spheres which microscopically are entire, grayish, and amorphous.

Gelatin stab. In depth, growth granular; on the surface a slow growth, which in 12 days is white and warty.

Agar colonies. Yellowish white, irregular, glistening, membranous, 1–2 mm.

Agar slant. Growth grayish — yellowish white, with a rough, finely cleft surface.

Bouillon. Clear, with a granular sediment and often a dirty gray membrane.

Milk. Not coagulated, reaction unchanged.

Potato. Slow growth, whitish yellow, dull; surface squamose.

Pathogenesis. Intraperitoneal inoculation of guinea pigs causes in 9–20 days a pseudotuberculosis of the abdominal viscera, with the fungus within the tubercles. By subcutaneous inoculation only an affection at the point of inoculation or neighboring lymph glands. Intravenous inoculations of cattle and sheep cause a slowly progressive pseudotuberculosis.

Habitat. Associated with a chronic form of tuberculosis in the subcutis of the intestines and of the internal organs.

15. **Streptothrix aurantiaca** Rossi-Doria

l.c., 1891.

Morphology. Branched filaments with aerial hyphæ and spores. Grows at 20°, but not with the exclusion of air.

Gelatin colonies. Waxy, yellow, becoming orange, with a whitish bloom; surface colonies not colored.

Agar slant. Colonies, becoming a warty membrane.

Potato. Growth thin, membranous, becoming orange.

Milk. Growth as orange-colored flecks, unchanged. Non-pathogenic.

Habitat. Air.

16. Streptothrix asteroides (Eppinger)

Cladothrix asteroides Eppinger; Ziegler's Beiträge, IX.

Oospora asteroides Sauvageau-Radais: Ann. Pasteur Inst., VI, 242.

Morphology. Branched filaments 0.2 μ thick; filaments break up into short quadrangular-coccoid segments, which, by the rupture of the wall of the filament at the apex, allow the latter to escape.

Agar colonies. Round, yellowish white, with a finely granular centre and a pale concentric border. Microscopically, delicate, stellately branched, becoming opaque in centre, with a delicate branched border.

Glucose agar slant. Firm whitish warts, becoming larger and rugose, ochre-yellow.

Blood serum. As before.

Gelatin slant. Growth orange-yellow, rough, rugose.

Potato. Growth slow, white, becoming brick-red warts; later pulverent on the surface due to aerial hyphæ.

Bouillon. Clear; on the surface, white disks which fall to the bottom.

17. Streptothrix carnea Rossi-Doria

l.c., 1891.

Morphology. Mycelium branched with aerial hyphæ and spores. Colonies characterized on all media by their minuteness and rosy color.

Habitat. Air (rare).

18. Streptothrix maduræ Vincent

Ann. Pasteur Inst., 1894.

Morphology. Branched filaments 1–1.5 μ thick, with aerial hyphæ and spores. Stained by Gram's method. Optimum temperature 37°.

Agar slant. Slow growth, colonies firm, warty, yellowish white, becoming reddish — bright red. No growth on *blood serum.*

Bouillon. Growth scanty, as granules.

Potato. Growth warty, white, becoming red orange, with whitish aerial hyphæ.

Milk. Slowly peptonized. Good growth in vegetable infusions, of a slightly acid reaction, as hay, potato. Media clear with brownish flocculi, which on the surface are red, often forming a membrane; the reaction becomes alkaline.

Pathogenesis. At most only a slight local reaction in rabbits and guinea pigs.

Habitat. Associated with a warty ulcerative affection of the feet, and rarely of the hands.

Madura foot. In the lesions no nodular swellings, but in the periphery of the colony a zone concentrically arranged, spindle-shaped elements, probably degeneration forms of the filaments.

CHLAMYDOBACTERIACEÆ Migula

Filamentous bacteria composed of rod-shaped cells, and surrounded by a distinct sheath. Division of the cells at right angles to the axis of the filaments. In Phragmidiothrix and Crenothrix, however, in the formation of gonidia, a division of the cells in three directions of space takes place. Reproduction by means of gonidia, which are either motile or non-motile.

I. Cell contents without sulphur granules.
  A. Filaments unbranched.
    1. Cell division takes place only in one direction of space. *Leptothrix*.
    2. Cell division, before the formation of gonidia, takes place in three directions of space.
      a. Filaments surrounded by a delicate, scarcely discernible sheath. *Phragmidiothrix*.
      b. Filaments surrounded by a plainly discernible sheath. *Crenothrix*.
  B. Filaments show false branchings. *Cladothrix*.
II. Cell contents contain sulphur granules. *Thiothrix*.

LEPTOTHRIX Kützing

Phycologia Generalis, 1843, 198.
Streptothrix Migula: Die Natürlichen Pflanzenfam., 1895.

Filaments unbranched, non-motile, enclosed in very delicate or rather thick sheaths, either fixed or associated in slimy masses. The separation of the filaments and the presence of a sheath is demonstrated by special methods of staining. The contents of the filaments become segmented, forming round or ovoid gonidia, which escape from the sheath and develop into filaments. Gonidia non-motile.

2 B

The above characters as given by Migula are referred to his genus *Strepto-thrix*. The Streptothrix of Cohn includes branched forms entirely distinct from the Streptothrix of Migula. Migula's genus is certainly more closely related to the Leptothrix of Kütsing, and later described by Cohn. Furthermore, the rule of priority would direct that the earlier name, *Leptothrix*, should be retained.

    I. Filaments not fixed, but associated in tangled masses.
        1. *Leptothrix hyalina* (Migula).

    II. Filaments fixed on a substratum.
        A. Filaments short.
          2. *Leptothrix epiphytica* (Migula).
        B. Filaments long.
          3. *Leptothrix fluitans* (Migula).
          4. *Leptothrix gigantea* Miller.
          5. *Leptothrix innominata* Miller.
          6. *Leptothrix buccalis* Miller.

1. Leptothrix hyalina (Migula)

Streptothrix hyalina Migula: Die Natürlichen Pflanzenfam., 1895.

Filaments 0.6 μ in diameter, forming tangled masses. A sheath is demonstrated with iodine.
Habitat. Water.

2. Leptothrix epiphytica (Migula)

Streptothrix epiphytica Migula: l.c.

Filaments short, colorless, fixed to algæ, etc., with a thick gelatinous sheath. Produce ovoid gonidia.

3. Leptothrix fluitans (Migula)

Streptothrix fluitans Migula: l.c.

Very slender filaments 10 mm. long, with a delicate sheath. Produce spherical gonidia, which generally remain glued to the filaments. Found attached to the stems of water plants and wet wood.

The following species are not sufficiently described as regards their morphology to determine whether they are true species of this genus, or whether they belong to the Bacteriaceæ.

4. Leptothrix gigantea Miller

Ber. deutsche bot. Gesellsch., 1883, Heft 5.

Occurs as tufts of filaments of considerable length and of variable thickness, either straight, irregular, or spirally bent; composed of long and short rods and cocci. In a certain stage a sheath is observed, from which the rods and coccoid forms escape.

Habitat. In mucus from human teeth.

5. Leptothrix innominata Miller

Die Bakt. der Mundhöhle, II Auf., Berlin, 1894.
Leptothrix buccalis (?) Robin: Histoire Naturelle des Végétaux parasites, 1853.

Filaments 0.5–0.8 μ in diameter, unsegmented, somewhat undulate. Stain a faint yellow with iodine. Not cultivated.

Habitat. From the teeth of man.

6. Leptothrix buccalis Miller

Leptothrix buccalis-maximus Miller: l.c.

Filaments 1–1.3 : 30–150 μ; occur singly, or in parallel bundles. Stain brown —violet with iodine, and composed of segmented rods. Not cultivated.

Habitat. From the teeth of man.

PHRAGMIDIOTHRIX Engler

Bot. Verein der Provinz Brandenburg, 1882, 19.

Filaments with a very delicate sheath, only visible in old filaments. The filaments consist at first of groups of cells in one plane, which later divide in three directions of space, forming Sarcina-like packets. Later, the single cells assume a spherical form, and become free.

1. Phragmidiothrix multiseptata Engler

l.c.

Filaments 3–12 μ broad by 100 μ long. Found attached to the bodies of crustaceæ, *Gammarus locusta.*

CRENOTHRIX Cohn

Beiträge Biol., I, 1875, 130.

Filaments fixed to a substratum, usually thinner at the base than at the apex, with thick sheaths. Cells cylindrical to flat, one-half the breadth of the filament. Gonidia of two kinds : *microgonidia*, formed by a segmentation of the vegetating cells, producing small spherical elements; and *macrogonidia*, produced by the vegetating cells in the neighborhood of the apex of the filament breaking up into larger oval elements. The gonidia may either escape or germinate within the filaments.

1. Crenothrix polyspora Cohn

l.c.

Long, stiff, unbranched filaments, composed of cells; in the young filaments a thin, and in the older filaments a thick sheath. There is often a deposition of oxide of iron in the sheath, which stains the latter brown. Filaments 1.5–5.2 μ broad. Vegetating cells one-half to four times the breadth of the filaments. Gonidia formation as already described. Cultures on artificial media not successful. According to Rössler, organisms grow in spring water containing fragments of sterilized brick, previously boiled in water containing sulphate of iron.

Habitat. Found in stagnant and running water, containing organic matter and iron salts, as thick masses of a brownish or greenish color.

CLADOTHRIX Cohn

Beiträge Biol., I, 1875.

Filaments generally with delicate sheaths, often fixed and forming tufts. Cells cylindrical. By intercalary growth a cell may break through the sheath laterally, and by continuous growth produce a false dichotomous branching. Reproduction by motile gonidia (swarm spores), which bear a little laterally to a pole, a bundle of flagella.

I. Filaments surrounded by thick gelatinous sheaths.

1. *Cladothrix natans* (Kützing) Migula.

II. Filaments with a delicate, or with a scarcely evident sheath.

A. A ferruginous species, with an accumulation of hydrated oxide of iron in the sheath. Do not grow in ordinary culture media.

2. *Cladothrix ochracea* (Kützing) Winogradsky.

B. Not as above specified.   Grow more or less readily on culture media.
  1. Gelatin·liquefied.
      a. Growth on gelatin or agar pinkish — reddish.
              3. *Cladothrix rufula* Wright.
      b. Growth on gelatin whitish.
          * Gelatin stained brownish by the growth of the organism.
              4. *Cladothrix dichotoma* Cohn.
              5. *Cladothrix profundus* Ravenel.
              6. *Cladothrix intestinalis* Ravenel.
        ** Gelatin not stained brownish.
            † Colonies on agar distinctly stellate in form.
              7. *Cladothrix invulnerabilis* Acosta-Grande-Rossi.
            †† Colonies on gelatin floccose — filamentous.
              8. *Cladothrix fungiformis* Ravenel.
              9. *Cladothrix intrica* Russell.
  2. Gelatin not liquefied.
              10. *Cladothrix non-liquefaciens* Ravenel.

1. Cladothrix natans (Kützing) Migula

Sphærotilus natans Kützing: Linnæa. VIII, 1833, 385.
Cladothrix natans Migula: Die Natürlichen Pflanzenfam., 1895, 46.

Filaments composed of rod-like elements, surrounded by a thick gelatinous
  sheath. Show false branching like *Cladothrix dichotoma*. Eventually
  round strongly refracting bodies form within the rods, which germinate
  either outside or within the mother cells.
Habitat. Found as slimy particles in factory water.

2. Cladothrix ochracea (Kützing) Winogradsky

Leptothrix ochracea Kützing: Species Algarum, 147.
Cladothrix ochracea Winogradsky.

In morphology like *Cladothrix dichotoma* ; distinguished from the latter by the
  accumulation in the sheath of hydrated oxide of iron. Does not grow in
  ordinary culture media.

3. Cladothrix rufula Wright

l.c., 433.

Filaments long slender branched, which in stained preparations are seen to
  be composed of rather long undulate segments, separated by a clear
  interval. In the mature condition, deeply stained bodies of a diameter
  2–3 times the width of the filament are seen.

Gelatin colonies. Deep: round, reddish, granular, border slightly uneven, dense or densely floccose. *Surface:* in 4 days, round, with pale ill-defined margins, about 2 mm., slightly sunken; microscopically, densely floccose in centre, and of a reddish color, thinner and lighter in color toward the border.

Gelatin slant. A narrow dense reddish stripe, beneath which the gelatin is slowly liquefied.

Gelatin stab. In depth, a slight growth; on the surface, growth round thin flat pinkish, below which is a slowly liquefying funnel.

Acid gelatin. No growth.

Agar slant. Growth thin, translucent, glistening, limited, slightly pinkish, usually of discrete colonies.

Bouillon. Medium clear; no membrane, and with a pinkish sediment.

Potato. Growth elevated, pink — pale reddish, limited, rather rough.

Litmus milk. No change in 2 weeks.

Glucose gelatin stab. No growth in depth.

Pepton rosolic acid. Unchanged. *Indol* slight. Grows at 36° C.

Habitat. Water.

Houston, 27th Report Loc. Gov. Board, England, Supplement, 1897-98, 289, notes a species of Cladothrix which liquefies gelatin and produces an orange-pink pigment.

Habitat. Soil.

4. **Cladothrix dichotoma** Cohn

Beiträge Biol., I, 3, 1875, 185.

Filaments 0.4 μ thick, composed of rod-like segments surrounded by a delicate sheath. Short motile gonidia are set free at the apex of a filament. Filaments may assume spiral forms and exhibit the false dichotomous branching. According to Macé (Compt. rend. CVI, 1888, 1622) the organism shows the following cultural characters : —

Gelatin colonies. In 4-5 days, small yellowish dots; later a brownish button with a whitish bloom; later depressed, due to a slow liquefaction of the surrounding gelatin; medium stained brown.

Gelatin stab. On the surface growth thin, grayish; gelatin slowly liquefied; medium remains clear, but is stained a deep brown.

Agar slant. At 35°, a thick glistening layer, very adherent, with often a whitish bloom. The agar is stained brown.

Bouillon. Whitish radiate flakes; medium clear, but stained brown. All cultures have a strong mouldy odor. A species named *Cl. dichotoma* was isolated by Ravenel (l.c., 15), with the following characters: forms long chains and filaments, with characteristic false branching; rods of variable length; many clavate and spiral forms.

Gelatin colonies. In 2 days, minute, white, 0.5 mm.; microscopically, dense in centre, brown, with a similar staining of the surrounding gelatin. In 5 days the colonies may be 3 mm.; the brownish discoloration extends far into the gelatin. Each colony lies in a crateriform depression, with a thick brown skin on the surface of the liquefied gelatin; microscopically, with dense brownish centres and filamentous borders.

Agar slant. A whitish line; agar stained brown; growth becomes wrinkled, gristly, and adherent.

Gelatin stab. In depth a slight growth, and later outgrowths; on the surface, a button, which rapidly sinks; gelatin a clear brown; in 7 days a crateriform liquefaction, becoming stratiform.

Potato. Growth thick, rough, grayish, wrinkled; medium stained a deep brown.

Bouillon. Growth at the bottom as dirty white flocculi; the medium becomes the color of brandy.

Litmus milk. Becomes a deeper blue, and in 6 weeks cherry-red; apparently peptonized.

Glucose gelatin stab. No gas or growth in depth. *Indol* produced. Grow at 36°.

Habitat. Water and soil.

The *Bismarck brown Cladothrix* Houston (l.c.) is probably a variety of the above.

Habitat. Soil.

5. Cladothrix profundus Ravenel

l.c., 17.

Long chains and filaments with false branching.

Gelatin colonies. *Deep:* in 2 days, radiate, filamentous. *Surface:* in 2 days, minute round whitish dots in saucers of liquefied gelatin, brownish; microscopically with brownish centres and radiately filamentous borders, becoming concentric. A whitish, radially folded membrane on the surface of the liquefied gelatin; the medium is stained brownish.

Agar slant. A tenacious leathery membrane, finely wrinkled, with a thinner undulate edge; bluish gray, with a metallic lustre; growth adherent.

Gelatin stab. In depth, a slightly arborescent growth; on the surface liquefaction crateriform, with vertical sides, becoming stratiform; liquefaction slow; liquefied gelatin a clear brown.

Potato. A finely wrinkled colorless stripe, becoming grayish, spreading, rough, rugose, becoming thin, dry, brownish, and less rugose; medium stained brownish.

Bouillon. Growth mostly at the bottom, with a few grayish flakes at the surface and around the edges. The medium becomes a sherry wine color.

Litmus milk. Becomes violet — plum-colored, alkaline, and is apparently peptonized.

Glucose gelatin stab. No gas; growth in the upper part of the stab and on the surface. *Indol* negative. Grows at 36°.

Habitat. Soil.

6. Cladothrix intestinalis Ravenel

l.c., 18.

Morphology. Long filaments, with false branching, which become segmented into rods of various lengths. The filaments have buds on them here and there which are almost spherical.

Gelatin colonies. Deep: radiately arborescent. *Surface:* minute white dots; microscopically, they show grayish densely floccose centres, outside of which is a filamentous zone and outside of this a corona of coarse spear points or arborescent outgrowths. The surface growth becomes white and mouldy; the gelatin is slowly liquefied and takes on a brownish color.

Agar slant. A narrow rough wrinkled stripe, adherent, friable; agar but little discolored.

Gelatin stab. In depth growth arborescent; on the surface a whitish mouldy growth, radially folded; gelatin slowly liquefied to a depth of 20 mm.; a slight brownish discoloration of the medium.

Potato. Growth thin, white, wrinkled, becoming thicker and more rugose and finally dry, white, mouldy; medium, but slightly discolored.

Bouillon. Whitish flocculi at the bottom; the medium the color of dark sherry wine.

Litmus milk. A lighter blue; a thick pellicle on the surface and a dirty brown ring around the tube. In 7 days medium violet, color slowly discharged, alkaline. and apparently peptonized.

Glucose gelatin stab. No gas, growth as in plain gelatin. *Indol* probably present; discoloration of medium interferes with the reaction. Grows at 36°.

Habitat. Soil.

7. Cladothrix invulnerabilis Acosta-Grande-Rossi

Centralblatt f. Bakteriol., XIV, 1893, 14.

Morphology. Not described.

On Agar. Small round dirty white colonies, very adherent, becoming silvery white, later yellowish, and in 14 days distinctly stellate in form, umbilicate in centre.

Gelatin stab. On the surface a slimy white colony, umbilicate in centre; gelatin slowly liquefied.

Potato. A broad band composed of confluent chalk-like colonies, with the odor of damp soil. The medium becomes blackish.

Bouillon. Turbid. In sterilized water an abundant cloudy growth. Grows in the absence of air.

Habitat. Water.

8. Cladothrix fungiformis Ravenel

l.c., 19.

Morphology. Long chains and filaments with false branching, not always easily made out.

Gelatin colonies. *Deep:* radiately arborescent. *Surface:* white, minute, which microscopically were densely floccose in the centre with gray, radiately arborescent borders. In 7 days the colonies are mouldy and sunken in the gelatin. Medium not discolored.

Agar slant. Growth white, mouldy, adherent, becoming fissured, and showing a dark brown substratum. Gives off a strong odor of rotten wood. In 3-4 weeks the agar becomes slightly darker.

Gelatin stab. In depth, growth arborescent; on the surface, a white button, under which a slow crateriform liquefaction takes place; sides of crater vertical. In 8 days the liquefaction has reached the sides of the tube. There is no discoloration of the medium.

Potato. A scanty, nearly colorless, streak; potato distinctly whiter along each border, becoming pale yellowish.

Habitat. Soil.

9. Cladothrix intrica Russell

Morphology. On gelatin, long slender cells united into long filaments. On potato, cells shorter with rounded ends, becoming segmented into short, plump individuals containing spores. A false branching is seen in gelatin cultures.

Gelatin colonies. In 24-36 hours small white dull mould-like colonies which rapidly liquefy the gelatin and which microscopically are floccose in centre and filamentous at the border.

Gelatin stab. In depth, growth arborescent; gelatin rapidly liquefied.

Agar slant. Growth thin, dull, white, from which filaments extend into the medium.

Potato. Growth irregular, dull, white.

Bouillon (made with sea water). An abundant friable sediment.

Habitat. Isolated from sea water and sea mud.

10. Cladothrix non-liquefaciens Ravenel

l.c., 16.

Morphology. Long chains and filaments with false branching. No spiral or coccoid forms. Filaments segmented into rods of various lengths which have square ends.

Gelatin colonies. Deep: in 2 days arborescent — radiate; in 7 days colonies show reddish yellow centres and radiately arborescent borders. *Surface:* in 2 days minute, whitish — yellowish; centres grayish, dense, opaque, yellow, borders of interlacing coarse filaments and spear points.

Agar slant. Growth thin, whitish. with smooth edges marked by yellowish points; later an elevated yellowish salmon-colored band, with a whitish bloom on the surface. Many offshoots from the lower side of the growth into the agar.

Gelatin stab. In depth, an arborescent growth; on the surface, after 7 days, a white dry button. A slight brownish discoloration of the medium at the surface.

Potato. Growth yellowish, rough, dry, rugose, later with a pinkish tinge; not abundant.

Bouillon. On the surface a thin pulverent film; medium clear; a whitish, flocculent sediment.

Habitat. Soil.

THIOTHRIX Winogradsky

Bot. Zeitung, 1887.

Filaments fixed, of unequal diameter, surrounded by a delicate difficultly discernible sheath, non-motile. Sulphur granules enclosed in the plasma. The filaments produce at their ends rod-shaped gonidia, which become detached, and, fixing themselves to some substratum, develop into new filaments.

        I. Filaments very slender.

1. Thiothrix tenuissima Winogradsky

l.c.

Filaments not exceeding 0.5 μ broad.
Habitat. Found in sulphur water.

        II. Filaments thicker.

2. **Thiothrix nivea** (Rabenhorst) Winogradsky

Beggiatoa nivea Rabenhorst: Kryptogamen flora, I.
Thiothrix nivea Winogradsky: l.c.

Filaments with a thin sheath, 2–2.5 μ broad at the base, and 1.4–1.5 μ broad at the apex; often 100 μ long, segmented at the apex, producing motile gonidia 8–9 μ long.
Habitat. Found in sulphur and stagnant water.

3. **Thiothrix tenuis** Winogradsky

Filaments very long, and about 1.0 in diameter.
Habitat. Found in sulphur water.

BEGGIATOACEÆ Migula

Filamentous bacteria. Filaments without sheaths, but with motility like *Oscillatoria* by means of an undulating membrane. The cell contents show the presence of sulphur granules. The formation of gonidia not known.

BEGGIATOA Trevisan

Prospetto della Flora Euganea, 1842, 76.

Filaments apparently not segmented except when stained with iodine. Colorless, or faintly rose-colored.
I. Filaments colorless.
 A. Filaments 3–4 microns thick.
  1. *Beggiatoa alba* (Vaucher) Trevisan.
 B. Filaments 7 microns thick.
  2. *Beggiatoa arachnoidea* (Agardh) Regensburger.
 C. Filaments 16 microns thick.
  3. *Beggiatoa mirabilis* Cohn.
II. Filaments colored reddish-violet.
  4. *Beggiatoa roseopersicina* Cohn.

1. **Beggiatoa alba** (Vaucher) Trevisan

Flora Euganea, 1842.

Filaments long, 3–4 μ thick, containing numerous strongly refracting granules of sulphur. Filaments break up into short segments, which then grow out into longer threads. Sulphates reduced to H_2S and to free sulphur.

Habitat. Found in dirty water, drain water from sugar factories, sulphur springs, etc. The filaments are attached to decayed plants, etc., producing slimy flakes.

2. Beggiatoa arachnoidea (Agardh) Rabenhorst

Flora, 1827, 634.

Found in swamp water, also in sea water. Filaments 7 microns thick.

3. Beggiatoa mirabilis Cohn

Hedwigia, 1865, 81.

Filaments 16 μ thick. Found in sea water, forming white growths on dead algæ, etc.

4. Beggiatoa roseopersicina Cohn

Beiträge Biol., I, 3. 1875. 157.

Filaments like (1), but colored as above. Found in stagnant water, forming a surface growth of a red violet color.

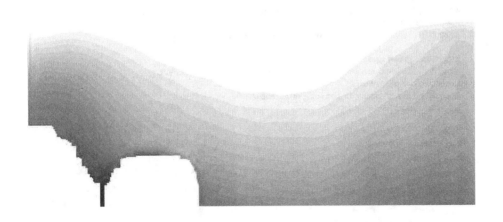

GLOSSARY OF TERMS USED OR USEFUL IN DESCRIPTIVE BACTERIOLOGY

Abnormal, differing from usual form.

Aborescent, branched, or treelike.

Achromatic, not readily colored by the usual staining methods.

Acrogenous, produced at the summit.

Actinomorphic, symmetrical; capable of bisection by planes into halves identical in form.

Aculeate, beset with sharp points or prickles.

Acute, ending in a distinct angle.

Aerobia, organisms which grow in the presence of air or free oxygen.

Agglomerated, } clustered or growing together, but not cohering.
Aggregated,

Alveolate, pitted so as to resemble a honey comb. Fig. 12, C.

Ameboid, assuming various shapes, like Amœba. Fig. 10, B.

Amorphous, without any definite structure.

Anaerobia, organisms which will not grow in the presence of air or free oxygen.

Anaerobiotic, the property of not growing in the presence of air or free oxygen.

Anastomosing, connected by transverse branches, forming a more or less perfect net work.

Antizymotic, preventing or checking fermentation.

Applanate, flattened, or horizontally expanded.

Arcuate, curved like a bow.

Areola (pl. areolæ), an area or small space with more or less definite boundaries.

Areolate, divided into areolæ.

Aromatic, having a pleasant odor, spicy, alcoholic.

Aseptic, not liable to putrefaction.

Auriculate, having auricles, earlike lobes or appendages. Fig. 13, e.

Biogenous, growing, living organisms.

Bion, an individual morphologically and physiologically independent.

Brunneus, deep brown.

Bullate, blistered, rising in convex prominences.

Butyrous, translucent and yellow.

Calcareous, of a dull, chalk-white color.

Canescent, hoary or gray.

Capitate, furnished with a head ; a semi-spherical colony.

Capsule, the gelatinous envelope surrounding a bacterium.   Fig. 2.

Carneus, flesh colored, pale red.

Cartilaginous, firm and tough, like cartilage.

Ceraceous, waxy, waxlike in appearance.

Chartaceous, of the texture of paper.

Chromatic, capable of being colored by staining agents.

Cineraceous, a little paler than cinerous.

Cinerous, ash gray.

Clavate, club-shaped.

Clouded, having a pale ground, with ill-defined patches of a darker tint gradually shading into it.   Fig. 12, f.

Cochleate, shaped like a shell, spirally turbinate.   Fig. 10, A.

Coerulescent, bluish, lighter than cœruleus.

Coeruleus, light blue, sky blue.

Concatenate, linked together in chains.

Concentric, having a common centre, the ringed structure of many colonies.

Conglobate, clustered together into a ball.

Conglomerate, clustered together.   Fig. 12, A.

Conidium, a propagative cell, naked or with a membrane, produced asexually, separating from the parent and capable of developing into a bion.

Continuous, not divided into septa.

Contoured, an irregular but smoothly undulating surface.

Convex, surface the segment of a circle, but flatly convex.

Coriaceous, leathery in texture.

Corrugated, in long folds or wrinkles.

Crateriform, disk-shaped.   Fig. 9, 1.

Crenate, edged with rounded teeth.

Cretaceous, opaque and white, chalky.

Cuneate, wedge-shaped.

Cyaneus, pure blue.

Dentate, with broad acute teeth.

Dichotomous, forked, bifurcated.

Discrete, separate, not confluent.

Echinate, beset with short prickles.

Ectogenic, capable of living outside the animal body, said of disease producing organisms.

Effused, spread out over the medium as a thin, veily layer.

Elliptical, shape of an ellipse.

Endemic, occurring in the one limited locality or region.

Endogenous, produced within another body.

Endophytal, growing within plants.

Endophyte, a plant which grows within another.

Entire, having a margin destitute of teeth or notches.

Epizoic, growing upon living animals, either parasitic or not.

Erose, gnawed as if bitten irregularly. Fig. 13, i.

Excretion, the separation of unassimilable matter from an organism.

Facultative, occasional, incidental.

False dichotomy, any dichotomous appearance which does not arise from a terminal division of the main axis.

Farinose, as if covered with a white, mealy powder, mealy.

Ferruginous, color of iron rust, brownish red.

Filamentous, threadlike, or composed of filaments. Fig. 11, E.

Filiform, threadlike.

Fimbriate, fringed, bordered by slender processes larger than hairs. Fig. 13, G.

Flavus, pure, pale yellow, lemon yellow.

Flexuose, wavy, winding.

Floccose, composed of matted, woolly hairs, denser than filamentous.

Floculent, as light feathery particles.

Fuliginous, dark brown, sooty or smoky.

Fuscescent, slightly brown.

Fuscous, grayish brown.

Fusiform, spindle-shaped, tapering towards each end.

Gelatinous, consistency of jelly.

Globose, spherical, shape of a globe.

Gregarious, growing in groups or clusters.

Grumous, clotted in clustered grains. Fig. 12, D.

Gyrose, marked with wavy lines. Fig. 12, I.

Habitat, the situation, locus, or mode of occurrence of an organism.

Heterosporous, having asexually produced spores of more than one kind.

Homochromous, of uniform color.

Homogeneous, of a uniform structure throughout.

Hyaline, clear or colorless, like glass or water.

Immarginate, without a distinctly defined rim or border.

Infectious, includes contagious as already defined, but applies also to diseases originating from germs which are to vegetate for a time at least outside of the affected animal or plant. In a strict sense infectious applies only to diseases produced by organisms which have their natural home outside the infected body.

Infundibuliform, shape of a funnel conical.

Lacerate, having the margin deeply cut into irregular segments as if torn. Fig. 13, f.

Laciniate, deeply cut into incisions or lobes, slashed; more irregular and coarser than fimbriate.

Lenticular, in the form of a double convex lens.

Ligulate, with narrow, tongue-shaped extensions.

Linear, long and narrow, with the sides parallel.

Lobate, border broadly rounded with equally broad sinuses.

Lobulate, having small lobes.

Marginate, with a distinct margin.

Marmorate, marbled, covered with faint, irregular stripes, or traced with vein-like markings like marble.

Membranous, thin and soft and usually translucent, like a membrane.

Moniliform, neckless-shaped, cylindrical and contracted at regular intervals so as to resemble a string of beads.

Moruloid, like a morula, segmented.

Mycelioid, referring to colonies with the radiate filamentous appearance of mould colonies.

Nacreous, translucent, grayish white, with pearly lustre.

Napiform, form of a turnip. Fig. 9, 2.

Nodose, knotted, swollen at intervals.

Nodulose, diminutive of nodose.

Ochraceous, brownish yellow.

Oleaginous, transparent and yellow, olive to linseed oil colored.

Olivaceous, dusky green.

Ovate, outline of an egg.

Papilla (pl. papillæ), a teat-shaped protuberance.

Papillate, having papillæ.

Parasite, an organism which grows upon or within another from which it derives its nourishment.

Patelliform, shape of a watch crystal or shallow saucer; more shallow than crateriform.

Pathogenic, producing disease.

Pellucid, translucent.

Penicillate, like a brush.

Plasmolysis, the contraction of the protoplasm under the influence of reagents.

Plumose, like a plume or feather, feathery.

Polymorphism, having a variety of forms under different conditions of growth or environment.

Pulverulent, as if covered with dust.

Pulvinate, in the form of a cushion, decidedly convex.
Punctate, dotted as if by punctures, like pin pricks.
Punctiform, in the form of a dot or point.
Pyriform, pear-shaped.

Radiate, spreading from centre, with irregular rays or lobes.
Raised, growth thick, with abrupt terraced edges.
Repand, like the border of an open umbrella.
Resinous, transparent and brown, varnish or resin-colored.
Reticulate, in the form of a network, as the veins of a leaf.
Rhizoid, of an irregular branched, root-like character.   Fig. 10, C.
Rimose, abounding in chinks, clefts, or cracks.
Rivulose, marked with lines like the rivers of a map.
Rosulate, shaped like a rosette, more regular than radiate.
Rudimentary, but slightly developed.
Rugose, irregularly wrinkled.

Saccate, shaped like an elongated sack, tubular, cylindrical.   Fig. 9, 3.
Saprophytic, living upon dead organic matter.
Sebaceous, translucent, yellowish or grayish white.
Sinuate, with a wavy outline.
uous, flexuose, curving back and forth.
ary, not closely associated with others.
:nous, producing spores.
scaly, covered with scales.
, diminutive of squamose.
, a layer, with upper and lower sides parallel.

group of four cells.
budding like the yeast plant.
, swollen at intervals.
trophism, variations produced in the organism through the influence of
: chemical nature of the medium.
ite, terminating abruptly as if cut off at the end, flattened.

icate, having an umbilicus or central depression.
aate, having a central projecting elevation.
aulate, diminutive of umbonate.
ate, hooked or abruptly curved at the end.

Vernicose, with a varnish-like lustre.
Verrucose, wart-like, or bearing wart-like prominences.
Verruculose, slightly verrucose.
Versicolor, changeable in color or appearance of different colors from different
points of view.
2 C

Vesicular, bearing or containing numerous vesicles.
Viable, capable of growing or manifesting life.
Villous, beset with hair-like extensions.
Virescent, greenish, or becoming green.
Viscid, sticky, adhesive, viscous.
Vitreous, transparent and colorless.

LIST OF SOME OF THE MOST IMPORTANT WORKS ON DESCRIPTIVE BACTERIOLOGY, ALSO REFERRED TO IN THE TEXT.

Adametz. Die Bakterien der Nutz, u. Trinkwässer, Vienna, 1888.

Billroth. Untersuchungen über die Vegetationsformen der Coccobacteria septica, Berlin, 1874.

Bütschli. Uber den Bau der Bakterien, Heidelberg, 1890.

Cohn. Untersuchungen über Bakterien; in Beiträge zur Biologie d. Pflanzen. Bd. I, 1872-75; Bd. II, 1877; Bd. III, 1879-80.

Conn. Descriptions of bacteria isolated from milk. Reports of the Connecticut (Storr's) Ag. Expt. Sta., 1893 and 1894.

Davaine. Article Bactéries, in Dictionnaire Encyclop des Sciences Médicales, 1868.

De Bary. Gen. Morphology of the Fungii, Myxomycetes, and Bacteria.

De Bary. Lectures on Bacteria, translated by Garnsey and Balfour, Oxford, 1887.

Duclaux. Mémoire sur le lait, in Ann. de l'Institut Agronomique, 1882.

Dujardin. Hist. Nat. des Zoophytes infusoires, Paris, 1841.

Dyar. On certain bacteria from the Air of New York city; in Report of the New York Acad. of Sci., VIII, 1895, 322-380.

Ehrenberg. Die Infusionsthierchen als Vollkommene Organismen, Leipzig, 1838.

Eisenberg. Bakteriologische Diagnostik, 1888 and 1891.

Fischer. Untersuchungen über Bakterien, Berlin, 1894.

Flügge. Die Mikroorganismen, 1st ed. 1886, 3d ed. 1885, Leipzig.

Frankland. Microorganisms in Water, London, 1894.

v. Freudenreich, E. Die Bacteriologie in der Milchwirthschaft, Jena, 1898.

Gessard. De la pyocyanine et son Microbe. Thèse inaugurale de la Faculté, de Médécine de Paris, 1882.

Hauser. Ueber Fäulniss-Bakterien, Leipzig, 1885.

Hoffmann. Mémoire sur les Bactéries. Ann. des Sci. Nat. bot., XI, 1869.

Jörgensen. Die Mikroorganismen der Gährungsindustrie. Berlin. 1898.

Kramer. Die Bakteriologie Landwirthschaft u. Landw.-Technischen Gewerben. Wien, 1890.

Lehmann-Neumann. Atlas u. Grundriss der Bakteriologie, II Aufl., 1899.

Lustig. Diagnostik der Bakterien des Wassers, 1893.

Macé. Atlas de Microbiologie, Paris, 1898.

MacFarland. Text-book upon the Pathogenic Bacteria.

Migula. Schizophyta. Lieferung 129, in Engler and Prantl. Die Natürlichen Pflanzfamilien, 1895.

Migula. System der Bakterien, 2 vols., Jena, 1897-1900.

Müller. Vermium terrestrium et fluviatilium historia, 1773.

Passet. Ueber Mikroorganismen der Eitrigen Zellgewebsentzündung des Menschen. Fortschritte d. Medezin, 1885, No. 2.

Prazmowski. Untersuchungen über die Entwickelungsgeschichte u. Fermentwirkung einiger Bakteriumarten, Leipzig, 1880.

Rabenhorst. Kryptogamenflora von Deutschland, Oesterreich u. der Schweiz,. Erster Band, Pilze, von Dr. Geo. Winter, Leipzig, 1884.

Ravenel. Memoires National Academy of Science, VIII, 1896.

Rosenbach. Mikroorganismen bei d. Wundinfectionskrankheiten des Menschen, Wiesbaden, 1884.

Schroeter. Die Pilze von Schlesien, in Cohn's Kryptogamenflora von Schlesien, 1885.

Sternberg. A Manual of Bacteriology, 1st ed. 1893, 2d ed. 1896. New York.

Trevisan-de-Toni. Schizomycetaceæ in Sylloge Fungorum by P. A. Saccardo. Vol. VIII, 1889.

Winter. Die Pilze, in Rabenhorst's Kryptogamenflora von Deutschland,. Oesterreich u. des Schweiz, Bd. I, 1 Abth. Leipzig, 1884.

Wright. Report on the Results of an Examination of the Water Supply of Philadelphia. Memoires of the National Acad. of Sciences, VII, 1895.

Zimmerman. Die Bakterien unserer Trink und Nutzwässer insbes. des Wassers der Chemnitzer Wasserleitung, Chemnitz, 1891.

Zopf. Die Spaltpilze. 1st ed. Leipzig, 1882, 2d ed. 1884, 3d ed. 1885.

INDEX

Micrococcus
botryogenus, 80.
bovis, 73.
candicans, 90.
candidus, 85.
carneus, 108.
catarrhalis, 73.
cerasinus, 108.
cerasinus-siccus, 108.
cereus, 204.
cereus-flavus, 104.
cerevisiæ, 88.
cinnabareus, 107.
citreus, 94.
citreus, 103.
citreus-conglomeratus, 100.
citreus-liquefaciens, 101.
concentricus, 86.
conglomeratus, 100.
coralinus, 106.
coronatus, 77.
cremfarbiger Micrococcus, 94.
cremoides, 93.
cremoides-aureus, 99.
cumulatus, 87.
cumulatus-tenuis, 87.
cyaneus, 109.
Dantecii, 106.
Demmei, 74.
descidens, 93.
dissimilis, 80.
eczemæ, 86.
endocarditis, 74.
endocarditis-rugatus, 74.
epidermis, 97.
epidermis-albus, 76.
expositionis, 92.
fervidosus, 89.
Finlayensis, 92.
Fischel No. 2, 59.
flava-varians, 102.
flavus, 99, 100.
flavus-liquefaciens, 97, 99, 100.
fœtidus, 81.
Freire, Micrococcus of, 79.
Freudenreichii, 79.
fuscus, 108.
gonococcus, 72.
gonorrhœæ, 72.
Gray white micrococcus Bumm, 80.
haemorrhagicus, 91.
Heydenreichii, 91.
Jongdii, 95.
Kefersteinii, 107.
lactericeus, 107.
lactis, 90.

Micrococcus
lactis-viscosus, 62.
liquefaciens, 78.
luteus, 99.
magnus, 85.
mastitis, 76.
Mendozæ, 84.
Milk-white Micrococcus Bumm, 85.
nivalis, 90.
ovalis, 89.
orbicularis, 101.
orbicularis-flavus, 101.
orbiculatus, 104.
plumosus, 94.
polymyositis, 77.
pyogenes var albus, 75.
pyogenes var aureus, 98.
pyogenes-bovis, 95.
pyrosepticus, 75.
quadrigeminus, 98.
radiatus, 79.
Rheni, 82.
Rhine water Micrococcus Burri, 82.
rosettaceus, 89.
roseus, 105.
rubescens, 105.
rugosus, 101.
salivarius, 97.
salivarius-pyogenes, 98.
salivarius-septicus, 87.
simplex, 78.
Sornthali, 86.
stellatus, 84.
subflavus, 96.
tardigradus, 95.
tardissimus, 85.
tenacatis, 88.
tetragenus, 84.
tetragenus-mobilis, 84.
tetragenus-mobilis-ventriculi, 115.
tetragenus-pallidus, 93.
tetragenus-subflavus, 96.
tetragenus-versatilis, 102.
tetragenus-vividus, 102.
Tommasoli, 101.
ureæ-liquefaciens, 77.
Uruguæ, 100.
varians, 102.
versatilis, 102.
versicolor, 95.
Vincenzii, 103.
viridis, 95.
viticulosus, 83.
xanthogenicus, 79.
Microspira
aquatilis, 339.

Typhus exanthematique, B. of, Babes-
  Oprescu, 229.

Vibrio
  bugula, 2.
  choleræ, 335.
  choleroides, 337.
  helicogenes, 339.
  Ivanoff, 337.
  Lissabonensis, 339.
  saprophiles, Weibel, 340, 341.

Vignal Bacillus j, 234.

Whooping Cough Bacillus, Afanassieff, 283.
Whooping Cough Bacillus, Czaplewski, 153.
Wild u. Rinderseuche, B. of, 137.

Yellow atrophy of the liver, B. of, Guarnieri,
  208.

Xerose Bacillus, Neisser, 355.

2 D

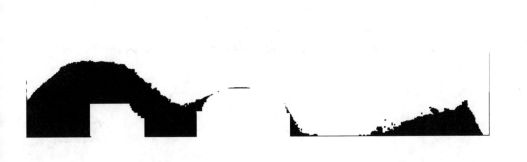

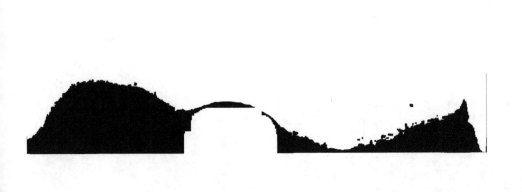

www.ingramcontent.com/pod-product-compliance
Lightning Source LLC
LaVergne TN
LVHW050151060326
832904LV00003B/110